Scientific Colonialism

A CROSS-CULTURAL COMPARISON

Papers from a Conference at

Melbourne, Australia

25–30 May 1981

Edited by

Nathan Reingold

and

Marc Rothenberg

SMITHSONIAN INSTITUTION PRESS

Washington, D.C.   London

Library of Congress Cataloging-in-Publication Data

Scientific colonialism.

"Papers from a conference at Melbourne,
Australia, 25–30 May 1981."
Includes index.
1. Science—History—Congresses. 2. Science—
United States—History—Congresses. 3. Science—
Australia—History—Congresses. I. Reingold,
Nathan, 1927– . II. Rothenberg, Marc, 1949–
Q124.6.S39   1986   509   85-42238
ISBN 0-87474-785-6 (alk. paper)

The paper used in this publication meets
the minimum requirements of
the American National Standard
for Permanence of Paper for
Printed Library Materials Z39.48-1984

We wish to thank the University of Melbourne
for its hospitality and particularly
the efforts of Professor R. W. Home.
The conference was made possible by
the generosity of the U.S. National Science
Foundation and the Department of
Science and Technology of Australia.

Contents

Introduction

NATHAN REINGOLD

MARC ROTHENBERG

Historians still argue about the origins, attributes, and consequences of the historical events and processes that occurred in Western Europe during the period, roughly speaking, A.D. 1300–1800.
There is no argument about the importance of what was taking
place. In many ways—political, economic, intellectual—the countries of the region began assuming an importance in world history
that prevailed in various senses until the middle of this century. It
was the age of European expansion overseas and of the flowering of
a distinctive cultural style referred to today as Western civilization.

The ships leaving Western Europe in what is often referred to by
westerners as the early modern period were technological achievements of that civilization, as were their armaments. Navigation
across the oceans represented a culmination of developments that
went back to Babylon, Egypt, and other ancient civilizations. Back
on shore, in the homelands, were countless individuals engaged in
art, in manufactures, in abstract speculation, and, even a few, in systematic probing of the natural world. By 1800 something had occurred in Western Europe that we now designate as the Scientific
Revolution and the Industrial Revolution. Historians still argue
about those two revolutions and will undoubtedly do so for a long
time. Whatever advantages the Western Europeans had in their earlier contacts with indigenous populations in the age of overseas expansion were enhanced by these later scientific, technological, and
economic developments. In time Western Europe had gained political and cultural hegemony over a large portion of the globe. How

that hegemony was accomplished and its persisting consequences are major issues to this day.

In general, the Western Europeans encountered two different situations as they spread overseas, although a precise line is hard to draw in a number of cases. One situation was the interaction of Western civilization with the ancient complex cultures of the relatively populous lands of Islam, the Indian subcontinent, and the Orient. Quite different were the cases in which the indigenous populations were relatively small, and they and their cultures were extirpated or pushed aside by European migrants. Australia and North America were the obvious examples. Latin America and Sub-Saharan Africa represent intermediate situations in which the native populations and their cultures persisted in varying degrees in symbiosis with European settlers and their ways of life.

What follows is an attempt at comparative history for the cases in which European migrants established Western societies on other continents. Specifically, these papers, given at a conference at the University of Melbourne, May 25–30, 1981, originated in an attempt at a juxtaposition of studies of events in Australia and the United States that bear on the development of disciplines and institutions of the pure and applied sciences. To these were added a small number of additional papers from other regions to provide different perspectives and to act, as it were, as a form of intellectual control over any tendencies at overhasty extrapolation. Pairing topics was not possible—to study geophysics in both countries, for example. That reflects in part the existing patterns of scholarship in the two countries. After the conference that lack of congruence or a mirror-image situation emerged both as a virtue and as a vindication of this comparative exercise. A wider range of situations was presented, but in such a manner as to give rise to a richer discussion of similarities and of distinctions.

The decision not to consider the colonial cases involving the cultures of Asia did not reflect any downgrading of their importance. On the contrary, the two organizers, R. W. Home of the University of Melbourne and Nathan Reingold of the Smithsonian Institution, were initially interested in expanding the scope of the meeting in that direction if funding permitted. There was an early realization of how historically far-reaching such cases were, coupled with a sense of their greater complexity. In addition, even a cursory survey of the bibliographic sources disclosed how much remained undone in the history of the sciences in Islam, India, China, and Japan. In the near future, however, there should be a carefully constructed meeting to explore systematically in a comparative mode what happened when Western science and technology interacted with these ancient, sophisticated cultures.

For purposes of this exercise in comparative history, Australia and the United States were simpler cases. The large-scale development of the study of the history of the pure and applied sciences in the latter country attracted the attention of Australian scholars embarking on equivalent investigations. For the participants from the North American republic, Australia was an unknown that presented novel possibilities for comparison and analysis. Both countries are large continental land masses with populations largely of European extraction. The original inhabitants survived but in a position of limited power and influence on the cultures of their two nations. Both were former colonies of Great Britain. The impress of the language and political traditions of the mother country are quite evident.

So are the differences between the two countries. One was remarked upon from the first contacts of the organizers and reappeared many times in papers and in oral discussions. The early independence of the United States enabled its citizens to experiment, to pick, and to choose among various alternatives presented by European models. In contrast, the continued imperial connection produced a lasting imprint on the sciences, as well as on other aspects of Australian life. Although the connection was sometimes regarded as irksome and even demeaning, the papers at the conference indicated that the relationship was often mutually beneficial. Feelings of responsibility for their overseas possession made the mother country both solicitous and supportive. On their part, many Australians readily turned to Great Britain in many situations, often with a sense of belonging to a great transnational entity, the Commonwealth.

A second difference not remarked upon explicitly was the effects of the sizes of the two populations. In a few of the papers on Australia were references to the physical differences of the two great land masses. That undoubtedly had an effect on the growth of population. The North American republic simply had more water to support a larger population. Distance from Europe provided a slower rate of migration to Australia. To the American participants, the numbers given for the science professoriate, for example, were reminiscent of the early nineteenth century in the United States. Even allowing for the differences in the internal structure of universities—that is, the multiprofessorial department in the United States compared to the single professor in Australia, as in Western Europe—the size of the Australian population created a distinctly different situation. For that reason, there was no tendency to apply an evolutionary model—that the United States represented a later stage which Australia would inevitably attain, barring some national disaster. Australia was now and would continue to be different from both Great Britain and from the United States.

What was common to the experiences of both the United States and Australia was the question of cultural hegemony. *Colonialism* in the context of this conference meant cultural hegemony, not political sovereignty. The early political independence did not immediately bring about a cultural parity with Western Europe in the sciences. In both countries there were attempts, often quite explicit, to attain a degree of self-sufficiency in the sciences and recognition as a peer from the older centers of research. Much of the history of the sciences in the United States is explicable in part as a succession of moves to catch up with and even to surpass the Western Europeans. Analogous efforts in Australia were more modest and were limited in goals, both by the persisting imperial or commonwealth link and the recognition of limited resources. As in the Canadian case, World War II signaled the start of a determined effort at cultural parity in the sciences and elsewhere.

For both the United States and Australia, the sciences constitute a case significantly different from politics and economics. Both political independence and economic independence were worthwhile goals, even though the two countries remained within particular international systems. Independence circumscribed, not absolute, was nevertheless a real condition, with genuine options available. In the sciences, the game was quite different. Parity and recognition did not imply independence. Except for a few bizarre cases, the scientists in both Australia and the United States were not seeking to found or to develop inherently distinctive national fields of science and technology. If investigators and their institutions were sometimes belligerently national in some characteristics, knowledge, in their view, was—or should be—transnational. National characteristics might influence or determine the origins of a particular piece of science or produce different rates and patterns of transmissions and acceptances of new knowledge. Once accepted, presumably by a worldwide community of specialists, knowledge, by conventional definition, was international. Its genetic origins and its passage into various national research communities ceased to count as factors bearing on validity—or so the conventional belief asserted, despite the strange history, for example, of the reception of Darwin's work in France. When examined in relation to cultural hegemony, the movement from colonial status is best seen as an attempt to contribute to and validate proposed additions to knowledge on a par with any of the older research communities in Western Europe. Looked at from an Australian-American perspective, the history of science is less the story of the spread of an international style or a universal body of knowledge than the emergence of new, additional centers of research and the fabrication of a net of linkages to the older centers.

In an effort to develop a more general view of the process of the diffusion of Western science, the participants in the conference considered, sometimes explicitly, sometimes implicitly, a number of descriptive models, among which George Basalla's three-stage model attracted the most attention.[1] Basalla saw a distinct pattern in the introduction of Western European science into non-Western European lands. Initially, the non-Western society or country served as the subject of scientific research. This phase was succeeded by the colonial phase, in which scientific activity in the new land is dependent upon the institutions and traditions of Western Europe. Finally, in the third phase, there occurs a struggle to establish an independent scientific culture. While no one would question that such phases did occur, the model was rejected as inadequate. It simply did not capture the richness and complexity of the diffusion of Western scientific culture. Nor did it take into account the fact that some of its processes also occurred within Western Europe.

Another way to regard the diffusion was to consider science as simply an intellectual process, with techniques and results that freely cross international boundaries. Such a vision divorces the content of science from the cultural context within which it developed. But this is unrealistic. History demonstrates that science develops in particular environmental settings; to understand science fully requires an understanding of the ecology of its environment.

Sociology offers a third alternative, that of the center and the periphery. This model can be offered in the form of either geographical or social distinctions. It is sometimes presented in the guise of the metropolis versus the provinces. In any event, this model presents the relationship of Western Europe to the rest of the world as an asymmetric one; those at the center make the decisions, establish the priorities, and award the laurels. While useful, this model is unnecessarily mechanical and rigid. In practice, moreover, it is sometimes difficult to decide who is at the center and who is at the periphery. The model downplays the initiatives and contributions that occur at the supposed periphery.

What emerges as a heuristic solution is to regard science as polycentric. The international scientific community could be envisioned as a group of cells, each cell linked to one or more of the other cells. These cells represent the center or focus of a particular national science, discipline, theoretical or experimental orientation, or just the scientific activity in a colony. A given scientist would have ties to a number of centers, both intranational and international, reflecting intellectual, political, and social linkages. Some of the relationships between centers would be hierarchical, others would be on a peer level. Questions of metropolis and province are answered in the context of the specific centers under examination and the scale of the inquiry. On the most global scale, for example, it might make sense to speak of the United Kingdom as the center of the British scientific empire and lump together the commonwealth activities as peripheral. More detailed and careful investigations, however, would demand that distinctions be made among such British centers as Cambridge, Edinburgh, and the scientific societies of London. Nor could the differences and complex relationships among, for example, the universities in New Zealand, Australia, and Canada be ignored. The existence of many centers promoted genuine internationalism while fostering the illusion of internationalism by masking contrary actions and viewpoints that arose from particulars of each center.

In the papers that follow there is relatively little consideration of the differences among Western European centers of scientific activity, but much of the differences among non-Western Europeans. It quickly becomes evident that the diffusion of Western science to Australasia, North America, and South America did not mean that the institutional forms and scientific content of Western science were simply transplanted. Even the most successful of the new centers—the United States, the Soviet Union, and Japan—did not simply become sites for new Göttingens, Parises, or Cambridges. The new communities of investigators did not merely replicate Western European science. Nor was there an effort at homogenization. Rather, in both large ways and small, the particular political, intellectual, social, and economic environments acted upon institutions and scientists, molding them into different, albeit related forms. Understanding how and why some elements of the Western European scientific culture were transformed, while others were transferred essentially unchanged, is essential to an understanding of the actual process of cultural diffusion.

1. "The Spread of Western Science," *Science*, n.s. 156 (5 May 1967): 611–22.

The Australian Experience

The Beginnings of
an Australian Physics
Community

R. W. HOME

The discovery, settlement, and geographical exploration of Australia
by Europeans was accompanied by its scientific exploration, both by
naturalists intrigued by its unfamiliar flora and fauna and by geologists in search of both minerals and understanding. Later, as the European settlements grew, a few chemists also found employment—
in the mining industry, for example, in tanning, brewing, and the
like, and in monitoring food and water standards. Physics, however,
was a much more abstract science, for the most part much less publicly accessible and with much less popular appeal than the relatively straightforward observational and collecting activities that
dominated the life sciences in Australia throughout the nineteenth
century. At the same time, it lacked the obvious practical utility
that gave sciences such as geology and chemistry a place in the colonial scheme of things. Not until well into the twentieth century,
in fact, was physics perceived to have a potentially significant part
in the economic life of the nation. Before that, it was generally regarded as a purely intellectual discipline, an appropriate part of university work but equally appropriately confined to the universities,
where it was sometimes still labeled "natural philosophy" instead
of "physics," as if to emphasize the point.

A few exceptions to this generalization need, however, to be
noted. During the course of the nineteenth century, most of the
Australian states established observatories in their capital cities
and, in addition to purely astronomical work and the maintenance

of the local time service, these undertook work more directly connected with physics. They were, for example, charged with collecting meteorological data and issuing weather forecasts, a task that involved coordinating large numbers of at best semiskilled observers throughout the region for which they were responsible.

In addition, there were some specialized geomagnetic observatories, the first of which was established in Hobart in 1840 by Captain James Ross during his four-year exploration of the Antarctic region in the ships *Erebus* and *Terror*. This station, which was supported by the British Admiralty and the Royal Artillery, was set up as part of a worldwide chain of observatories whose results were transmitted to London for reduction by Edward Sabine and his staff at the Woolwich Arsenal. The system lasted only a few years, however, after which the Hobart observatory, along with most of the others, was closed. Envisaged from the beginning as a temporary institution and staffed for almost all its existence by a serving British naval officer who was recalled shortly before it was closed, it had little direct effect on indigenous Australian science.[1]

Whether, in fact, the Hobart geomagnetic observatory warrants a place at all in a history of Australian physics might at first seem debatable. In effect, apart from functioning as a supply base, Australia was, from the point of view of the organizers of the expedition, merely a convenient platform on which Ross could erect a fixed observing station. Its contribution was surely seen as analogous to that of St. Helena to Edmond Halley's seventeenth-century survey of the Southern skies or of the Cape of Good Hope to John Herschel's astronomical investigations. Australians were not expected to make any contribution to the scientific work of the Ross expedition or of the Hobart observatory in particular; the expedition brought with it all the scientific knowledge and skills it was expected to need. Certainly the arrival of the *Erebus* and *Terror* in the Derwent estuary gave a significant, though temporary, fillip to Hobart's intellectual life, and a number of local residents contributed to the observing program. When the observatory was closed again, however, and the bulk of its equipment shipped back to England, it left in Hobart little trace of scientific progeny. There simply was no local physics community whose work could have profited from the establishment of the observatory in Australia or which could have carried on the work once the initial British support was withdrawn; neither, for that matter, does there appear to have been much local interest in carrying on the work. A few of the astronomical and meteorological observations were continued on a voluntary basis from his own home by a local resident, Francis Abbott, who had been trained at the observatory. For the rest, the observatory was but a short-lived alien intrusion on the developing Australian intellectual landscape of Tasmania.

By a curious quirk of fate, however, Ross's observatory did have consequences for Australian physics. Inspired by its work and saddened by its discontinuance, the young German geophysicist Georg von Neumayer persuaded King Maximilian II of Bavaria to supply him with an extensive collection of magnetic and meteorological instruments with which to establish a new geophysical observatory in southeastern Australia. The first site granted him was on Flagstaff Hill in Melbourne, where he set up his instruments in 1857; later, the equipment was transferred to the new, permanent site of the Melbourne Observatory. The program of systematic observing of the earth's magnetic field that Neumayer established during his five years in Victoria became a regular part of the Melbourne Observatory's activities and has been continued to the present day, despite the closing of the observatory itself in 1944. Some of Neumayer's results were published in Australia, but most were brought out in Germany following his return to Europe.[2]

This history can be contrasted with that of another major geomagnetic observatory established, three quarters of a century later, on Australian soil. Again the initiative came from outside Australia, on this occasion from the United States, from the Department of Terrestrial Magnetism of the Carnegie Institution of Washington, in Washington, D.C. Once more the necessary staff and equipment were imported, so Australia's role in the establishment of the observatory was again merely that of a conveniently located fixed platform on which to set up the measuring instruments. Again the observations recorded were reduced and published elsewhere, in reports issued from time to time in Washington.[3] Once set up, however, the Carnegie Institution observatory, located at Watheroo, Western Australia, became integrated into the scientific life of the host country as Neumayer's had but in a way that the earlier Hobart observatory had never been. The imported staff developed close and friendly ties with the physics department of the by then established

University of Western Australia, through the long-serving professor there, A. D. Ross. Through the years Ross encouraged the people from Watheroo to participate in Perth's scientific life, to such good effect indeed that when in 1942 a separate Western Australian Division was organized within the Australian Branch of the Institute of Physics, its first chairman was Watheroo's then observer-in-charge, W. D. Parkinson. At the same time, a number of research students from Ross's department gained experience by working for a period at Watheroo. The importance of the observatory's work came to be widely enough understood within the Australian community that when the Carnegie Institution at last decided, in 1946, to withdraw its support, the Australian government was prepared to take over and was able, furthermore, to find suitable trained staff in Australia to continue the work.

The formal teaching of physics in Australia dates from the establishment, in the early 1850s, of the first two state universities, one in Sydney and one in Melbourne. In both, there was from the start a strong emphasis on the physical sciences. At Sydney, three professors were appointed initially, two of whom, Morris Birkbeck Pell and John Smith, were to teach the scientific disciplines. Both in the titles of their chairs and in their teaching, Pell and Smith reflected the traditional division between the mathematical and experimental parts of physics. Pell, senior wrangler and second Smith's prizeman at Cambridge in 1849, was appointed professor of mathematics and natural philosophy. His outstanding undergraduate record shows him to have had a first-rate mathematical mind. In Australia, however, he confined his mathematical skills for the most part to the classroom and to actuarial consultancy work. His few original papers dealt exclusively with actuarial matters. So far as natural philosophy was concerned he restricted himself largely to the classroom and to the traditionally mathematized branches of the subject, optics and astronomy, which he evidently taught, as did many at this time, as branches of mathematics. Smith's appointment, on the other hand, was as professor of chemistry and experimental physics. His principal expertise, however, and for many years his chief interest, was in chemistry, and he gave elementary lectures on experimental physics only every other year.

At Melbourne, two of the four professors appointed initially were appointed to scientific chairs. The division of responsibilities between them was somewhat different, however, from the Sydney pattern. Frederick McCoy, whose chair was designated as natural science, made the field sciences—geology, mineralogy, and natural history—his particular domain, together with some chemistry. W. P. Wilson was appointed professor of mathematics but in fact also took responsibility for all teaching of physics within the university. Like Pell, Wilson had had an outstanding undergraduate career at Cambridge. Senior wrangler and first Smith's prizeman in 1847, he had published *A Treatise on Dynamics* in 1850 and served as foundation professor of mathematics at Queen's College, Belfast, before taking up his Melbourne appointment. At Melbourne his lectures in natural philosophy extended over a two-year course and were "illustrated by models and experiments," on the apparatus for which he invested no less than £500 during his first year. Wilson was active in the scientific life of the infant colony in which he found himself. He read papers to the Royal Society of Victoria from time to time and served as a committeeman involved in the work of the Melbourne Observatory, his efforts and those of the government astronomer, R. L. J. Ellery, being crowned in 1868 with the commissioning of the Great Melbourne Telescope with its four-foot reflector. He also took part in 1871 in an important expedition to north Queensland organized by the Melbourne Observatory in order to observe a solar eclipse. So far as physics was concerned, however, Wilson's only contribution was his teaching.

For all four of Australia's first science professors and for a number of their immediate successors, their appointments meant lifelong exile from the scientific centers of Europe. The journey to Australia was still in those days arduous as well as long—it was, indeed, too much for Melbourne's first professor of classics, who never recovered from the voyage and expired shortly after his arrival—and was not to be undertaken lightly by anyone. Communication with England was slow and intellectual contacts accordingly difficult if not impossible to maintain. All four appointees were young and had yet to make names for themselves in the wider scientific world. Once in Australia, their chances of doing so were virtually nil—and so, too, therefore, were their chances of subsequently returning to university posts in Britain. All this they must have recognized at the time they accepted their appointments. Yet accept them they did, attracted no doubt by the very high salaries being offered, as well, perhaps, as by a spirit of adventure and the challenge of establishing an institution of higher learning in one of Her Majesty's most distant dominions.

So far as the biological sciences were concerned, Australian natural history offered its own peculiar attractions. In physics there were no such compensations. Yet the early physical science professors would not necessarily have seen the likelihood of soon falling out of touch with current research in their field as a serious drawback. In the English universities of the time, research was not yet seen as a normal part of a professor's function, and in this the new Australian institutions duly followed suit. Contributing to the general liberal education of undergraduates was what an Australian professor was paid to do, and any research he chose to undertake was regarded more as a hobby than anything else. Furthermore, whatever the merits of the Cambridge mathematical tripos examinations, one thing they did not do was instill a research ethos in the students subjected to them. Memory, speed, and accuracy were tested, not originality. Outstanding graduates such as Pell and Wilson emerged from the system well trained for careers as mathematics teachers, but not trained at all to pursue physical investigations on their own. They undoubtedly saw their Australian chairs as precisely the kind of job for which they had been trained, and they filled them competently.

The same may be said of Horace Lamb and W. H. Bragg, the first two professors of mathematics appointed by the University of Adelaide following its establishment in the 1870s, despite the eminence to which they subsequently rose as research physicists. Bragg in particular fits the pattern very well. Appointed to the Adelaide chair in 1885, not long after finishing as third wrangler in the Cambridge tripos examinations, he had had no training in physics before the reading he did on board ship en route to take up his new position other than his two terms of experimental classes in the Cavendish Laboratory following his graduation. For a period of nineteen years thereafter he taught his classes conscientiously without ever undertaking a piece of major research. His eventually becoming involved in the investigations into radioactivity that were to carry him to the Nobel Prize was a new departure for him. On the other hand, Bragg's predecessor, Lamb, spent several years in Cambridge after completing the tripos examinations, and during that period he lectured and read widely on current research in mathematical physics. He therefore arrived in Australia much better equipped than the others to undertake research and much more accustomed to thinking about doing so. Furthermore, he had a mind to take on such work. During his ten years in Adelaide he devoted most of his spare time to writing a number of substantial papers on various topics in applied mathematics and also his well-known *Treatise on the Motion of Fluids* (1879). Significantly, Lamb's work was in a field in which neither equipment nor assistance was required. He had no research students, and he worked entirely on his own. Even he, however, felt the want of a suitably stocked library, as he indicated in the preface to his *Treatise:*

I have endeavoured throughout the book, to attribute to their proper authors the various steps in the development of the subject. The list of Memoirs and Treatises at the end of the book has no pretensions to completeness, and as it is to a great extent based on MS notes which I have no present means of verifying, some of the references may possibly be inexact.

For many years, the numbers of students attending university in Australia, in all courses, remained minuscule, and whatever physics was taught in either Sydney, Melbourne, or Adelaide was taught as part of a general program of liberal education leading to the B.A. degree—or, at Melbourne, as part of the country's only medical and engineering courses. Not until separate B.Sc. programs were established at all three universities during the 1880s did the opportunity arise for specialized training in the sciences. Still, however, the numbers undertaking advanced courses in physics remained small, and it was not unusual for Bragg in Adelaide to have no students at all in his final-year class.[4] T. R. Lyle at Melbourne had the largest first-year enrollment on account of the presence of the medical students and also students taking courses in civil and mechanical engineering. Even so, during the 1890s his first-year numbers averaged only 126, dropping to 25–30 in the second year, most of these being engineering students, and a mere 5–6 in the third year.[5] Only with the introduction at around the turn of the century of the first courses in electrical engineering did the numbers of students taking advanced physics courses begin to rise significantly. Shortly before this the first graduate students appeared, who would typically spend an additional year or perhaps two at the university doing research after completing their bachelor's degrees before either finding employment or going overseas in search of further research experience. Until the turn of the century, in fact, those actively engaged in physics in Australia can almost be counted off one by one. There were the professors in the four universities (the University of Tasmania having been added to the list in 1890) and, toward the end of the period, occasional assistant lecturers or demonstrators and a few research students. In addition, there were the official government astronomers in Sydney, Melbourne, Adelaide, and Perth, each of whom was still also responsible for his state's meteorological service, and government meteorologists in Queensland and Tasmania. There were also a number of enthusiastic amateur astronomers and meteorological observers scattered across the country. The Astronomical Society of South Australia was formed in Adelaide in 1892, and branches of the British Astronomical Association in Sydney and Melbourne soon afterward, the latter of which, however, survived for only a brief period. State survey departments also did some routine astronomical work.[6] A few telegraph engineers, largely self-taught, carried their interests in electricity beyond their places of employment, in the case of William Filmer of Maitland, N.S.W., to such good effect that he was one of the first people in Australia to generate X rays successfully, early in 1896. Another early X-ray investigator was a school teacher in Bathurst, N.S.W., Father Slattery.[7]

Such a small and disparate group cannot be seen as constituting even the beginnings of an Australian physics community. Indeed, it cannot even properly be described as a group, since hundreds of miles of territory separated one state capital from the next, and in physics as in so many other things, those living in the different states long remained in almost total isolation from their fellows elsewhere in the country. At least until the formation of the Australasian Association for the Advancement of Science (AAAS) in 1888, their links, tenuous as these were, with physicists "at home" in Britain were probably closer than any they maintained with others elsewhere in the Australian colonies. After 1888, the AAAS meetings provided opportunities that were eagerly grasped for discussions with other Australian physicists, but the Association met, on the average, only every second year, and between meetings the tiny collections of individuals in each state were once more thrown back almost entirely upon their own resources.

The largest group was in Melbourne, where even physics had been able to profit from the economic boom of the 1880s. The Melbourne Observatory, directed by Ellery until his retirement in 1895 and then by Pietro Baracchi, was by far the best equipped and staffed in the country.[8] At the university, a splendid new set of laboratories was presided over by Lyle, a former student of G. F. FitzGerald's at Trinity College, Dublin, and professor of natural philosophy since 1889. Lyle's chair was not a continuation of Wilson's under another name, but had been newly created earlier in the same decade. The chair of mathematics continued as a separate position, occupied throughout this period by E. J. Nanson. Mathematics was also strongly represented in one of the residential colleges attached to the university, Ormond College, where the first master, J. H. MacFarland, himself a Cambridge graduate and competent mathematician, built up a fine library of mathematical books and a strong tutorial program.[9] As part of the latter, the college employed Alexander McAulay as a lecturer in mathematics for several years around 1890, during which time, despite his isolation from Europe, he managed to play a significant role in the creation of the vector calculus.[10] Subsequently, in 1893, McAulay was appointed the first professor of mathematics at the infant University of Tasmania. Another college, Trinity, also had a mathematics lecturer on its staff in most years. Also in Melbourne at this time, living frugally on earnings from occasional part-time employment which sometimes included lecturing at the university, was William Sutherland, whose publications on molecular physics were sufficiently important to earn him a place in the *Dictionary of Scientific Biography*, one of the few Australian scientists to attain that honor.[11] Another Cambridge graduate, E. F. J. Love, held a continuing appointment at the University as Lyle's assistant lecturer.

Lyle and his opposite number at Sydney University, Richard Threlfall, who had previously been J. J. Thomson's demonstrator at the Cavendish Laboratory, were the first Australian physics professors to enjoy the luxury of having research students working under them. Threlfall was the most experienced of the Australian physics professors. He, too, soon had a splendid new building erected, and he equipped it superbly. He and later Lyle were much more active re-

searchers than their predecessors had been, presenting original papers on physical subjects from time to time to their local Royal
Societies or at AAAS meetings and usually sending copies simultaneously to London for publication in one or another of the journals
there. Threlfall, however, returned to England in 1898, to be succeeded by the competent but less distinguished J. A. Pollock. Thereafter, except for the brief period of Bragg's flowering at Adelaide in
the first decade of this century, Melbourne was to remain the paramount center for physics in Australia until at least the onset of
World War II.

Two other developments during the 1890s were to prove of lasting
significance for Australian physics. As if to emphasize the entirely
colonial standing of Australian science at the time, both occurred in
England rather than in Australia. First, in 1891 the Royal Commissioners for the Exhibition of 1851 established their celebrated system of science research scholarships, "devised to enable selected
students from University institutions throughout the Empire, who
had given 'distinct evidence of capacity for original research' to continue their work for two or three years elsewhere in leading laboratories in this country or abroad."[12] These scholarships were highly
unusual for that period in that they were not open to students who
wished to work for ordinary university degrees, but were awarded
purely for research. Australian students were not slow to take advantage of them, and as a result a steady flow set in of the best
science graduates of the Australian universities, including of course
a number who majored in physics, toward England.

Second, in 1896 the University of Cambridge instituted a new
degree of B.A. by Research and thus gave official status to a new
category of student, the research student.[13] Physics students in particular took advantage of the new regulation and flocked to the Cavendish Laboratory to work under the direction of J. J. Thomson and
his staff. Most notably, the 1851 Exhibition scholars who wished to
do research in physics almost invariably went there.[14] These two
innovations, taken together, greatly increased the opportunities for
the very best Australian physics graduates to become engaged in research and also considerably strengthened the already close connections between Australian physics and Cambridge. In the event,
these ties were to grow even stronger in the years that followed.

A few years later, of course, the Rhodes scholarship scheme was also instituted, which provided six scholarships each year, one for each state, to take outstanding Australian students to England. The Rhodes awards never placed the same emphasis on research, however, as did the 1851 Exhibition awards. Furthermore, Rhodes scholars were required to go to Oxford, and at Oxford physics was not strong during this period. For a would-be physicist, therefore, a Rhodes scholarship was very much second best, so the Rhodes scheme plays a much less significant part in our story than the 1851s. From the point of view of physics graduates, in fact, the various endowed traveling scholarships that began to be awarded by the Australian universities at about this same period were also more important than the Rhodes. Such awards enabled two Sydney graduates, for example, E. M. Wellisch and S. G. Lusby, to go to the Cavendish before 1910, and several others followed in their footsteps later.[15]

The pattern thus established persisted, with only minor variations—something of a tendency to go to Manchester to work with Rutherford, for example, or to Leeds to work with Bragg following his return to England, instead of going to the Cavendish—until the outbreak of war in 1914. Not all the outstanding students went to England, of course, and some of those who remained at home subsequently became important figures in the Australian scientific community.[16] All those who could go, however, did. Many, including some of the best of them, never returned.

To sum up: During the period until 1914, physics in Australia was largely a university-based discipline. Within the six widely separated universities—the Universities of Queensland and Western Australia having been added to the list in 1910 and 1911, respectively—the subject was taught largely as a service course for medical and, particularly, engineering students and for students majoring in the other sciences. Only a handful of students completed a major in physics, and an even smaller number proceeded to Honours. Apart from secondary science teaching, there were virtually no jobs outside the universities for which a background in physics would have been specified as a prerequisite, except perhaps that of assistant at one of the larger state observatories. Inside the universities, vacancies appeared only infrequently. Teaching loads were heavy. Most professors did some research, though all except Threlfall found it hard to do so regularly. They published their papers locally and also sent the best of them to England, usually to the *Philosophical Magazine*, for parallel publication there, and they encouraged their occasional research students to do likewise. The most productive professors were rewarded in due course with an F.R.S. There was never any suggestion that a professor in Australia might build up a research school: on the contrary, the best of his Honours students were expected to proceed to England, usually to the Cavendish, to pursue their research there.

Bragg's last five years at Adelaide—he left early in 1909—show that it was not impossible under these conditions to do first-rate work. Bragg's mechanic proved perfectly capable of producing the quite complex pieces of apparatus he required. Obtaining the use of some radium was at first a difficulty but was soon overcome. So, too, was a later problem concerning the availability of satisfactory samples of very thin metal foils. Delays in communication with England were a nuisance. The advent of regular steamer services had reduced these considerably, however, and they no longer loomed as an insuperable problem. The local library evidently subscribed to the leading French and German periodicals as well as to those published in England, for Bragg regularly referred to them in his own papers. As collaborators, Bragg recruited such Honours students as he had—two of whom, R. D. Kleeman and J. L. Glasson, were awarded 1851 Exhibition scholarships on the basis of their contributions to the work, while the third, H. J. Priest, subsequently took up a lectureship at the University of Queensland—and also his colleagues, E. H. Rennie and W. T. Cooke from the university's chemistry department and J. P. V. Madsen, the lecturer in electrical engineering. It was, however, for the sake of his research that Bragg eventually decided to return to England, chiefly, it appears, because he felt intellectually isolated in Adelaide, so far from the main centers of activity in his field. Bragg at first intended to go to McGill University in Montreal rather than to Leeds; apparently he did not regard Canada as suffering from the same kind of isolation from the rest of the world as did Australia. Furthermore, it seems that it took a Bragg to cope with the problem of isolation in the way that he did while still in Adelaide, because once he left, physics research there virtually ceased. Not even Madsen, who under Bragg's stimulating guidance had done some highly significant work on the scattering of β and γ rays, was able to continue successfully on his own.[17]

In a number of respects, therefore, Australian physics before World War I was paradigmatically colonial, dependent on Britain for the higher training of its best graduates and for the validation of its work through publication in London. There was never any suggestion that material should be sent elsewhere in the first instance than to Britain for publication, and neither did Australian graduates in physics—unlike their colleagues in chemistry—ever follow the American pattern of going to Germany rather than Britain for their research experience. On the other hand, the discipline in Australia was gradually becoming self-sustaining as more and more positions were filled, not by expatriate Britishers, but by Australians who had had at least their basic training in Australia.

Unlike physicists in the United States, however, Australian physicists were not consciously building up an independent national scientific base.[18] On the contrary, the feeling was very strong among them, as among the bulk of the Australian population until relatively recent times, that Australia was a part of a unified imperial system under the British crown. There was little or no feeling among physicists of "us" in Australia as against "them" in England: rather, it was simply taken for granted that "we" were all part of a single cultural and political system. We can see today, for example, that while the 1851 Exhibition scholarship scheme certainly fostered scientific research in Britain, it also served systematically to siphon off some of the best Australian scientific talent at its initial flowering and thus may actually have inhibited the development of an indigenous program of scientific research. Such thinking would have been entirely alien to an earlier generation of Australians. It was long widely, though not of course universally, accepted that Australia should rest content with its traditional function in the imperial economy of exporting food and wool and importing the manufactured goods it needed from England. Just so, it was generally agreed that England was the appropriate place within the Empire for doing the bulk of the empire's physical research—whereas Australia was perhaps a more appropriate place to undertake certain classes of agricultural research. The 1851 Exhibition scholarship scheme was therefore not seen as exploitative. It could not weaken Australian science, since "Australian science" was not a perceived categorial possibility. "British science" benefited from the scheme and therefore, as a matter of course, it was in Australia's interest.

The First World War marks a watershed in the history of Australian physics as in so much else of Australian life. While hostilities lasted, significant Australian work in physics ceased. As many Australian physicists as could do so went to England to put their talents directly to work for the nation's—that is, the Empire's—benefit. A few remained at home, charged with the task of maintaining some kind of scientific training in the universities. With the signing of the peace, however, things did not return to their prewar state. For a time the numbers of physicists remained small, partly, as A. D. Ross noted, because during the war few students had proceeded to Honours, and partly because of "many qualified persons who went to England during the war on special service accepting permanent posts there."[19] Thereafter, however, the numbers began slowly but steadily to grow, and for the first time those involved formed themselves into a national organization. The center of gravity of these developments lay in Melbourne, where the university's Department of Natural Philosophy under its new professor, Thomas Howell Laby, developed into a much larger and more securely established center for physical research than Australia had ever seen before.

Laby had been one of the first Australian students to go to the Cavendish Laboratory as an 1851 Exhibition scholar. He was also one of the rather smaller number who returned to the antipodes, at first to the chair of physics at Victoria University College in Wellington, New Zealand, and then, in 1915 to the Melbourne chair in succession to Lyle. At the Cavendish, Laby had become immersed in the leading problem of early twentieth-century experimental physics, the study of the new ionizing radiations and the ionization they produced. He had published several useful papers that arose out of his researches, and he had also become active in the social fabric of the laboratory; he was, for example, secretary of the organizing committee for the celebrations that marked the twenty-fifty anniversary of J. J. Thomson's accession to the directorship. The contacts and friendships Laby thus established later contributed greatly, as we shall see, to the success of the Melbourne School under his direction. While at the Cavendish Laby also began the collaboration with G. W. C. Kaye that led to the publication in 1911 of the first edition of the well known Kaye and Laby *Tables of Physical and Chemical Constants and Some Mathematical Functions*. Later editions of this work were to follow at regular intervals through the years, the tenth and final edition prepared during the lifetime of either author being published posthumously in 1947.

Laby's research interests carried the Melbourne department into what were for it, quite new directions, especially into work on X rays, on precision measurement, and later on the propagation of radio waves in the atmosphere. To all these areas he brought a spirit of critical exactitude that earned international respect for the work of the department. The extent to which he emphasized research in the training of physicists was new to Australia. "My experience with these men," he wrote to A. D. Ross in 1926, "makes me consider that experience of research work and of thesis writing is an essential part of the training of anyone who calls himself a physicist. Lectures and examinations do not train originality of mind, independence of thought, capacity to take responsibility, or (in Australia) the candidate's power of expression."[20] To a much greater extent than his colleagues in the other state universities, he was careful to build up particular areas of research concentration in his department and to channel his research students into these. In addition, he made it clear to them that they were expected to publish the fruits of their work, and he was highly critical of colleagues elsewhere who did not do likewise: "There are already too many temptations in Australia", he complained, "to men merely to do work which interests without bringing it to a conclusion and in a finished condition for publication."[21]

During the 1920s, Laby was able gradually to expand the lecturing staff of his department, and he encouraged his lecturers, too, to keep up their research. Several of these men were former students of his who had subsequently spent some time at the Cavendish Laboratory, and during the 1930s this pattern continued. Extraordinarily close links developed, in fact, between Laby's department and the Cavendish—by now under Rutherford's supervision—during this period. Laby and Rutherford had come to know each other during Laby's own period at the Cavendish, and through the years their acquaintance ripened into warm friendship. The two men corresponded regularly, and whenever Laby visited England (which he did regularly), he and his family spent some time at the Rutherfords'.[22] Likewise, Rutherford stayed with the Labys when he visited Melbourne. The emphasis on research in Laby's department left his students exceptionally well placed in the competition for 1851 Exhibition scholarships, and no fewer than twelve of them gained the award during the interwar period. All went to the Cavendish, as did Laby's favorite student H. S. W. Massey, who took up a more remunerative Melbourne University traveling scholarship instead of the 1851. Rutherford trusted Laby's judgment and always found a place in his laboratory for any student he recommended. Laby, for his part, was evidently careful to recommend only those students in whom he had complete confidence: the others had to rest content with such experience as they could acquire in Australia. When it came to filling a lectureship, only a Cavendish man would do, and Laby relied heavily on Rutherford's advice in making his choice.[23]

Laby was never content to rely solely on university funds to support his department's activities. In 1918, for example, when H. J. Grayson died and Lyle, retired from his chair but still active, bought Grayson's superb ruling engine from his widow, Laby assisted Lyle in finding laboratory space at the university for the equipment and obtained a research fellowship which, supplemented by Lyle's own funds, enabled him to employ an assistant, Z. A. Merfield, to maintain and develop it. Though Merfield's work was later clouded by controversy, he did during the 1920s produce a number of extremely high-quality ruled gratings with Grayson's machine, some of which were used by Laby's students in their investigation of X rays.[24] Again, Laby's advice was sought when the Australian government decided in 1926 to buy a large quantity of radium for medical purposes, and when, in due course, the Commonwealth Radium Laboratory (later the Commonwealth X-ray and Radium Laboratory) was established, it was located in the Melbourne University grounds and for many years worked in close association with Laby's department.[25] Similarly, Laby was active in the establishment of radio research in Australia. In 1926 he persuaded the Broadcasting Company of Australia to fund a three-year research fellowship in his department and when, soon afterward, an Australian Radio Research Board was set up under the Council of Scientific and Industrial Research (CSIR), the Melbourne department was chosen as one of its two research bases, the other being Madsen's Department of Electrical Engineering at the University of Sydney. Throughout the 1930s, radio research was an important part of the department's work in which a number of postgraduate students became engaged and to which a number of men who subsequently became leaders in

Australian physics contributed, including D. F. Martyn and J. L. Pawsey.[26]

Melbourne's preeminence was not based just on the university and its associated units. The Melbourne Observatory, now directed by J. M. Baldwin, remained the major astronomical institution in the country, though the new Commonwealth Solar Observatory on Mt. Stromlo, near Canberra, grew steadily in importance as the states handed over more and more of their traditional astronomical work to it.[27] In addition, both the research laboratories of the Postmaster General's Department and the Research Division of the Commonwealth Meteorological Service were based in Melbourne.[28] So, too, at Maribyrnong, were the Defence Department's Munitions Supply Laboratories, long the chief center for metrological work in Australia. Small wonder, then, that when A. D. Ross was organizing an Australian Branch of the Institute of Physics, he was convinced that "the majority of the Corporate Members is likely to be always in Victoria."[29] This must, in fact, have appeared inevitable to most Australian physicists at that time.

Yet, curiously, it was just at this period that the first stirrings of new life were being felt in Sydney, to such effect that by the early years of World War II, there were as many physicists active there as there were in Melbourne. Their distribution, however, was rather different. The university's physics department, though about as large as Melbourne's, did not place so much emphasis on research, and its output of published work was much smaller. Furthermore, the Sydney Observatory, after stagnating for several years, underwent a drastic reduction in 1926 at the hands of the New South Wales government and became an institution of negligible scientific significance.[30] The new vigor appeared elsewhere and was chiefly connected with the burgeoning field of radio research. Madsen, professor of electrical engineering at the University, was chairman of the Radio Research Board and a driving force behind its work. A flourishing research group, attached to his department, soon became established and began to specialize in problems in ionospheric physics. D. F. Martyn became the leading figure in the group following his transfer from Melbourne in 1932, and his work, some of it done in collaboration with V. A. Bailey of the university's physics department, was of great significance in the development of the field. Simultaneously, a substantial research group was built up at the Syd-

ney headquarters of Amalgamated Wireless (Australasia) Ltd. (AWA),
Australia's principal manufacturer of electrical and electronic components, 50 percent of which was owned by the Australian government until its sale by the Menzies government during the 1950s.
Many of the AWA research staff had their introduction to radio research as postgraduate students working on problems initiated by
the Radio Research Board. At the outbreak of war in 1939, several of
them were recruited, along with Radio Research Board personnel, to
work on a highly secret project, namely radar, in a new Radiophysics
Laboratory established in the grounds of the University of Sydney.
Others stayed at AWA to oversee the company's contribution to the
war effort.[31]

Two other developments at Sydney during the interwar period
need to be mentioned. First, a Cancer Research Committee, which
sponsored research into, among other things, X-ray and radium therapy and which thus helped to promote a certain amount of physical
research, was set up within the university in 1921. Unfortunately,
though generously funded, the committee and its work became em-
broiled in controversy and scandal, and, compared to the amount of
money invested, little of any value was achieved.[32] Second, and
much more significant, in 1939 the CSIR established, again within
the grounds of the University of Sydney, a National Standards Laboratory that quickly became one of the country's principal employers
of physicists. The laboratory, which marked CSIR's first venture into
research that would benefit manufacturers rather than primary producers, filled a long-felt need. There had been many calls for the
creation of such a body, and its formation had been expressly provided for in the 1926 act establishing the CSIR. Thereafter, however,
planning proceeded slowly and was then overtaken by the onset of
the financial depression. The driving force behind the revival of the
idea in the mid 1930s was G. A. Julius, chairman of CSIR, but Madsen was again deeply involved and supervised the initial construction and staffing of the laboratory once the decision was made to go
ahead with it.[33]

Sydney thus became an important center of Australian physics
rather later than Melbourne, and its rise was closely connected with
the growing importance of physics to Australian industry as this
became more sophisticated. With the AWA laboratory Sydney was

the home of what was probably the most technically advanced industrial undertaking in Australia. Melbourne's strength, on the other hand, remained firmly rooted in the more traditional areas of "pure" research, and its university department remained preeminent, "the only worthwhile school of physics in the Dominions," as Oliphant was later to describe it.[34] The Watheroo Observatory apart, in none of the other states was a significant amount of physics being done; such physics as there was remained confined to the universities and in them chiefly to teaching.

No longer, however, did the distance between the state capitals loom as so great a barrier as it did before 1914. Beginning soon after the end of World War I, and doubtless connected with the general rise in national pride and consciousness that the war engendered throughout the Australian community, Australian physicists began for perhaps the first time to see themselves as distinctively Australian as well as part of an Empire-wide community of British physicists. Not that they denied their Britishness—far from it—but they now also perceived that they had something important in common, as Australians, with their fellows in the other Australian states, and they began to organize themselves accordingly.

Laby, Empire-minded though he was, was also perhaps the first to conceive of a national organization of Australian physicists.[35] He evidently wrote to Rutherford in 1920 or early 1921 to tell him what he had in mind, but did not succeed at that time in taking the idea any further.[36] He revived it three years later, however, when A. D. Ross sought his views, advocating to Ross the formation of an Australian Physical Society, at meetings of which papers on physical subjects could be presented and which might also sponsor a journal in which to publish such contributions.

Ross, though, had different ideas. He believed that if such a society were formed, except when its meetings were held in conjunction with an AAAS meeting, "they will be attended merely by local members, and the meetings will convert the organization into a State society."[37] Furthermore, he thought there were still far too few physicists in Australia to sustain an organization of that kind. Finally, he was very much opposed to creating another new and inevitably minor (not to mention expensive) journal: "I deprecate the idea of having any more journals in which physical papers can be hidden away. If a journal is to be issued we should require so large a subscription that we should frighten away all the juniors on whom we must be dependent for a considerable share of the work."[38]

Ross's alternative was to make use of the recently formed London-based Institute of Physics, taking advantage of a clause in the rules of that body that permitted the formation of local branches of the institute in different parts of the country by setting up a local Australian Branch: In part, no doubt, Ross's preference was yet another manifestation of imperial sentiment in Australian public life, something perhaps to be expected of someone who had himself emigrated from Britain only a few years earlier to take up his chair.[39] He also, however, was able to marshal strong objective arguments to support his case. Furthermore, he clearly regarded the arrangement he was suggesting as merely temporary: "I am quite in agreement," he wrote, "with the idea of having local institutions in Australia when the time is ripe, but it appears to me that the profession of the physicist in Australia is still to[o] young to stand by itself, and that we shall need all the help we can get from home to secure recognition for Physics."[40]

What Ross had in mind, therefore, was not at all another scientific society for the hearing of learned papers, but a professional organization devoted to improving the status of the physicist in Australia. He advocated a British connection not for its own sake but for the prestige which would then accrue to the Australian offspring in an environment where the "colonial cringe" was still very much in evidence: "Connection with the British Institute," he wrote, "will insure standing and encouragement which would be lacking in a society having no such connection."[41] In other words, by attaching themselves to the British Institute's coattails, Australian physicists could gain a level of community respect that would not be forthcoming for a purely local organization. The pattern is a familiar one and is still to some extent observed today. Furthermore, the British connection could conveniently be used to justify setting higher standards for admission to the profession of physics than could perhaps otherwise plausibly be maintained in a community still at that time largely dominated by egalitarian sympathies and in which the number of people who would initially be qualified to join would be small indeed.[42]

Not content with mere advocacy of an idea, Ross wrote to the

institute in London to seek authority to act as its Australian organizer, at the same time that he sought the views of those in charge
of the teaching of physics in the various state universities.[43] The
officers of the institute were somewhat startled by Ross's audacity
in proposing the formation of an Australian branch at a time when
there were only three members in the whole of the country—
namely, Laby, Ross, and, as an associate, N. A. Esserman from the
Defence Department laboratories at Maribyrnong. They were happy,
however, to encourage Ross's efforts to expand the Australian membership and agreed to grant him semiofficial status as the institute's
"Hon. Local Secretary for Australia."

Except for Laby, who preferred his own scheme, the other Australian professors gave cautious approval to Ross's proposal, while an
enthusiastic A. L. McAulay, lecturer-in-charge at the University of
Tasmania, applied at once to join the institute. So, too, did Edward
Kidson, chief of the Research Division of the Commonwealth Meteorological Service, a New Zealander whom Ross had got to know
some years earlier as observer-in-charge at Watheroo and on whose
encouragement and advice Ross soon came to rely quite heavily.

Ross was a persuasive letter writer, and during the months that
followed he persuaded a number of others to apply for membership.
He was also an adroit manipulator of the institute's regulations.
Seizing upon the requirement that applicants for membership be
sponsored by at least two people who were already members, he
quickly established a de facto procedure whereby all Australian applications were circulated for approval to all fellows of the institute
then resident in Australia—which for some considerable time
meant, on account of the departure of Laby and McAulay temporarily for England and long delays in London in processing other applications, just himself and Kidson. Then, when the number of Australian fellows began to increase, he persuaded London to formalize the
arrangement by authorizing the establishment of a small Australian
committee, with himself as secretary, to advise on all future Australian applications.

Ross was also true to his word and seized every opportunity to use
his position and the institute's name to advance the status of physics in the country. An early occasion arose in 1925 when a position
in Kidson's department in the Meteorological Service was filled by
an untrained person promoted from within the service rather than

by someone with an adequate background in physics. Kidson was enraged and sought Ross's help, who promptly lobbied members of the government and also Sir Brudenell White, chairman of the Commonwealth Public Service Board. Though when it came White's reply was not completely satisfactory, it did at least go some way toward recognizing special qualifications for such positions in the future.

Ross called meetings of the Australian Corporate Members—that is, fellows and associates—of the institute to coincide with the AAAS meetings at Adelaide in 1924 and at Perth in 1926. Several matters of concern at the second of these gatherings illustrate the growing corporate feeling among Australian physicists that was being engendered by Ross's never-ending flow of letters. First, those present welcomed the news that the University of Tasmania had advertised a chair of physics and agreed in the light of this that a letter should be sent to the University of Western Australia to point out that it was now the only Australian university that did not have a separate chair of physics—Ross's appointment was as professor of mathematics and physics—and urging it, successfully in this instance, to create one.

Second, there was a lengthy discussion at the meeting concerning the comparative standing of degrees in physics from the various Australian universities. Some Australian degrees had already been recognized by the board of the institute in London as a prima facie qualification for membership, before ever Ross began his letter-writing campaign, and one of Ross's first successes had been to persuade the institute to recognize degrees from other Australian universities as well. The existence of his committee, however, and its duty to make recommendations to London on applications from all parts of the country soon forced those concerned to compare the training provided for physicists in the various states. In other words, here, too, we find a growing concern with professional questions, indeed an explicit awareness for perhaps the first time by a body of Australian physicists that physics constituted a profession whose standards needed to be carefully guarded. The particular point at issue at the 1926 meeting was McAulay's success during his visit to England in persuading the London board to recognize his university's three-year pass degree, with a major sequence in physics, as an appropriate qualification for membership in the institute. Ross was furious, since his committee had in the meantime, he thought with McAulay's concurrence, settled on a four-year honours degree as the standard Australian qualification that should be recognized. Melbourne, however, was an exception here; its more specialized three-year honours degree was, it was agreed, acceptable. As a result of the board's action, Ross argued, three-year pass degrees from the other Australian universities would also have to be recognized. Yet this would in his opinion lower the standing of the fellowship of the institute to an unacceptable level. Far better, he thought, for the board to reverse its decision on the Tasmanian degree. The meeting agreed; the Melbourne members even offered to waive the recognition of their own three-year honours degree rather than see the recognition of the Tasmanian degree maintained. With one voice, therefore, Australia's physicists urged McAulay to withdraw his three-year degree from the institute's list, and he quickly fell into line. The incident, while thus happily resolved, also gave Ross an opportunity to enhance the standing of his committee by pressing upon the London board the notion "that the Board should, before arriving at a decision, refer all purely Australian matters to the Australian Local Committee for report."[44]

Finally, the meeting resolved that a formal protest should be lodged with the powers that be over the absence of a physicist from the just announced Council for Scientific and Industrial Research. In this instance, however, they were less successful, despite Ross's lobbying even the Prime Minister, S. M. Bruce, on the matter. A. C. D. Rivett wrote a soothing reply on behalf of the CSIR executive, stressing his sympathy with the physicists' concern but pointing out that "for the present the need for a physicist is not so great as it will later become, because our work for some time will be almost solely on the side of primary industry."[45] In fact the difficulty appears to have been one of personality rather than principle, of a kind that was to recur on a number of occasions in the years that followed. The problem was simple. Laby was generally acknowledged to be the most distinguished physicist in the country and the obvious person to represent the discipline on bodies such as the CSIR. At the same time, he was a notoriously difficult man to work with on committees, and on this occasion he appears to have talked his way off the council before ever being appointed to it. As Ross reported to V. A. Bailey, who had been unable to get to the Perth meeting, "Rivett . . . tried to get the Executive of the Council to put Laby on the Council, and got them to meet him, but I believe that Laby rather tried to push some pet schemes of his with them and so got their backs up."[46] A letter from Laby to Ross in the same file reveals what these pet schemes were. Concerning the CSIR Laby wrote, "I should have liked to have seen the Section represent to the Commonwealth Government that if Australian manufacturers are to be efficient then research, standardising, and testing work in physics needs to be done in the Universities. I would suggest a joint memorandum to this effect should be prepared."[47] Such an idea, while sensible in itself and also one that would have appealed both to the Chairman, Julius, and to Rivett, ran quite counter to the emphasis on research for primary industry on which the CSIR's political support was based. Presumably it was Laby's enthusiasm for pursuing his scheme at once, despite the political pitfalls, that frightened off his listeners and led to his exclusion from the council. Yet the small size of the Australian physics community and Laby's standing in it were such that if he were not appointed, neither could any other physicist be appointed instead; and so the discipline went unrepresented altogether on the council for some considerable time.[48]

The number of professionally qualified physicists in Australia in fact remained very small. Despite Ross's best efforts, by mid 1926 the Australian membership of the institute had risen to only seventeen, and he could see only about another half dozen potential recruits "and probably a few other younger men" in the entire country.[49] Such figures surely vindicate Ross's judgment that Laby's proposal for an Australian Physical Society was premature. In fact, not until Oliphant's triumphal return to Australia in 1949 as head of the School of Physics in the newly established Australian National University was the idea of an Australian Physical Society again seriously canvassed, and even then it drew little support and came to nothing. In the meantime, attempts were made to meet Laby's objective at least in part through the formation of a separate section for physics and mathematics in some of the State Royal Societies. Only in New South Wales, however, did even this succeed; by early 1927, Kidson in Melbourne was complaining that "one cannot, somehow, get enthusiasm and cooperation in this city,"[50] as the proposed section within the Royal Society of Victoria sputtered to a halt despite the greater number of physicists in that state.

Somewhat to Ross's surprise, however, the infant and not yet formally constituted Australian branch that he had brought into being itself began to take on some of the trappings of a physical society, in a manner very different from anything envisaged by the institute itself in England.[51] At first, meetings of the branch were held only in conjunction with the regular AAAS meetings and were then confined strictly to branch business and matters of professional concern. Beginning in 1928, however, the branch also began to sponsor conferences at which research reports were presented and discussed.

The first such conference was held in Canberra in August 1928 following a decision made by the corporate members of the Institute of Physics at their meeting during the AAAS conference in Hobart earlier that year. The host was W. G. Duffield, director of the Mt. Stromlo Observatory. Thirty-four people attended, and another eight, though unable to attend, contributed papers. The meeting was most successful, a highlight being a discussion of the new quantum mechanics introduced by Mrs. G. H. Briggs, who with her husband, a lecturer in physics at Sydney University, had returned to Australia not long before after spending an extended period in Cambridge, and H. S. W. Massey, who was at the time in the throes of preparing an outstanding M.Sc. thesis on the quantum theory at Melbourne University. For many of those attending, the discussion would have constituted their first real exposure to the power of the new physics. Another major discussion, introduced by Laby, dealt with "Radio Research in Australia." A third, "Geophysical Prospecting," was prompted by the activities of the Imperial Geophysical Experimental Survey, then in full swing in Australia.[52]

A second conference, held in Melbourne twelve months later, was attended by fifty-two people. Once again a session was devoted to geophysical prospecting, and several papers dealt with radio research and ionospherics. Massey and his fellow Melbourne M.Sc. student C. B. O. Mohr were, however, the only ones to contribute papers on quantum mechanical topics.[53] At this meeting it was decided thereafter to hold similar conferences in approximately alternate years, between AAAS meetings, and in fact four more were held before the outbreak of World War II, in 1931, 1933, 1936, and 1939. All were

Table 1.   Australian Corporate Members of the Institute of Physics, 1923–50

Year	Number	Year	Number	Year	Number	Year	Number
1923	3	1930	n.a.	1937	n.a.	1944	134
1924	7	1931	n.a.	1938	n.a.	1945	137
1925	n.a.	1932	21	1939	55	1946	143
1926	17	1933	20	1940	70	1947	158
1927	18	1934	21	1941	82	1948	197
1928	n.a.	1935	n.a.	1942	83	1949	204
1929	18	1936	30	1943	115	1950	246[a]

n.a. Not available.
a. Includes new membership grade of graduate.

highly successful, with many papers being presented and fifty or more people in attendance on each occasion. By the 1936 meeting, in fact, the number of papers being offered was such that it was necessary to give serious consideration to running parallel sections for at least part of the time, though in the end the problem was avoided on this occasion by a very tight timetable.[54]

These meetings undoubtedly helped consolidate the still informally constituted Australian branch of the London Institute as the authoritative voice of Australian physics. The number of Australian corporate members of the institute remained almost constant until the early 1930s, then slowly began to rise as the steadily increasing number of honours graduates from Melbourne and Sydney swelled the ranks. At Melbourne, 1928 was seen at the time as a vintage year, because no fewer than two students, Massey and Mohr, began work on the M.Sc. degree.[55] Within a few years after this, however, several new students were starting every year. Then with the establishment of the National Standards Laboratory in 1939 and another sudden demand for physicists occasioned by the outbreak of war soon afterwards, the numbers began to increase much more rapidly (see table 1).

The year 1939, in fact, marked another watershed in the history of Australian physics. I have mentioned already the establishment of the National Standards Laboratory and the secret Radiophysics Laboratory at the University of Sydney. In the same year, the CSIR established a Division of Aeronautics in Melbourne and, with war looming, the Defence Department began taking on new staff at the

Munitions Supply Laboratories at Maribyrnong as well. At about the same time, E. H. S. Burhop at Melbourne University took Australian physics research into a new era with the commissioning of a neutron generator and the successful completion of Australia's first experiments on the atomic nucleus. The year was important from an organizational point of view as well, for it saw the rather belated formal confirmation of the arrangements which had gradually evolved out of Ross's inspired letter writing during the preceding decade and a half. An Australian branch of the Institute of Physics was at last formally established with its own constitution and elected office-bearers and committee. With the onset of war the board in London delegated to the branch complete control of its own affairs, including, for the duration of hostilities, full power of decision on Australian applications for membership of the institute. It was to be a further twenty years, however, before an independent Australian Institute of Physics was launched. In the meantime the Australian physics community continued to cling firmly to Britain's coattails.

But that is another story. I shall bring my tale here to a close at the point we have now reached, with Australia's physicists girding themselves for a war in which their discipline was to come very much into its own, even in their small and distant corner of the world.

How does the development of Australian physics, as I have described it, compare with that elsewhere? Perhaps the most striking feature of the Australian experience is the very small number of people engaged in the discipline throughout the period I have discussed. So small are they that they make significant comparisons with what was happening in the United States, for example, where large numbers of physics Ph.D.s were being graduated each year, extremely problematic. Parallels which, on the basis of size alone, seem much more likely are with Canada, South Africa, and perhaps Argentina. Nevertheless, even in the United States, a lone physicist in a small-town university, for example, must have suffered from intellectual isolation. What counts here, surely, is not the absolute size of a country's population of physicists, but its distribution. In Melbourne in the 1890s, with half a dozen or so physicists active, there was probably more scope for stimulating discussion than in many cities in more populous parts of the globe. What was perhaps peculiar to the Australian (and New Zealand) environment was how difficult it was to get beyond that local half dozen to meet and exchange ideas with the wider world community of physicists.

Connected with but different from the isolation problem engendered by a small and scattered population is the difficulty that arises in these circumstances in establishing satisfactory career patterns for would-be practitioners of the discipline. Pyenson has identified this as a key problem in early twentieth-century Canadian physics,[56] and the same problem certainly existed in Australia. Except for the openings that appeared, with exceeding rarity, in the country's universities and state observatories, there were for a long time literally no openings whatever for physicists in Australia. In fact, the established career pattern was, as we have seen, for the aspiring physicist to go to Britain to practice his craft. Furthermore, given the nature of Australia's links with the mother country this was seen not as a nationally enervating brain drain, but as something inevitable and perfectly in accord with the natural order of things. Not until the crisis of World War II brought home to Australians just how remote they really were from Britain was this national attitude of mind to change significantly. Employment opportunities for physicists in Australia had begun to expand shortly before this, and thereafter they increased dramatically. So, too, somewhat more slowly, did the number of trained physicists available to fill them.

My final point concerns the way in which physics in Australia quickly became indigenous, more quickly perhaps than in comparable countries elsewhere. While roughly the first two generations of physics professors were imported from Britain, most appointments thereafter were of local men, albeit many of them had received their advanced training abroad, preferably at the Cavendish Laboratory. Only for a brief period after World War II, when there were many more positions open for physicists than qualified Australians to fill them, did Australia once more—but now in a different context—depend on other countries to meet its needs. At no stage was there any idea, as there appears to have been in Argentina, of building up the country's physics by staffing the universities with foreign experts on short-term appointments.[57] In the Australian universities, the fixed-term contract is a recent innovation indeed!

Research for this essay was supported by a grant under the Australian Research Grants Scheme. The author is indebted to the Australian Institute of Physics and the Basser Library, Australian Academy of Science, for allowing him to consult the Institute's archives, and to Misses Jean and Betty Laby for granting him access to their father's papers. He is also grateful to the Misses Laby and to Sir Mark Oliphant, Professor C. B. O. Mohr, Dr. Kenneth Rivett, and Mrs. Verna Rowbotham (née Ross) for allowing him, as appropriate, to include quotations from various documents. A shorter version of the essay appeared in *Historical Studies* 20 (1982–83): 383–400.

1. A. Savours and A McConnell, "The History of the Rossbank Observatory, Tasmania", *Annals of Science* 39 (1982): 527–64. The background to the establishment of the observatory and its effect on intellectual circles in Hobart are described by John Cawood, "The Magnetic Crusade: Science and Politics in Early Victorian Britain," *Isis* 70 (1979): 493–518, and Michael E. Hoare, "'All Things are Queer and Opposite': Scientific Societies in Tasmania in the 1840s," *Isis* 60 (1969): 198–209.

2. *Australian Dictionary of Biography*, s.v. "Neumayer"; G. von Neumayer, "Description and System of Working of the Flagstaff Observatory," *Transactions of the Philosophical Institute of Victoria* 3 (1859): 94–103; Neumayer, *Results of the Meteorological Observations Taken in the Colony of Victoria, during the Years 1859–1862; and of the Nautical Observations Collected and Discussed at the Flagstaff Observatory, Melbourne, during the Years 1858–1862* (Melbourne, 1864); Neumayer, *Discussion of the Meteorological and Magnetical Observations Made at Melbourne during the Years 1858–63* (Mannheim, 1867); Neumayer, *Results of the Magnetic Survey of the Colony of Victoria Executed during the Years 1858–64* (Mannheim, 1869).

3. Carnegie Institution of Washington, *Researches of the Department of Terrestrial Magnetism*, vols. 7A, 7B, 13 (Washington, D.C., 1947–48).

4. G. M. Caroe, *William Henry Bragg, 1862–1942: Man and Scientist* (Cambridge: Cambridge University Press, 1978), p. 49.

5. T. R. Lyle, in *Royal Commission on the University of Melbourne. Part II: Minutes of Evidence on Administration, Teaching Work, and Government of the University of Melbourne* (Melbourne, 1903), p. 46.

6. These various astronomical activities are usefully summarized in the article "astronomy" in *The Australian Encyclopaedia* (Sydney, 1958), vol. 1, pp. 278–86.

7. Hugh Hamersley, "Radiation Science and Australian Medicine, 1896–1914," *Historical Records of Australian Science* 5, no. 3 (1982): 41–63.

8. In addition to the great 4-foot reflector, the Observatory's equipment included at this period a Dallmeyer photoheliograph of 4-inch aperture and 60-inch equivalent focal length, two equatorial telescopes—a 4½-inch aperture, 60-inch focal length achromatic instrument by Cooke & Sons, and one of 8-inch aperture and 100-inch focal length by Troughton & Simms—and a large transit circle, 8-inch aperture, 108-inch focal length, circles 3 feet in diameter, also by Troughton & Simms; see *Australian Encyclopaedia*, s.v. "astronomy."

9. Unfortunately the library was recently broken up, but I have in my possession a rough catalogue of its contents, which reveals a strong emphasis on the applied branches of the subject, very similar to that which long dominated the Cambridge tripos examinations. Many of the books acquired were, in fact, standard Cambridge texts.

10. Michael J. Crowe, *A History of Vector Analysis: The Evolution of the Idea of a Vectorial System* (Notre Dame, Indiana: Notre Dame University Press, 1967), chap. 6.

11. See also W. A. Osborne, *William Sutherland: A Biography* (Melbourne, 1920).

12. *Ninth Report of the Commissioners for the Exhibition of 1851* (Cambridge, 1896), p. 7.

13. *Cambridge University Calendar for the Year 1896–1897* (Cambridge, 1896), pp. 120–23.

14. *Record of the Science Research Scholars of the Royal Commission for the Exhibition of 1851, 1891–1960* (London: for the Commissioners, 1961), passim.

15. *Calendar of the University of Sydney*, 1908, p. 441; ibid., 1909, p. 237.

16. Kerr Grant, who studied under Lyle at Melbourne, succeeded Bragg at Adelaide and remained in that chair until 1948. J. P. V. Madsen was graduated from Sydney University in 1900 and was then appointed to a lectureship at Adelaide, where he did some important work on radioactivity with Bragg. Following Bragg's departure for England, Madsen returned to Sydney, where he rose to become the country's first professor of electrical engineering, also retiring in 1948. O. U. Vonwiller was appointed an assistant lecturer at Sydney University immediately following his graduation in 1902. Thereafter he was promoted regularly through the ranks until he eventually succeeded Pollock as professor in 1922. He retired in 1946.

17. R. W. Home, "W. H. Bragg and J. P. V. Madsen: Collaboration and Correspondence, 1905–1911," *Historical Records of Australian Science* 5, no. 2 (1981): 1–29; Home, "The Problem of Intellectual Isolation in Scientific Life: W. H. Bragg and the Australian Scientific Community, 1886–1909," *Historical Records of Australian Science* 6, no. 1 (1984): 19–30; Caroe, *William Henry Bragg*, chap. 2.

18. Daniel J. Kevles, *The Physicists: The History of a Scientific Community in Modern America* (New York: Alfred A. Knopf, 1978).

19. A. D. Ross to F. S. Spiers, 17 November 1923, in box 86/13, Australian Institute of Physics files, Basser Library, Australian Academy of Science, hereafter referred to as "AIP files".

20. T. H. Laby to A. D. Ross, 29 September 1926; box 86/13, AIP files.

21. Laby to Ross, 26 April 1926, ibid.

22. Rutherford's contribution to the correspondence, together with drafts of some of Laby's letters to him, are preserved among Laby's personal papers recently deposited in the University of Melbourne Archives.

23. Melbourne was not alone in setting such a requirement, as is made clear by C. B. O. Mohr's description of the selection procedure for a lectureship at the University of Cape Town: "It was made clear to me that a Cavendish man was required, and that if I applied, I would be practically certain to obtain the job." And obtain it, he did! Mohr to Laby, 25 October 1935, Laby papers.

24. Interview with C. B. O. Mohr, 7 March 1980. On Grayson and his ruling engine and Lyle's fostering of Grayson's work, see J. J. McNeill, "Diffraction Grating Ruling in Australia", *Records of the Australian Academy of Science* 2(3) (1972): 18–38.

25. J. F. Richardson, *The Australian Radiation Laboratory: A Concise History, 1929–1979* (Canberra: A.G.P.S., 1981).

26. W. F. Evans, *History of the Radio Research Board, 1926–1945* (Melbourne: CSIRO, 1973).

27. Eventually, in 1944, even the Melbourne Observatory was closed and much of its equipment, including the large reflector, transferred to Canberra.

28. On the research work of the meteorological service, see Isobel M. Kidson, compiler, *Edward Kidson* (Christchurch: Whitcombe & Tombs, n.d. [1941?]).

29. Ross to E. Kidson, 25 June 1926, box 86/13, AIP files.

30. David Hutchison, "William Ernest Cooke, Astronomer, 1863–1947," *Historical Records of Australian Science* 5(2) (1981): 58–77, especially pp. 71–74.

31. W. F. Evans, *History of the Radiophysics Advisory Board, 1939–1945* (Melbourne: CSIRO, 1970); D. P. Mellor, *The Role of Science and Industry* (Canberra: Australian War Memorial, 1958), chap. 19.

32. O. U. Vonwiller, "Cancer Research in the University of Sydney," *Journal of the Cancer Research Committee,* 1 October 1938, pp. 52–73.

33. G. Currie and J. Graham, "G. A. Julius and Research for Secondary Industry," *Records of the Australian Academy of Science* 2, no. 1 (1971): 10–28; see also F. W. G. White, "John Percival Vissing Madsen," ibid. 51–65.

34. M. L. E. Oliphant to Mrs. B. Laby, 24 June 1946, Laby papers.

35. Laby was for many years an active member of the Round Table, an Empire-wide organization devoted to promoting Britain's influence in the world; see D. K. Picken, "Thomas Howell Laby," *Obituary Notices of Fellows of the Royal Society* 5 (1948), 733–55.

36. Rutherford's encouraging response has survived: Rutherford to Laby, 23 April 1921, Laby papers.

37. Ross to Laby, 16 November 1923, box 86/13, AIP files.

38. Ross to Kerr Grant, 11 December 1923, ibid.

39. Ross was a Scot, a graduate of the University of Glasgow, who had also spent some time as a research fellow at Göttingen before becoming the first professor of mathematics and physics at the University of Western Australia in 1912.

40. Ross to A. L. McAulay, 26 December 1923, box 86/13, AIP files.

41. Ross to T. Parnell, 16 November 1923, ibid.

42. Ross to Laby, 16 November 1923, ibid.

43. The account that follows is based on Ross's voluminous correspondence from this period, now preserved in the AIP files, boxes 86/13 and 86/14.

44. Ross to Institute of Physics, 18 September 1926, AIP files, box 86/13.

45. A. C. D. Rivett to Ross, 21 September 1926, ibid.

46. Ross to Bailey, 10 September 1926, ibid.

47. Laby to Ross, 29 September 1926, ibid.

48. Kidson's view probably reflected that of most Australian physicists at the time. "Professor Laby should be on, undoubtedly, though I can quite understand that they don't want him at meetings. He would certainly be a joy forever to them. Still he has the ability and his advice is not to be ignored." Kidson to Ross, 4 July 1926, ibid.

49. Ross to Vonwiller, 6 June 1926, ibid.

50. Kidson to Ross, 28 February 1927, ibid.

51. See Ross to Kidson, 25 June 1926: "It seems to me that the matter of the publication of research papers, the holding of meetings to discuss researches, and such subjects are not among the true functions of the Institute," ibid.

52. *Conference of Australian Physicists, Canberra, 15th to 18th August, 1928: Proceedings and Abstracts of Papers,* AIP files. On the IGES, see A. B. Broughton Edge and T. H. Laby, eds., *The Principles and Practice of Geophysical Prospecting, being the Report of the Imperial Geophysical Experimental Survey* (Cambridge, 1931); Barry W. Butcher, "Science and the Imperial Vision: The Imperial Geophysical Experimental Survey, 1928–1930," *Historical Records of Australian Science* 6, no. 1 (1984): 31–43.

53. *Conference of Australian Physicists, Melbourne, 20th to 23rd August, 1929: Proceedings and Abstracts of Papers,* AIP files.

54. Laby to E. H. Booth, 10 April 1936, copy in Laby papers; see also *Fifth Conference of Physicists and Astronomers, Sydney, 25th to 28th May, 1936: Proceedings and Abstracts of Papers*, AIP files.

55. Interview with C. B. O. Mohr, 7th March 1980.

56. Lewis Pyenson, "The Incomplete Transmission of a European Image: Physics at Greater Buenos Aires and Montreal, 1890–1920," *Proceedings of the American Philosophical Society* 122 (1978): 92–114.

57. Ibid.

The History of Telecommunication in Australia: Aspects of the Technological Experience, 1854–1930

ANN MOYAL

Australian history has been conspicuously short on the examination of the history of technical subjects and of the place of technological development in the country's evolution. As early as 1961, Geoffrey Blainey observed: "We historians are uneasy outside the old triad of political, social and religious history; we are inclined to avoid the history of technical subjects even more than did the historians of the last century with their narrower compass of history."[1] The comment remains valid today. The history of technology in Australia stands as a broad and nearly empty canvas on which to depict the major underpinnings—and their social interconnections—of an increasingly industrialized society.

This paper, preliminary and exploratory, is focused on three central issues:

• the emergence in a country governed or, as Blainey suggests, "tyrannised" by distance, of a vital telegraph and telephonic network:
• the importance of Australia as an importer and advanced adapter of telecommunication technology transferred from Britain, the United States, and Europe;

• the emergence of a federally institutionalized framework for the spread of telecommunication and, with it, the development of an engineering culture in Australia.

The Plunge to Interconnectivity

It was nine years after Samuel Morse tapped his famous message, "What Hath God Wrought," from Washington to Baltimore in May 1844, that Australians responded to the prospects of the new telegraph technology. It was brought to Australia by Samuel McGowan (1829–1887), a Canadian trained in the law, who had been acquainted with Morse and directly associated with Morse's colleague, Ezra Cornell, inventor of the first telegraph insulators and a principal of the early telegraph industry.[2] McGowan, who had managed the New York–Buffalo Line, was one of a bevy of entrepreneurs who canvassed subscriptions for America's spreading private enterprise lines. He combined zeal with technical knowledge and, in 1853, attracted by reports of Victorian gold, emigrated to Melbourne, bringing with him several sets of the new Morse instruments, batteries, insulators, and other necessary materials with the express intention of developing a private company to establish and work telegraph lines between Melbourne, Sydney, Adelaide, and the goldfields at Ballarat and Bendigo. It marked the first transfer of electric telegraph technology to Australia.

In the event, government intervention forestalled the private development of telegraphy in Victoria. McGowan's demonstration of his apparatus in Melbourne stirred excited praise. "To us, old Colonials who have left Britain years ago," the *Argus* editorial reflected the following day, "there is something very delightful in the actual contemplation of this, the most perfect of modern inventions. . . . We call the electric telegraph the most perfect invention of modern times . . . as anything more perfect than this is scarcely conceivable, and we really begin to wonder what will be left for the next generation, upon which to expend the restless energies of the human mind. Let us set about electric telegraphy at once."

By September 1853, tenders were invited by the Victorian government for the construction of a line between Melbourne and Williamstown. The contract went to McGowan. By March 1854, the short line was in operation, and within three years, the colony of Victoria was webbed by a network of telegraph lines stretching from Melbourne to Ballarat, toward the South Australian border at Portland, and north and northwest to the Murray River border. The closed-circuit system of Morse transmission was employed and the received signals were recorded on an embossing recorder. While much of the construction work was empirical, ad hoc, and derivative of the example of the Melbourne to Williamstown line, it was more than a mechanical craftsman system; yet little has come to light on the contractors who tendered for and successfully completed the extension of the wires.

There were some hitches in operating skills: telegraphers had to be trained. In New South Wales, where Parliament voted £38,000 for building a telegraph line to link Sydney with the Victorian border at Albury in 1857, the first completed telegraph line "for the business of transmitting intelligence" from Sydney eighteen miles to Liverpool hiccoughed into operation in December that year. Under the watchful eye of Sir William Denison, governor of New South Wales, an engineer by training himself, the first message ran, "Can you read my writing?" No answer was received for several minutes, the *Sydney Morning Herald* reported next day, but on the question being repeated "an answer arrived that the pen of the instrument at Liverpool did not mark and was out of order." Secondary messages like "Have you got my writing plain" met silence and at last a terse reply from the Liverpool telegraphist that he "did not get it plain enough." "Whether he was unable to rectify the defect in his instrument," the *Herald* mused, "or believed from not being able to read the writing that all queries were at an end, we are unable to say."[3]

Despite human inexperience, lines were soon operating south from Sydney more than 200 miles to Yass and Gundagai. The Tasmanian government erected a line from Launceston to Hobart in 1857; a private line linking Adelaide and its port, erected by James Macgeorge, appeared early in 1856, while the South Australian government, averse to private enterprise, in the same year erected a parallel line from Adelaide to Port Adelaide and on to Semaphore in St Vincent's Gulf. The job cost £6,048. In 1861 the first messages were telegraphed in Queensland from Brisbane to Ipswich, and eight years later the line had extended up the long north coast via Bowen and on to Cardwell. Only Western Australia lagged behind, its first telegraph being erected by private enterprise between Perth and Fremantle, aided by government-supplied convict labor in 1869. The network was acquired by the government in 1871.

Within a decade of McGowan's initiative, some thousands of miles of Morse's Lightening Lines silhouetted the Australian countryside. Over the long distances "repeaters" were installed, initially human operators who read the incoming Morse signals and transmitted them on to their destination. Much of the growing expertise was of American derivation, though Britain's Wheatstone automatic instrumentation for passing on messages in minor offices in city and suburbs—and its evolutionary improvements—became widely diffused through Australia from the early 1860s.

It was Charles Todd (1826–1910), appointed from England in 1855 as superintendent of telegraphs and simultaneously observer (later astronomer) of South Australia who brought British knowledge, experience, materials, and scientific underpinnings to telecommunication development in Australia and became a prime mover in Australian telegraph technology. The British telegraph, launched somewhat earlier than Morse's, differed from the American system, though integration and refinements toward internationalization of the Morse system were increasing.[4] Todd's early career spanned a period as supernumerary computer at Greenwich Royal Observatory. With skills in mathematics and astronomical observation, he became junior assistant to Professor Challis at Cambridge University, 1848–54, assisting in the determination of longitude by telegraphic means between Cambridge and Greenwich. In 1854 he was back at Greenwich as superintendent of the galvanic apparatus for the transmission of time signals.

This involved him in close cooperation with the Electric Telegraph Company and with C. V. Walker one of the pioneer experimenters with submarine cables. Not surprisingly, Todd became steeped in and fascinated by telecommunication technology and was a prime candidate, supported by the Astronomer Royal Sir George Airy, for the South Australian post. Todd had never built a telegraph line, but under Airy's guidance he supervised the acquisition of equipment and stores, made important contacts with the telegraph world, and assimilated British telegraphic methods and skills. While Airy wondered in private whether Todd "had the boldness and independence of character which may be required in an

Australian establishment," Todd dreamed of large-scale telegraphic ventures, writing to the explorer Captain Sturt in 1854, "I look forward to . . . the time when the telegraph system will be extended to join the several seats of commerce in Australia and also, it is no idle dream in the present age of wonders, when I shall be able to meet Dr. O'Shaughnessey by connecting Asia by submarine cable thence via Calcutta to London."[5]

Todd arrived in Australia with British telegraph plant and one technical assistant, E. C. Cracknell. He declined the offer of bringing operators, determined to rear his own corps of telegraphers. The British technique, which he transferred to Australia, was the undergrounding of telegraph cable, a method adopted disastrously by Morse in his initial attempt on his Baltimore-Washington experimental line.[6] Acting on British specifications and instructions, Todd completed the underground line from Port Adelaide to Semaphore. But he quickly regretted this uncritical dependence on Imperial technology, which, he wrote, "added very greatly to the cost and was very soon an endless source of trouble as indeed were all underground wires laid in England . . . owing to the insulation becoming defective."[7]

It fell to Todd to negotiate for the government with Samuel McGowan for the construction of an intercolonial line between Melbourne and Adelaide, which, strung securely above ground, was completed in August 1858 at a cost of £39,000. Three months later the line was joined from Sydney to Melbourne on a uniform system and Australia's three mainland capitals were telegraphically united. Brisbane was linked to Sydney in 1861. British influence and experience was further extended by E. C. Cracknell's appointment as superintendent of telegraphs in New South Wales in 1861, and his brother , W. J. Cracknell, as superintendent of telegraphs in Queensland in 1863 (both men trained at Oxford); but attempts to make an intercapital join between Melbourne and Tasmania in 1859 foundered on the immense complexity and contemporary difficulties of submarine cabling. While a cable "perfectly made" and coiled in Britain was transshipped and dropped from the S.S. *Omeo* across Bass Straits from Williamtown to Cape Otway in 1859, under the troubled eye of Victoria's Superintendent of Telegraph Samuel McGowan, the £53,000 venture proved unsuccessful. Contact was intermittent, there were frequent breaks to mend and the cables were abandoned as hopeless in 1860. It was not until 1869 that a second attempt to establish a cable connection successfully linked Tasmania telegraphically with the mainland.

In most colonies, the telegraph brought promising returns to sponsoring governments. Telegrams were the new currency of communication. In 1857, three years after the beginning of telegraph operation in Victoria, 13,000 telegrams were transmitted that year, bringing in £5,648 to the Colonial Department. In South Australia a remarkable 35,792 messages were sent in 1856, yielding a financial return of £1,183, which went some way toward covering, and shortly exceeding, the cost of installation. Queensland, with its later investment and larger distances, had telegraph wires stretching over 1,131 miles in 1865; it dealt with 47,697 telegrams that year and realized £13,383 for the government at Brisbane. Internal charges were moderate. Between Adelaide and its port, as one example, the sender paid sixpence for the first twenty words exclusive of names and addresses, and thereafter threepence for each additional ten words. Costs rose considerably over longer distances and rocketed when overseas rates were introduced. But Australians took to telegrams like ducks to water. There was something about the laconic staccato style of the communication that suited their unloquacious temperaments. There were also more powerful impulses. The telegram emerged not only as a vital social link joining families and friends in joy and disaster. It became at once an important legal policy instrument, communicating sentences of law passed in remote outposts, tracking bushrangers and other criminals;[8] transmitting decisions; communicating appointments; assisting railway traffic control;[9] an economic agent diffusing commercial and price information, informing on the state of the local gold market; and in every sense transferring knowledge, instructions, and human feelings from city, country, suburbs, capital cities, and back again.

By the end of the century Australia had become one of the largest users of the telegraph. "In no country in the world," wrote the statistician Thomas Coghlan in 1901, "has the development of telegraphic communication been as rapid as in Australia, and in none has it been taken advantage of to anything like the same extent."[10] Thirty years later, Western Australians held the world lead. "The habit of sending telegrams had so captured Australians generally," the *West Australian* newspaper reported that year, "that Western Australians send more telegrams on average than any other people in the world," and Australia-wide, the ratio in relation to the population was higher than in any other country except New Zealand.[11]

A significant impetus to "this most perfect of modern inventions" derived undoubtedly in Australia from the striking feats of planners, contractors, and linemen who stretched the technology across the continent in the great Overland Telegraph exploit of 1871–72 and five years later built the almost equally challenging East-West Telegraph line from Albany via Eucla to Adelaide.

This is not the place for an analytical account of these epic ventures. Politics underscored the battle for the Overland Telegraph route. The inspiration of the plan to link Australia with a submarine cable brought ashore at Darwin from Java and uniting the isolated continent with an overland line across the center to Port Augusta in South Australia has commonly been attributed to Todd. Some doubt now exists on that point. There was keen, even bitter, rivalry between the Australian colonies to seize the advantage of securing one of the submarine cables then snaking their way around the oceans of the world, joining countries and continents, to terminate in their territory. Rivalry between the governments of Queensland, abetted by New South Wales, and South Australia sharpened significantly in 1870. Queensland backed a cable coming ashore from Java at Darwin and thence by landline to Bourketown. South Australia, whose governance controlled Northern Australia and hence Darwin, wanted the Java submarine cable landed and connected from Darwin by a direct overland route—already demonstrated as a land passage by John McDouall Stuart's overland exploration of 1861–62—to Port Augusta and Adelaide. By swift political negotiation and some sleight of hand, South Australia won through with a contract signed with the British Australian Telegraph Company in June 1870. Contractual conditions were stringent. The overland line of 2,000 miles across desert and semidesert country was to be ready for operation in eighteen months, the deadline the 1st of January, 1872. The colonial government voted an immediate £120,000; Todd was put in charge. "Then, perhaps for the first time," he reflected, "I fully realised the vastness of the undertaking I had pledged myself to carry out."[12]

In 1870, the technology underlying the Overland Telegraph construction was rudimentary. The project, subdivided into three parts
of Southern, Central, and Northern Sections, was contracted out,
the Southern and Northern parts to private entrepreneurs, while
government surveyors and workmen completed the Central Section
of 550 miles. Overseers and suboverseers were involved in close supervision of the work. Todd, ubiquitous supremo, issued oral and
detailed written instructions and maintained a restless overview-
ing of the work. Technicians, telegraphers, and linemen were self-
taught. Transport was crucial. Horse-drawn vehicles carried men
and materials to the sites. Instructions conveyed that wooden poles,
tapered to specifications and not less than nine inches in diameter
at the butt, were to be hoisted and fixed "in a most substantial manner" and positioned at not fewer than twenty to the mile. Three
thousand wrought iron poles, along with the galvanized iron wire,
were imported from England and deployed in sectors where suitable
wood was unavailable, though their late arrival delayed their fullest
use. In softened ground each pole was strutted with a sixteen-foot
pole to prevent the line wire pulling the pole over, and, in a region
of the highest incidence of lightning in Australia, each second pole
was fitted with a lightning rod. The bare wire line rested on insulators, fastened to spindles attached to pole or crossarm, and overseers
kept written checks of the daily work of each man, the poles cut
and erected, and the length of wire put up each week into the empty
sky.[13] Most careful instructions were spelled out for the conservation of that most special technological resource, the horses, and
some attempt was made to instruct the working parties to deal moderately with the Aborigines, firing only "in the last extremity," as
the technological cavalcade plunged unwittingly through sacred
sites and tribal lands. In their turn, the Aboriginals found their rewards in aiming their spears at the white porcelain insulators "of
the best approved pattern" made by the Berlin Imperial Pottery, to
claim sharp new cutting shards.

In this human and technological experiment, ingenuity was
pressed to its straining point. The early telegraph "engineers" knew
little of theory. In any event, the conditions were novel and unre-
mitting. The constant enemy was leakage, and the insulators—
60,000 in all across the 2,000 miles of open wire—were challenged

by conditions that began in temperate zones and experienced every variation from misty rain, dry aridity, dust, monsoonal rain, sun temperatures as high as anywhere in the world, and savage assaults from spiders and other insect invaders of the semienclosed undersides of the insulators. White ants were to remain a permanent scourge of telecommunication connections in Australia. In 1871 they gnawed their way through wooden poles and were barely daunted by the steel poles imported from the firms of Siemens and Oppenheimer.[14] For these and other reasons repeaters were essential, and maintenance provisions and testing made constant demands on the manpower of the work. With difficult country, engineers used the minimum number of hand repeating stations—eleven were installed on the route—scattered near water holes at intervals of 300–500 miles. With the restricted knowledge of the period, strange false currents wreaked havoc on the line. Slowly it was learned that these were related to weather conditions and sunspot activity, and the construction proved instructive to research.[15]

For the technicians and linemen, groping with diverse obstructions, demands on their inventiveness were constant and profound. Anxiety steams off the pages of the overseers' diaries. Sunstroke, death, and exhaustion played their part. In November 1871, the cable from London, via India and Java, was landed at Darwin, but monsoons intervened. The Northern Section was not completed until May 1872, and the final join was made at Frews Ironstone Pond on August 22, 1872. Eight months behind schedule, the Overland Telegraph had been completed in two unforgettable years. Uninterrupted communication from Britain now extended across the length of the fifth continent. Lodged at a campsite at Central Mount Stuart that day and connected by a portable telegraph set, Todd was deluged with congratulations from all parts of the colony. Astronomer and meteorologist as well as an intrepid superintendent of telegraphs, Todd was later to make sound scientific use of the arduously installed technology by deploying his repeater outposts and the telegraph system to build a national system of meteorological and climatological intelligence.[16]

The construction of the East-West Telegraph to bring Western Australia into telegraphic contact with Adelaide and the rest of the world was launched from Albany in April 1875. A collaborative enterprise between the governments of Western and South Australia, it called on much of the expertise and ingenuity demonstrated on the Overland Route but took more than two and a half years to complete. Winding from its starting point at Albany around the great Australian Bight largely within sight of the sea, the line was erected on jarrah wood poles, most ferried by ship around the Bight, cast off in bundles, rafted ashore, and hoisted by derrick up perpendicular cliffs to be moved by horse or camel cart to the sites inshore. The border town of Eucla became the important, if bizarre, repeater station for the line. Here telegraphists, seated on their respective sides of the electrical apparatus, took down the Morse signals from their capitals and handed them across to their opposite numbers for transmission in different colonial codes. "Saturday 7 pm" (8 December 1877), ran the first message, "Eucla line opened. Hurrah."[17]

The skills and adaptiveness illustrated by Australia's early longline telegraph construction were notable in any terms and indelibly stamped Australia's attitude and policies toward future telecommunication techniques. Australia would pioneer many applications of telecommunication technology in the century ahead and would accommodate the applications of an evolving and widely derived range of network systems to suit her particular geographical and human needs. Yet in the field of early telegraphy, Australians failed to contribute to original inventions or to add to evolving theory of electrical science. The country's remoteness, indeed, fostered negative effects. Dr. Edward Dey and Benjamin Babbage (1815–78), two noted telegraph experimenters in England, both forsook their enquiries on emigrating to South Australia, though Babbage, scientist, explorer, surveyor, and engineering son of the more famous Charles, served Todd in planning the Overland Telegraph, supervised the Southern Sector contractors, and introduced the concept of concrete telegraph posts into South Australia.[18] In nineteenth-century Australia, electrical science was not expounded in the universities but found its expression exclusively and empirically in the rugged but often creative application of technology and materials imported from overseas. Of stimulus to a local manufacturing industry there was none. The cutting and end trimming of wooden poles; some hastily devised insulators for the first telegraph line, and Babbage's concrete posts, were the only known contributions to the early public telegraph communication network in Australia.[19]

Telephony: The Prolific Experience

In telephony, the Australian experience proved different. Within months of Alexander Graham Bell's patenting of his telephone invention in 1876, Australian experimenters were widely engaged in turning out variants of Bell's instrument in all the Australian colonies. The "carrier" was the *Scientific American* of 6 October 1876, in which Bell's experiments were detailed and diagrammed, which reached Australia early in 1877.

Australian experimenters moved alongside Bell's improvements in carrying the invention further. W. J. Thomas, working from homemade telephone instruments in Geelong, managed early in 1877 to link houses in his locality both by music and conversation and, more ambitious, to use the telegraph line between Geelong and Queenscliff and Queenscliff and Melbourne to convey catches of song. In Tasmania, Alfred Biggs, an astronomer and schoolteacher at Campbell Town Public School, simultaneously made three hand telephones out of Tasmania's beautiful Huon Pine and, after telephoning over short distances, communicated with Launceston from Campbell Town. In Brisbane, in February 1878, a medical practitioner, Dr. Severn, experimented with homemade telephones over a mile of telegraph wire, and other researchers carried a telephone call from Ipswich over fifty miles of wire. In northern Queensland, the inland coastal telegraph routes were also used to transmit voice communication over as much as 1,000 miles of line. In Perth, newspapers reported successful experiments with telephones over Perth–Fremantle telegraph lines in March 1878.

But by far the most fertile and significant telephone researcher in Australia was Henry Sutton (1856–1912) of Ballarat. Self-taught, an inventor of eclectic talent and range, Sutton from the earliest age was spawning models and machines from continuous-current electric dynamos, electric motors, vacuum pumps, heavier-than-air flight machines, and a television apparatus for transmitting the running of the Melbourne Cup to Ballarat, now recognized as some seventy years ahead of the technological backup of his time. A distinguished modeler and draftsman, Sutton devised and constructed some twenty telephone designs but resisted the practice of patenting on the grounds that discovery "should benefit fellow workers in science."[20] It did. Sixteen of his telephones were subsequently patented by others overseas.

During this aptly named period of invention, the colonial superintendents of telegraphs also joined the active process of experimentation and research. E. C. and W. J. Cracknell carried out experiments in late 1877 and January 1878, respectively, in telephonic communication by voice across 140 miles from Maitland to Sydney and in the first experiments at the Brisbane Telegraph Exchange. Queensland's Superintendent, W.J., was clearly dissatisfied with his crude implements and, on securing Bell's promise of improved models, informed the Queensland Parliament that "further experiments with this scientific wonder will be made."[21] In South Australia, Todd, too, used Bell-type instruments in January 1878 to achieve telephonic voice communication across telegraph lines between Adelaide and Semaphore, and similar experiments were conducted on parts of the Darwin line between Beltana and Strangways, a distance of 200 miles.

Where did this local ingenuity lead? Certainly some of Sutton's inventions passed by others' patenting into telephonic communication overseas. But it has yet to be discovered whether the labors of indigenous inventors were absorbed by the embryonic telephonic programs in the Australian states. Historiography at present sheds no light.[22] Reports of the Commonwealth Postmaster-General's Department suggest that Australia looked to Britain for standardized telephone equipment for distribution throughout the country from 1901 until local production of handsets began in 1933. Innovation traditionally proved a poor handmaiden to inventive enterprise in the colonies. Bell, however, recognized the singular ingenuity of Sutton's work, and on a journey to Australia in 1910 he visited to see the complete telephone system installed by Sutton in the family warehouse at Ballarat.[23]

Telephony, in fact, spread slowly in Australia. The first installation of a regular service, thanks to private enterprise and skills, was brought into operation in two offices of the Melbourne hardware importers Messrs. McLaren Bros. and Rigg in January 1878, with telephones designed by J. S. Edwards of Melbourne.[24] Government telephone exchanges, however, moved slowly. Melbourne installed its first exchange with seventeen subscribers in 1880; Sydney had a government-operated exchange in 1882; Hobart and Adelaide in

1883; Perth in 1887; and exchanges began to open in provincial towns by the late 1880s. In 1887, the decision to place all telephones in each colony under government control in association with the colonial post offices was made, a policy that predated Britain's move to consolidate its posts and telephone networks by twenty-two years.

By federation, in 1901, there were 33,000 telephones operating in Australia, 100,000 in 1910. With the gradual development of trunk line services between the capital cities—Melbourne–Sydney, 1907; Melbourne–Adelaide, 1914; Sydney–Brisbane, 1923; Perth–Adelaide, 1930; and Hobart–Melbourne, 1936—the figures leapt dramatically forward. Across the wide communication axes of Australia, apart from local calls, 32 million trunk calls were recorded in 1928, 35½ million in 1930, and 40 million in the post-depression year of 1938. By contrast, telegrams sent nationally in 1928 reached 16 million, sank to 13 million in 1938, and assumed decreasing importance in relation to telephone communication by the late 1930s, though the war brought a strong resuscitation.[25]

Australian telephony would absorb significant technological changes in the first quarter of the twentieth century. The importation from the United States in 1912 of the Strowger Automatic Telephone Exchange, a step-by-step automatic dialing system that accelerated connections by allowing dialing by subscribers without manual intervention, marked the first introduction of this exchange system to the Southern Hemisphere and only the second installation of the system after that in Great Britain, in the British Empire.[26] Installed initially at the Geelong Exchange, Strowger spread quickly to capital and suburban centers to become a foundation part of the Australian telephone network. The transfer of this technology ahead of many countries characterized the strategic forward planning that became a hallmark of Australian telecommunication policy.

At the same time, World War I isolated Australia from its sources of imported telephone copper wires and forced the growth of local manufacturing industry. Copper wires were fabricated for the Australian Post Office by E. R. and S. Smelter of Port Kembla in 1918. The Australian firm Metal Manufacturers was supplying paper-insulated, lead-covered cables by 1923 and, while imports persisted, continued to meet demands for all types of communication network cables, including dry-core lead, lead-antimony, covered cables, a single pair to a thousand pairs. It was amalgamated with British Standard Telephones and Cable Pty. Ltd. to form Austral Standard Cables Pty. Ltd. in 1947. By 1970, Australia had at last become self-sufficient in the manufacture of all types of telephone cable, including coaxial.[27]

Institutional Consolidation:
the Rise of an Engineering Culture

Federation brought marked changes to telecommunication organization. With the Act of Federation of January 1901, the separate administrations of posts and telegraphs in the six colonies—states as they now became—were consolidated into one Commonwealth bureaucracy, the Postmaster-General's (PMG's) department in March of that year. Confronted by "six different Acts, six different regulations and practices differing in every State," the new Commonwealth government organized a government postal and telecommunication monopoly that, in the postmaster-general's words, would be "of best service to the people of the Commonwealth." Introducing the Post and Telegraph Bill in the Commonwealth Senate in June 1901, the postmaster-general, Senator Drake, gave pertinent expression to a philosophy that would underlie telecommunication practice and development in this century. A telegraph monopoly, the postmaster-general told Parliament, "allows us to look ahead; take advantage of every innovation, and adapt it to the benefit of the public." If the telegraph system had been in the hands of a private proprietor, he reflected, "I am inclined to think that it would have fought against the introduction of the telephone service. . . . Instead, the Postal Department . . . welcomed the telephones and assisted to a great extent in making the innovation of practical value to the community."[28]

From its inception, the large new instrumentality was seriously handicapped in its progress by the shortage of trained staff. The largest employer of manpower in the Commonwealth, the PMG's department inherited the practices and deficiencies of colonial times. With growing traffic and a country flexing its sinews further for telephone and telegraph connections in the remoter rural parts, the department's greatest need was for qualified engineers, mechanics, and technical men. These were in short supply in Australia. Feats of nineteenth-century telegraphy and telephone connection had rested on British and American experience and empirical improvization by practical men. There was minimal formal training in Australia. The Mechanics Institutes, scattered in the capital cities and sometimes allied in provincial centers to the School of Mines, offered some instruction in mechanical electrical engineering. The inventive Henry Sutton taught electricity and applied magnetism to privileged students at Ballarat School of Mines from 1883 to 1887, but such knowledge and capacity were rare. Telegraphers, mechanics, and engineering staff were trained traditionally on the job.

The Australian universities were dauntingly slow in providing professional courses in electrical engineering. Tuition existed in engineering at Melbourne University from the mid 1860s, and William Kernot, M.A., earned the first certificate in engineering in 1866. Kernot in turn taught civil engineering, became first professor of engineering at Melbourne University in 1883, and took part in a Victorian inquiry into the undergrounding of telephone and telegraph wires, but he offered no instruction in electrical engineering fields.[29] His counterpart at Sydney, Professor William Warren, appointed in 1884, founded an engineering school, "built his courses on American rather than British models," and by 1910 had established three major teaching streams, which included electrical engineering.[30] But professional training in engineering with telecommunication components only became available much later, during the 1950s, despite the existence of engineering schools in Queensland, Adelaide, and Perth. The PMG's department trained most of its own engineers, building more and more on science degrees from the Australian universities, until 1950–55, when appropriate degree training in electrical engineering became available.

The situation in fact changed little until after World War II. Within the PMG's department, however, there were early pressures to fashion an organizational structure in which engineering was of central importance and to upgrade the efficiency and the economy of construction and maintenance.[31] From 1906 operation and planning for telephone communication was assigned to the chief electrical engineer, and in the first decade of the PMG, engineers were upgraded managerially in the states. But a shortage of trained engineers remained. Despite inducements to staff to "undertake an approved course of instruction in telegraph and telephone engineering at any recognised institution" pressed in 1911,[32] the chief electrical engineer, John Hesketh, reported in 1912 that, "the supply of staff for the engineering branch in various sections has been insufficient for the past six years, and there appears to be no sign of relief."[33] Harassed by shortage, the PMG's department turned to Britain, advertising the post of assistant electrical engineer and recruiting J. M. Crawford from the British Post Office. It was the first of many such tugs on the imperial apron string; recurring appeals for engineering reinforcement continued into the post–World War II period and, surprisingly, into the 1970s.[34]

Undoubtedly the universities' failure to grasp the significance of professionally trained manpower for Australia's telecommunication development had serious consequences for the development both of a trained civil service of engineers and for the promotion of local telecommunication research and development. Instructive contrasts clearly lie in the historical relations between professional engineering education and the development of telecommunications in countries as diversely productive as Britain, the United States, Canada, and Germany.

During the first two decades of the century, the PMG department initiated spasmodic efforts to reform its own house. In 1914 a committee was appointed to investigate the best means of recruiting and training both junior members and the technical staff, but the palliatives of a technical library, technical societies, and better-equipped telegraph training schools failed to shake the problem. By 1918, the harassed but determined Labor postmaster-general, William Webster, noting the deficiencies of funding and men, declaimed in his annual report that he had only "with great difficulty been able to keep the Service off the rocks—it is still in the breakers."[35] It was Webster's successor, W. G. Gibson, who found the passage through the shoals. Faced with mounting public demands for services, inadequate supply, and a demonstrable lack of skilled staff, Gibson turned for help to Britain and found it providentially in Harry Percy Brown, British Post Office engineer.

H. P. Brown (1878–1967) was to become one of Australia's public service giants. Son of a former superintendent of the London Telegraph Office, Brown had no degree but entered the British Post Office and, after rapid advancement, held responsibility for the technical planning and management of all telephone and telegraph plants in Britain. In 1913 he visited India for the East India Railway Company, acting there as adviser on telegraph, telephone, and railway traffic control, and in 1916 he was put in charge of "emergency communications" in British home defense, which called on all his power of improvization. On the Australian government's overture to the imperial government in 1922, Brown was chosen as the man "most capable in all England" for assisting the Australian government in an advisory capacity. He arrived in Melbourne late in January 1923 for one year's stay and remained for the rest of his life. In December 1923 he was appointed secretary of the PMG's department at a salary that soared above that of any other public servant.

Brown's appointment was part of a broad Commonwealth government approach to upgrade the national scientific and technological enterprise. Australia had already looked to Britain for advice, and a model for its Advisory Council on Science and Industry which would lead, with high-powered British expertise to guide it, to its permanent Council for Scientific and Industrial Research (CSIR) in 1926.[36] Brown's advent four years earlier was designed to pump knowledge, high engineering experience and, above all, British Post Office management skills into the confused and disorganised PMG's department. With his acceptance of the permanent executive leadership of that body, its future was assured.

From a debut of press and parliamentary outrage and shock, "Was there no Australian who could hold the post?" Brown, "the human whirlwind," went on to reorganize the Postmaster-General's department and to build it on a base of engineering management. Significantly, for the first time, there was an engineer at the top. It was an idea that charmed one professional engineering journal. Mr. Brown, the *Chronicle* reported, was appointed less because he was an acknowledged telephone and telegraph expert, but "because he was an engineer." He was, therefore, "a trained thinker . . . and preferable to anyone who has not been taught to think."

Brown's first five years in office were to demonstrate the relevance of his engineering management and planning approach. Extensive reconstruction and major works programs bore his stamp, and within four years he had established a research section at headquarters with a staff of eleven. Known widely in the press as Poohbah

Brown, for his trenchant approaches, his impact on postal and telecommunication technology was profound. He holds a pivotal position in the history of Australian telecommunication.[37] First and
foremost he built an engineering environment in which, in headquarters administration and in the states, he cultivated the development of advancing telecommunication around a blend of technical and managerial engineering skills. Though shortages of skilled
manpower did not disappear, he provided the encouragement by
which engineering staff could rise rapidly along administrative paths
and hence be increasingly visible in PMG management. His spur to
the PMG's Research Laboratories in 1924 provided the means
whereby extensive evaluations and testing of new technology in pilot stage could be performed and the adaptive skills of Australian
telecommunication staff marshaled to the sound scientific under-
pinning of national advancement.

The style and character of Brown's administration was already
deeply set when he left the PMG's department in October 1939 and
was linked with key new developments such as wireless broadcasting and advanced carrier wave telephony and telegraphy. A quarter
century later, the dominance of the telecommunication side of the
activities of the Australian Post Office was a central factor in leading
to the partition of functions between the historically linked postal,
telegraph, and telephone services and the creation, in 1975, of two
statutory bodies, the Australian Postal Commission and the Australian Telecommunications Commission (Telecom Australia).

This paper was part of a preliminary overview of a study commissioned by Telecom
Australia in 1981 and published as *Clear across Australia: A History of Telecommunications* (Melbourne: Thomas Nelson Australia, 1984, 437 pp). It is reproduced with
acknowledgment to Telecom Australia.

    1. G. Blainey, *Historical Studies of Australia and New Zealand*, May 1961, p. 338.
    2. R. Thompson, *Wiring a Continent* (Princeton, N.J.: Princeton University Press,
1947).
    3. *Sydney Morning Herald*, 31 December 1857.
    4. A uniform telegraph system, based on Morse, was established by the first Inter-
national Telegraph Convention and signed in Paris in 1865. The International Telegraph Union followed.
    5. G. W. Symes, "Australia and World Communications: A Summary of Develop-
ments, 1830–70," paper presented to a symposium on the occasion of the centenary
of the Adelaide-Darwin Overland Telegraph Line (Institution of Engineers, Australia,
and the Australian Post Office, August 1972), p. 3. O'Shaughnessey was a telegraph
experimenter in the East India Company.
    6. Undergrounding swallowed up two thirds of Morse's $30,000 congressional ap-

propriation; the insulated wire was painfully unraveled by Cornell and later erected on poles above ground.

7. South Australia, *Report of Superintendent of Telegraphs,* Tenth Report, 1884.

8. Ned Kelly, bushranger extraordinary and national folk hero, was successfully tracked by telegraph.

9. The alliance between the telegraph and the extending railways in the nineteenth century was important in most countries. In Australia railwaymen kept an informing watch in remote areas on needed repairs to the telegraph line.

10. Thomas Coghlan, *The Seven Colonies of Australia* (1901)

11. 1 January 1938, p. 12.

12. Charles Todd, Public lecture notes, 1873, South Australian Archives.

13. Instructions to overseers in charge of works, Overland Telegraph, Port Augusta to Port Darwin, copy in Telecom, Telecommunication Museum, Adelaide; M. J. Gooley, "The Construction of the Overland Telegraph, Port Augusta to Darwin," in *The Centenary of the Adelaide-Darwin Overland Telegraph Line* (Adelaide: Institution of Engineers, Australia, and Australian Post Office, August 1972).

14. F. P. O'Grady, "The Overland Telegraph Line Technology of the 1870's," ibid.

15. Ibid.

16. Ann Mozley Moyal, *Scientists in Nineteenth Century Australia* (Sydney: Cassell Australia, 1976), p. 154.

17. J. Moynihan, "In Touch with the World," *Western Australian,* 6 December 1977, p. 58.

18. *Australian Dictionary of Biography, 1851–1890* (Melbourne: Melbourne University Press, 1969), vol. 3, pp. 65–66.

19. McGowan had great trouble with the insulators. Melbourne had no glasshouse; the Sydney manufacturer botched the production, and McGowan himself had to resort to inventing and manufacturing a quantity made from shellac and tar; see *The Argus,* 11 March 1854, p. 5; see also Sir Samuel Jones, "The Development of the Australian Telecommunications Manufacturing Industry," in *The Centenary of the Adelaide-Darwin Telegraph Line.*

20. *Australian Dictionary of Biography, 1851–1890* (Melbourne: Melbourne University Press, 1976), vol. 6, pp. 226–27. In Sutton's "TV system," which he called "telephany," he used all the latest technology, such as the recently invented Kerr effect, the Nipkow disc—which Baird was to use in the 1920s—and the selenium photocell. But its weak link in the 1870s was that the signal had to be transmitted by telegraph lines, for radio had yet to arrive, and these were too slow to transmit the dashing horses of the Melbourne cup. Information from Dr. C. Coogan, CSIRO, Melbourne.

21. Superintendant of Telegraphs, Report to Parliament, Queensland, 11 March 1878.

22. G. Linge, *Industrial Awakening* (Canberra: Australian National University Press, 1980), offers a pioneering look at some other aspects of manufacturing invention and innovation.

23. Postmaster-General's Department, *Fourth Annual Report* (1913–14), p. 37. On this visit to the Royal Commission into Postal Services Bell gave riveting testimony in comparing American and Australian telephone development; see Postmaster General's Department, *First Annual Report* (Victoria: Government Printer, 1911), appendix 29, pp. 2417–26. (No annual reports were issued for the first nine years of the department's existence.)

24. *The Cyclopedia of Victoria,* vol. 1 (Melbourne, 1903), pp. 568–69.

25. *Postmaster-General's Department Twenty-fifth Annual Report* (1937) and *Twenty-ninth Annual Report* (1941).

26. The Strowger system demonstrates the converse maxim "in the midst of death we are in life." A. B. Strowger, a Kansas City undertaker, moved to experiment because he believed that the exchange girls were diverting to business elsewhere calls that were intended for him, designed the world's first automatic telephone system, by which each digit, step by step, activated and carried through the connection without the aid of the "hello girl." The system was patented in 1891.

27. Jones, "Australian Telecommunications Manufacturing Industry," p. 33. Other active local firms besides Austral Standard Cables Pty. were Olympic Cables Pty., Conqueror Cables Pty., and Beacon Cables.

28. Commonwealth of Australia, *Parliamentary Debates*, Senate, 6 June 1901, vol. 1, p. 749.

29. *Australian Dictionary of Biography, 1851–1900* (Melbourne: Melbourne University Press, 1974), vol. 5, pp. 20–22.

30. Ibid., vol. 6, p. 357.

31. Commonwealth of Australia, Posts and Telegraph Department, *Report of the Departmental Committee of Inquiry into the Telegraph and Telephone Systems of the State of the Commonwealth* (Adelaide: S.A. Government Printer, 1901), Recommendations, p. 37.

32. Postmaster-General's Department, *First Annual Report* (1911), pp. 13, 25.

33. Postmaster-General's Department, *Second Annual Report* (Victoria: Government Printer, 1912), p. 24.

34. A campaign during the early 1950s to recruit British engineers followed a notice in the annual report for 1948–49 of a "dearth of qualified men." The campaign of 1971 anticipated the addition of fifty engineering graduates from Britain.

35. Postmaster-General's Department, *Ninth Annual Report, 1918–19* (Victoria: Government Printer, 1920), p. 21.

36. Sir Frank Heath was lent from Britain's Department of Scientific and Industrial Research (DSIR) in 1925.

37. *Australian Dictionary of Biography, 1891–1939* (Melbourne: Melbourne University Press, 1979), vol. 7, pp. 437–39.

Patron and Client: The Web of Intellectual Kinship in Australian Anthropology

I have not at present any subject on which I want information from Australia, but I will not forget your obliging offer. If the suggestion has not already occurred to you, will you allow me to make one viz—to keep full & accurate notes of the mental powers, such as the capacity of abstract reasoning etc, etc. of the Australians. Also of their quasireligious beliefs,—their curious marriage laws & all other such points. With your means of obtaining information & with your scientific habits of thought, you could certainly write a very valuable memoir or book, in the course of a few years.

> Charles Darwin to A. W. Howitt
> September 1874

In Australia and Polynesia you are several strata below barbarism into savagism and nearer to the primitive condition of man than any other investigator.

> Lewis Henry Morgan to Lorimer Fison
> 8 May 1872

Anthropology is scarcely "practical" enough for the Australian mind in fact a preliminary lecture would be necessary to enlighten the general Melbourne public with regard to the meaning of the word & the aim of any individual calling himself an anthropologist.

W. Baldwin Spencer to E. B. Tylor
23 May 1889

Australia has at all events the great advantage that there is no end of pioneer work to be done and work which, in Anthropology at least, must be done soon if it is to be done at all. There is a great charm in pioneer work which quite compensates for the loss of many advantages which are of course absent in a new country though one cannot help at times feeling how much one misses in being so far away from the centre of civilization.

W. B. Spencer to H. Goulty
16 July 1900

The origins of Australian anthropology are exemplified by these observations involving eminent Victorians. There is an interplay between European patron and colonial client; a thread of social evolution connects assumed objective factual observation with the substantiation of models drawn from the milieu of social Darwinism. Reflected also are the tensions and frustrations of academic isolation, associated with elements of cultural cringe and the denigration of Australian pragmatism, but contrasting with the intellectual stimulation of discovery. This paper will offer reflections upon such factors, within the context of post-Darwinian science.

Anthropology of a sort is as old as the first European contacts with Australia. Even the records of the first substantial Dutch visit contain ethnographic observations made on a Cape York beach in 1623. Grimly symbolical of later race relations in this continent, unsuspecting Aborigines met violent deaths. The fragmented and credulous observations of early navigators is not our concern. Such ethnographic records were rarely anthropological in possessing any concern for the systematic study of social phenomena, and I have examined these fleeting observations elsewhere.[1]

Until well into the nineteenth century most accounts were uncritical or filtered through a mesh of prevailing philosophical notions and expectations, such as the idea of the noble savage or the depravity of unredeemed mankind. The short chronology of Genesis prevailed, while it was commonplace to "find" Israel's lost tribes or remnants of other ancient civilizations in unlikely antipodean places. There was little scope in this world of marvels or chaos for a scientific analysis of indigenous society. Even so, the knowledge that European scholars derived from Captain Cook's visit or from first fleet chroniclers, especially David Collins, qualifies Australian ethnography to be included tentatively within Basalla's first stage of scientific development, when "the nonscientific society . . . provides a source for European science."[2]

The growth of British humanitarianism and European utilitarianism influenced the course of anthropology from the 1830s and this impinged indirectly on Australia. It could be claimed that the Aborigines, along with other indigenous colonial races, were endowed with the status of humanity. It can be no accident that the decade which produced the abolition of the transport of convicts to New South Wales, the House of Commons Select Committee on the Condition of Aborigines in British Colonies, and the establishment of the Aborigines' Protection Society produced the first execution of Europeans for the murder of Aborigines at Myall Creek in 1838. Similarly, this period witnessed the advent of sympathetic native protectors, such as G. A. Robinson in Tasmania and William Thomas in Victoria, whose greatness as ethnographers has been appreciated only recently.

The most significant trend in European anthropology was the formation of learned ethnological societies. Their membership overlapped with the activities mentioned above, and their pragmatic purpose was to stimulate colonists to record data systematically. They assumed that any educated observer would be enabled to record native customs in meticulous and routine fashion by means of questionnaires and manuals of inquiry.[3] Presumably, following the Europocentric notion that all savages were alike, these guidelines had universal application.

In England during 1839, the British Association for the Advancement of Science established a committee to produce a set of questions and accompanying instructions. At the same time, in France, the Société Ethnologique de Paris issued its *Instruction générale adressée aux voyageurs, 1840.*[4] Much influenced by the French pamphlet, the British Association committee promptly framed 89 questions, which also reflected the humanitarian interests of its exponents. By 1844 these questionnaires had been circulated widely in colonies, while the 1852 British Association proudly exhibited copies of its *Manual of Ethnological Inquiry.* It now contained 103 questions, under the general headings physical character, language, grammar, individual and family life, buildings and monuments, geography and statistics, social relations, and religion and superstitions.

Objectivity was the keynote of this positivist document, which was intended "to induce Consuls, political and other residents and travellers to obtain precise knowledge. . . . Any amount of knowledge, however trifling it may appear in itself, may be of great value in connexion with other knowledge. . . . We are seeking Facts, and not influences, what is observed, and not what is thought."[5]

Unadorned fact, indeed, became the truth of nineteenth-century anthropology and, transferred into the context of the classification of society and its institutions, it became the evolutionists' lodestone. The use of questionnaires dominated the Australian anthropological scene during the period 1872–85, with Lewis Henry Morgan the prime instigator.

J. G. Frazer later exerted his influence. Frazer, who never set foot in any field or sighted a single savage during his long life, produced 507 *Questions on the Customs, Beliefs, and Languages of Savages,* for "the civilised enquirer."[6] As he sat in his study during 1898, cobbling "facts" into explanations, he urged Baldwin Spencer, then beginning his anthropological career, to eschew "general theories and discussions. What we want . . . is a clear and precise statement of facts . . . that and nothing else."[7]

It seems appropriate that the author of *The Golden Bough* later used a mining simile to praise Spencer, then nearing the end of his career, for opening up "a deeper mine into the past of human institutions than any one else. . . . I have worked at the products you have brought up from the mine."[8] These delusions of objective grandeur involving colonial fact-finder and imperial synthesiser, however, anticipate our discussion.

The factual questionnaire survey, as canvassed by the British Association, possibly influenced colonial authorities by 1845. In that year, the Legislative Council of New South Wales appointed a select committee to report upon the condition of Aborigines, although their basic concern lay in controlling their land rather than in welfare or academic research. Eighteen questions were distributed to clergymen, an initiative that attracted praise from the Aborigines' Protection Society.[9] Lieutenant-Governor Charles LaTrobe proved to be a consul sympathetic to the questionnaire method. In 1853 he framed his own set of questions for Victorian pioneers, which emphasised the importance of Aboriginal data. He received more than fifty written responses, which constitute a significant source for the contact period in Victoria.[10]

The 1858 Select Committee of the Legislative Council of Victoria on the Aborigines proved most receptive to the British Association initiative. It adopted the association's manual and distributed its 103 questions widely. In its report, it published all responses, in the expectation that the corpus would be "prized by the learned societies of Europe."[11] Published in this obscure colonial parliamentary paper, it was either unknown or ignored by those overseas for whom it was intended. Its contents reflect the weaknesses of amateur anthropology conducted by questionnaire. Its random data are only as reliable as the comprehension, critical ability, and diligence of the respondents allowed. Such "facts" prove subjective and disconnected. One of the longest comments was supplied by an assertive and racially arrogant Bourke street phrenologist in whom the committee placed its trust.[12] Given the ambient atmosphere of protestant Christianity, it hardly is surprising that Victoria's savages were stated to be devoid of any religious belief; evidently they also lacked sexual life.

It is a chronological quirk that this lengthy colonial document, published without any integrating body of theory, belongs to the year in which *The Origin of Species* transplanted time and society beyond Eden's confines. Of science and learning before Darwin, J. C. Greene has emphasized the influence of enlightened free enterprise, rather than state patronage, in Britain.[13] He found that private enterprise was also the primary influence in America, but developments were less centralized and less well endowed. They were hampered, further, "by public attitudes that placed no high value on intellectual achievement per se."

The parallel holds for Australia, but with important qualifications. Public priorities were comparable, and the decentralization of learned institutions and societies within state capital cities proved striking. (Sydney's *Australian* Museum and Melbourne's *National*

Museum still challenge any name proposed for the new Commonwealth museum in Canberra.)

Despite the minimal expenditure by colonial governments on science, however, particularly anthropology, it is difficult to separate
private and public activities in the tiny microcosms that constituted
the young colonies. Governors Macquarie, Brisbane, and Grey, for
example, and Lieutenant-Governors La Trobe, Arthur, and Franklin
all stimulated the search for knowledge and supported the founding
of scientific societies or learned institutions. Colonial legislatures
actively assembled data that they deemed to be research material. In
the anthropological arena, colonial administrations actively encouraged the acquisition of useful knowledge about Aboriginal society.
Indeed, such officials often showed greater interest in and solicitude
for Aborigines than the colonists, who despised those whom they
were supplanting on the land.

Some of the most percipient and detailed observations made during the period before 1859, however, owed little to the British Colonial Office and nothing to the British Association manuals. These
were the accounts written by educated explorers, often in colonial
government employ, but always acting under strong official patronage. A. P. Elkin called this phase one of "casual or incidental anthropology."[14] This is an appropriate attribution, although it over-
states the incidental nature of the better accounts.

John Oxley, Charles Sturt, Thomas L. Mitchell, Ludwig Leichhardt, Edward J. Eyre, and Governor George Grey, together with the
ethnographic artist, George French Angas, recorded objective data
whose quality has stimulated and assisted numerous research projects within recent years.[15] Perhaps the two most valuable sources
were never published and remained unread until recently. These
were the journals and papers of George Augustus Robinson and William Thomas, salaried public servants employed in Tasmania and
Victoria as Protectors of Aborigines.

Apart from their official duties or the encouragement of their seniors, it is evident that the interest of such observers in Aboriginal
society was stimulated by their personal experiences. While some of
these men held professional qualifications, their comprehension of
ecological or environmental controls was instilled in all of them
only during their harsh bush schooling. They recorded pragmati-
cally in their journals themes which foreshadowed many research

"discoveries" and models of the past two decades. These include the symbiotic relations between man, flora, and fauna in hunter-gatherer society; the ecological role of man as a fire agent; ceremonial exchange systems and seasonality as dynamic cultural forces. George Grey described Australian totemism half a century before it became a matter of deep anthropological controversy.[16]

During this period before evolutionary theory influenced attitudes or the formulation of models of social behavior, the nearest approach to systematic social theory developed incidental to the field experience of such discerning colonial public servants or adventurers. Few of these individuals could be typed under Reingold's American classification as either cultivators or practitioners.[17] The perceptive approach to Aboriginal society by Eyre, Grey, and Robinson, for example, is best explained as a combination of a spirit of pragmatic colonial inquiry and acquired bush lore, together with grudging comprehension of Aboriginal adaptive ingenuity. Whatever sponsorship was involved in such work was provided essentially from within colonial society rather than from Britain, although the fact that Grey and Eyre devoted more than a quarter of their popular books to aboriginal subjects suggests that European readers were interested in such matters.

During the later decades of the century a new factor intruded into ethnological investigations, involving the participation of distinguished overseas scholars as mentors and patrons. Before considering the resultant network of intellectual kinship, however, it is necessary to emphasise that colonial governments played a more positive role in the collection and dissemination of ethnographic data than their negative record of welfare and racial discrimination suggests. Their considerable indirect support merits closer investigation than has been attempted. The collaboration of some senior public servant usually provided the occasion for assistance, but it would have proved impossible without wider government assent.

R. Brough Smyth, secretary of the Victorian Board for the Protection of Aborigines from 1861, was the archetypal civil servant who used public funds to assemble and publish useful knowledge. His massive two-volume compilation, *The Aborigines of Victoria*, was published by the Government Printer in 1876.[18] Ten years later, E. M. Curr's four volumes, *The Australian Race*, were issued by the same printer.[19] Both Smyth and Curr acknowledged the assistance of prominent public servants in several colonies. Curr stated that, without the cooperation of every colonial secretary in Australia, "it would not have been possible to extend the inquiries."[20] Critical for the production of these source books was the collaboration of governments in providing the addresses of the police and other functionaries with whom the compilers corresponded and in facilitating the delivery of letters, often without postage charges.

As a senior Victorian public servant, A. W. Howitt used official channels to facilitate his research. Aboriginal informants traveled to his Bairnsdale home with free rail warrants. All the numerous questionnaires that he and Lorimer Fison designed during several years were printed by the government printer, with Brough Smyth's collaboration, and despatched around Australia through official channels. More than 500 copies of one circular were printed in 1874 and were posted over Smyth's signature.[21] Only one year after the opening of the Overland Telegraph to Darwin, A. W. Howitt optimistically told his mother that through the assistance of Todd, head of the postal department, he intended sending copies of his questionnaires so that "men all along the telegraph line [will] work for us."[22]

When Baldwin Spencer spent 1901 on his transcontinental expedition with F. J. Gillen, the Victorian government paid Melbourne university both his salary and that of his replacement. Gillen was treated similarly by the South Australian post office. In the course of several years, the South Australian government invested funds in supporting Spencer and Gillen with facilities such as rail and horse transport, free telegraph, and police guides. Such funds seem trifling today, but at that time men like Howitt and Spencer otherwise financed their own activities. It should be remembered that, while Tylor and Frazer showered praise and advice on these anthropologists and while Frazer successfully petitioned the two colonial governments to release Spencer and Gillen for the 1901 expedition, neither the British government nor the eager academics thought fit to provide a pennyworth of funding.

It is relevant that several other important late nineteenth-century anthropological studies were published under colonial government authority. These include three of the most important south Australian reference books, two by G. Taplin, *Folklore, Manners, Customs and Languages of the Aborigines of Australia* and his 1874 account of the Narrinyeri, and one by T. Worsnop, *The Prehistoric Arts,*

Manufactures, Works, Weapons, etc, of the Aborigines of Australia.[23]
The Queensland government published W. E. Roth's *Ethnological Studies among the North-West-Central Queensland Aborigines* in 1897, while he was a government medical officer. After Roth became Protector of Aborigines, the government printer issued his important *North Queensland Ethnography Bulletins*, the later numbers of which were published by the Australian Museum, Sydney. In Western Australia, Daisy Bates was employed initially in Perth in the Registrar-General's office to compile a volume on Aborigines. Despite the patronage of Andrew Lang in England, her book was never published. It gained anthropological notoriety, however, because she claimed later that Radcliffe-Brown plagiarized her manuscript.[24]

These activities were so characteristic of the later nineteenth century that Elkin designated the period the "compiling and collating phase."[25] It represented considerable indirect colonial government assistance, but there are no indications that the British Colonial Office encouraged such activities.

R. R. Marett, Rector of Exeter, Baldwin Spencer's Oxford college, characterized anthropology in a manner appropriate to the course of systematic research in Australia from 1870: "Anthropology is the child of Darwin. . . . Reject the Darwinian point of view, and you must reject anthropology also."[26] Social Darwinism elevated Aboriginal society to the forefront of academic disputation. As an explanatory model and possibly as an intellectual solace in a harsh racial contact situation, it justified its lowly place in both nature and in the state. Like the sociological systemists who espoused evolutionary theory overseas, Australian practitioners classed their investigations as a variety of natural science. Their findings proved basic to world systems. Thus A. W. Howitt and Lorimer Fison were woven into the fabric of Lewis Henry Morgan's *Ancient Society* and from there passed into Engels's *Origin of the Family*. E. B. Tylor, James Frazer, and Andrew Lang classed those two Australians, along with Baldwin Spencer and others, as friends or combatants; they all stand tall in the pages of Emile Durkheim's *Elementary Forms of the Religious Life*. Both Malinowski and Radcliffe-Brown commenced their careers with a diet of the same authors.[27]

Testimony to the rapidity with which the intellectual climate changed is provided by the first congress of the Australasian Association for the Advancement of Science in 1888. Anthropology achieved sectional status from the outset. The first lecture, "Outlines of Anthropology," was delivered by J. J. Wild, assistant-secretary of the Royal Society of Victoria and a former artist on the *Challenger* expedition. Wild defined the discipline "as the critical examination of the intellectual and material progress of man from the earliest ages down to the present."[28] He felt assured that the study of Aboriginal society would illumine "the moral and intellectual conditions which appertain to the earliest stages of the human race." Progress and evolution had become the integrating themes whereby to order social data.

Elsewhere I have traced the interconnection between overseas evolutionist patrons and the careers of Howitt, Fison, and Spencer: here I will emphasize only salient matters to illuminate the history of anthropology through its most distinguished figures.[29] The Australian connection with systematic research ostensibly began in 1872, when Fijian missionary Lorimer Fison, already a disciple of L. H. Morgan, advertised in the Australian press for regional correspondents. He recruited Gippsland magistrate A. W. Howitt, whose mind was receptive to theories of social evolution and whose abilities place him in the front rank of Australian intellects.

Howitt had read Lyell's *Elements of Geology* in 1868, then the *Origin of Species* and Lubbock's *Pre-Historic Times*. By 1869 he had formed remarkably challenging opinions concerning both the antiquity and the status of the Aboriginal race. It is significant that he reached these conclusions in his bush isolation, long before Morgan's influence. He wrote to his sister on 18 April 69:

I have come to the conclusion that the Australian black is a wild man by nature and you "cannot wash a blackamoor white." . . . These blacks have the minds of children and the bodies of adults. I think they are indigenous to the soil and date from a period anterior to the great physical changes in Australian Geology which prevented migration into Australia of the fauna of the later Tertiary period.[30]

Howitt later told his mother of the force of Darwin's writing:

I . . . see their force and value more every day. I think the Descent of Man a wonderfully reasoned and suggestive book. All my studies have tended to impress their views on me and not least the study of ethnological questions relating to these blacks.[31]

A respectful Howitt evidently wrote to Charles Darwin shortly afterward and received the reply quoted at the beginning or this essay. If he felt disappointment at his courteous dismissal, he soon consoled himself with letterbox anthropology involving numerous questionnaires, which he and Fison began distributing with Morgan's encouragement.

Inspired both by Morgan's unstinted and frequent praise and by his urgent sense of destiny that their discoveries of kinship systems produced vital data on the origins of society, Fison and Howitt commenced systematic classificatory studies. They adapted and elaborated Morgan's questionnaire technique; but this was no slavish imitation. Critics who infer that these Australian researchers were devoid of theory and simply accepted direction should read their correspondence. They read all available interpreters of social institutions such as McLennan, Tylor, and Lubbock and compared their theories both with Morgan's and with their own experiences.[32]

Over several years, Howitt wrestled with the problem of how to frame questionnaires, and his letters testify to his attempts to check information supplied by correspondents. It was Morgan, however, who supplied the initial map of human social development, advice on how to fill in the detail, and encouragement to complete the task. In a letter to Fison praising his Fijian research, Morgan clearly expressed his vision of social Darwinism and the psychic unity of mankind. It is understandable, consequently, that Fison and Howitt approached ethnology with the conviction that it was a branch of natural science:

I have never in my limited studies found such striking proof of the natural logic of the human mind as systems of consanguinity discloses. We are held to the logical results of the ideas born in our brains. We cannot overthrow them, escape them or transmute them; but are compelled to follow in the line of their growth and development. Evolution is the law of mind as well as of nature. These savages show the specific identity of their brains with our own by their mental work as revealed in their institutions.[33]

To appreciate the importance of Morgan's blueprint, compare the random jottings on the Coopers Creek Aborigines, which Howitt wrote in 1871 for Brough Smyth's *Aborigines of Victoria*, before he was in contact with Fison, with the classificatory essay on the Brabrolong tribe, compiled in 1873 for the same volume, in which Howitt took Morgan's schema as his organizational basis.[34] Morgan sent Howitt a copy of *Ancient Society* in 1877. In expressing his thanks and agreement with its thesis, Howitt observed, "I am now going to reread your book and note everything that requires investigation here. Several new directions for enquiry have suggested themselves. I have already commenced on the food question, and I find that in our tribe the distribution of food among the family group is regulated by strict laws, as are also the positions of the camps. . . . A wide vista has opened itself to me."[35] Four years later, Howitt commented that he was following "the path which you have laid out. The more I work and the more materials I collect the more clearly I see how wonderful a field of inquiry there is here."[36]

Others also began to take interest in postal ethnological inquiry. Morgan had proposed publication of the study by Fison and Howitt, *Kamilaroi and Kurnai* (1880), as a memoir of the Smithsonian Institution. He failed, partly because the authors were impatient to publish quickly, to ensure priority in the face of local competition. They sensed rivalry from other colonial ethnologists, who, they claimed, "were using the methods of enquiry we have invented and which they have appropriated from our circulars without the least acknowledgement."[37] In this letter Howitt referred to four plagiarists, but he did not name them. It is evident, however, that he considered E. M. Curr and G. Taplin to be prime villains. Taplin later wrote that his questionnaires were inspired by a letter received from a South African linguist.[38] Since there is no doubt that he had already known of Howitt and Fison, however, their suspicions had foundation.

Apart from the fashionable use of local correspondents prompted by Fison, however, it is possible that some compilers were influenced by a British connection. The old British Association ethnological questionnaire had been taken over by the Anthropological Institute, which in 1874 published the first edition of its influential *Notes and Queries on Anthropology.*

To judge Australian anthropology in the 1870s by its two chief practitioners, Fison and Howitt, it was partisan to Morgan's cause and held little respect for English social theorists. Morgan died in 1881, however, and, since his health had waned, the balance of intellectual influence swung towards Britain. It was Tylor's pen and other personal issues that determined this shift; government policies played no part.

Some years earlier Fison had posted an article from Fiji to the Anthropological Institute in London. It was lost in transit, but he deduced that the failure to acknowledge it was the result of his being snubbed for his association with Morgan. A personal note of enquiry from the Fijian governor, Sir Arthur Gordon, to his friend Lieutenant-General Pitt-Rivers eventually elicited a letter from Tylor inviting Fison's collaboration. About the same time, Howitt's sister met Tylor and Tylor expressed interest in Howitt's research. The contacts thus forged between Tylor and the two Australians developed into a relationship as warm as the Morgan link and had consequences both for the direction taken by their research and for much of Tylor's speculation.[39]

The nexus of friendship and philosophical kinship between these three social evolutionists proved a critical intellectual influence on the youthful Baldwin Spencer, who in 1887, at the age of twenty-six years, took up the foundation chair of biology at the University of Melbourne. Together with D. O. Masson, simultaneously appointed to Melbourne's chemistry chair, and Edgeworth David, Sydney University geologist, Spencer became one of Australia's most influential scientific spokesmen. Fortunately for the course of Australian Anthropology, the elderly Howitt and Fison now lived in Melbourne, and Spencer enjoyed their friendship.[40]

Son of a Congregationalist Manchester manufacturer, Spencer spent a period at Owens College, where the young biologist Milnes Marshall introduced him to evolutionary theory. From 1881 he studied biology at Oxford under H. N. Moseley, who had returned recently from the *Challenger* expedition with a keen interest in ethnography. Spencer was fortunate in his opportunities, but he seized them. He attended lecture courses by Tylor and Ruskin in company with Halford Mackinder, future doyen of British geography. He and Mackinder shared rooms and vacationed together, and Spencer's early appreciation of Australian biogeographical factors possibly owe much to this association. It is a reflection of Moseley's inspiration, however, that members of Spencer's class included Henry Balfour, the future curator of the Pitt Rivers Museum, and W. E. Roth, the ethnographer of northern Queensland.

Spencer was a student when Pitt-Rivers, the systematist of the evolution of technology, donated his vast ethnographic collection to Oxford. During 1884 Moseley and Tylor employed Spencer to assist with its transfer from London and to arrange displays in the new Pitt Rivers Museum. When Spencer later applied for the Melbourne post, Tylor, Moseley, and R. Lankester, a future director of the British Museum (Natural History), wrote him references. It was a useful background for Spencer, who, from 1899, began a twenty-eight-year term as honorary director of the National Museum of Victoria. It is relevant, also, that he immediately set about the systematic collection of ethnographic material from Australia and the Pacific region. His 1901 *Guide to the Australian Ethnographical Collection in the National Museum of Victoria* set a standard of guide seldom matched by any Australian museum until recently. It also echoed remarkably the evolutionary theories on society and material culture of Tylor and Pitt-Rivers.

Spencer and Mackinder established the Oxford Junior Science club, to which they invited leading evolutionary biologists. Some of them were members of the electoral committee that chose Spencer; T. H. Huxley, also, was a member. Runner-up to Spencer was A. C. Haddon, who subsequently achieved anthropological fame in Torres Strait. Spencer later sought his advice on anthropological field matters, including photography, before undertaking his 1901 expedition.

Spencer maintained a network of links with British scientists during his subsequent career. An important contact was the London biologist G. B. Howes, a close friend and an editorial contributor to *Nature.* Howes highlighted Spencer's activities and alerted readers to his proposed overland anthropological expedition with Gillen, in tone recalling the romantic approbation accorded explorers half a century earlier.[41] Howes was one of several scientists who promoted Spencer's fellowship in the Royal Society. He also urged his cause for chairs at Oxford and Manchester, although Spencer withdrew his candidacy in both cases.

In Melbourne, Spencer rapidly established an active teaching department, in which field excursions were an innovative feature, and he developed a significant taxonomic research center. Not surprisingly, he promoted the formation of a student science society. Although his first lecturer, A. Dendy, came from the Manchester department, all other staff appointments appear to have been Australian, mostly from within the department. One remarkable aspect of his vigorous department was its number of female research students and staff, including Australia's first female university lecturer and Melbourne's first two associate professors. In 1919, at the time of his retirement, his entire staff was female. It seems likely that this was a unique record in the history of Australian science.

Spencer brought to Melbourne the vigor and administrative ability that characterized his brief Oxford career. While he maintained a network of influential colleagues in Britain, he soon acquired an aura of Australian nationalism. His arrival in Melbourne coincided with preparations for the Centennial celebrations. Even so, this does not explain why he wrote some scientific articles for the *Centennial Magazine*, where his illustrations appeared in 1888 alongside those of Tom Roberts, Streeton, Conder, and McCubbin. With the professor of classics as coeditor, he also launched a cultural review, *The Australasian Critic*, which proved a victim of the 1890 depression. By 1919 Spencer owned one of the largest collections of Australian impressionist art.

Spencer was also largely responsible for the early nature conservation movement, being particularly active in the establishment of the Wilsons Promontory National Park. In a percipient article on the indiscriminate ringbarking of eucalypts, he predicted massive erosion and other ecological effects.[42] Not surprising, at the first meeting of the Australasian Society for the Advancement of Science (AAAS), Spencer proposed a Standing Committee on the Protection of Native Birds and Mammals. As director of the National Museum he was active in attempts to prevent the export of fauna, even opposing a British Museum collecting expedition. In 1920 he urged the creation of a systematic biological survey, deploring the fact that most research was undertaken by foreigners, because of the failure of wealthy Australians to contribute funds toward biological research. His plea on behalf of Australian museums must advance his claims for inclusion in Basalla's third phase, 1967, that of scientific independence. "I would like to enter a strong protest against the sending to Europe of collections that can well be described here," Spencer asserted.[43] Local museums, he added, "are quite as competent to describe and take care of Australian type specimens as are our own colleagues in Europe."

In 1913, Spencer advised the Commonwealth government that

stringent control over the export of Aboriginal ethnographic and
skeletal material was necessary. Atlee Hunt, Department of Home
and Territory Affairs secretary, agreed to study the question in the
light of the New Zealand Maori Antiquities Act. Spencer's opinions
were incorporated into a proclamation dated 22 November 1913,
under the Customs Act. After that, export of ethnographic material
was prohibited unless by accredited scientific institutions. Spencer
implemented it the following year, when he blocked the export of
sixty-eight Aboriginal crania to Edinburgh, on the grounds that
Edinburgh already owned a collection larger than that held by any
Australian institution.[44] The flaw in this progressive move lay in the
reality that Spencer was an accredited museum director himself.
Through the years he used his own rich ethnographic collections as
a medium of exchange. "Spencer and Gillen" items appear today in
museums around the world.

Spencer arrived in Victoria at the period when both biological and
anthropological studies attracted amateur interest and poorly managed learned societies struggled to cope with these demands. Spencer's genius was to capitalize on this interest and channel it into
research activities. For a decade from 1889 he acted as secretary and
editor to the Royal Society of Victoria. He developed more positive
programs and initiated an important memoir series, including anthropology. Membership overlapped with that of the Victorian Field
Naturalists' Club, in which Spencer became the dominant force.
Under his guidance a series of useful field surveys was undertaken.
He was associated closely with the Port Phillip Biological Survey,
one of the earliest systematic attempts to record and assess Australian biological resources.

Near the close of his life, Spencer wrote an obituary of his Sydney
biologist friend, J. J. Fletcher. Fletcher returned from studies in London the year in which Spencer went up to Oxford. His reflections
possess an authentic autobiographical nostalgia for that brave new
world of evolutionary discovery, now half a century away:

*In study, field and laboratory scores of eager students relieved from
the dead weight of the special creation theory, were working under
the stimulus of an entirely new outlook on the world of life. It is
difficult for the students of the present day to realize the excitement of those times when everything was new and stimulating and*

when, further still, it was possible for one man to have a good all round knowledge of ... the salient features of the different branches of Science.[45]

Spencer intruded upon the Australian scene at the critical time with appropriate skills and the virtue that, as a newcomer, he was unencumbered by inherited attitudes or alignments. The first heroic century of Australian exploration had discovered Australia for the white man. It was time for a new kind of exploration involving meticulous, detailed regional or thematic research on specific issues. Just as all the main geographic regions and features had been discovered, so too, the broad outlines of Australian flora and fauna were known. Spencer's own research, however, and that of his colleagues and students, added immeasurably to their detailed comprehension. A systematic approach to taxonomic and habitat descriptions was essential. Consequently theses and memoirs describing earthworms, land planarians, marine flatworms, leeches, sponges, frogs, lung fish, and the like flowed from the biology school; a bibliography published in 1904 lists almost a hundred kindred departmental publications.[46] Today these descriptive studies seem unsophisticated and tedious, but they represent a crucial stage in Australian biological research. They also constituted an orderly, not a magpielike, assemblage. Spencer set priorities and the work possessed theoretical value. Various species were selected because of their potential evolutionary significance or their biogeographical interest, and they were assessed within the conceptual framework of biological evolution. Amateur field naturalists and students alike must have sensed the purposeful intent of their disparate field collections or laboratory dissections, as they were shown to possess pattern and purpose. There was also the satisfaction of field observation being turned quickly into published note, with due acknowledgment for assistance rendered on excursions or in the laboratory.

Spencer's interests shifted toward anthropology after his involvement, as zoologist, in the 1894 Horn expedition to Central Australia, one of the few scientific expeditions financed by an Australian businessman. Howitt and Fison possibly exerted the greatest influence on his anthropological progress, however. Queens College opened the Melbourne campus in 1888, under the mastership of an Owens College man, E. H. Sugden. Aiming to replicate Oxbridge,

Sugden established College fellowships, which ensured that Howitt, Fison, and Spencer, three of the first fellows, met regularly under congenial circumstances.

Spencer always acknowledged the guidance of these two elders, and he and F. J. Gillen dedicated their first book to them. The fact that Spencer readily accepted their version of Morgan's social evolution must have been the result of his own training and familiarity with Tylor's theories. It was Tylor upon whom he called for advice, after his Alice Springs fieldwork in 1896. Actually, Tylor proved obstructive and wanted to censor Spencer's text of *The Native Tribes of Central Australia*. It was rescued from his puritanical control by James Frazer.

Spencer is known best for his collaboration with Frazer, and their correspondence has been edited.[47] Spencer had read and approved *The Golden Bough* but had not met its author, who heard of Spencer's research from Fison. (Fison met Frazer when attending a British Association meeting, his visit being engineered largely by Spencer and Howitt in contact with Tylor). Frazer wrote to Spencer, via Fison, expressing his interest. Within a few months Frazer had volunteered to act as Spencer's literary agent and editor, securing Macmillan as publisher to Spencer and Gillen.

This relationship proved mutually positive, like that between Howitt and Morgan. Although dependent upon Frazer for suggestions and encouragement, Spencer retained his intellectual integrity. Both men, however, possessed minds formed by the first generation of evolutionary theorists, and their concepts and their assumptions remained fixed. From publication of *Native Tribes* in 1899 to *The Arunta* in 1927, Spencer presented an implicit stadial social model in which Australian "savages" typified the lowest stage, with their childlike minds adapted to that stage of evolution. The physical unity of the minds of savage men—for women were never considered—was assumed. Indeed, it is possible that Spencer's fatal last expedition to Tierra del Fuego was prompted by his expectation that he would confirm his interpretation of Australian superstition and magic—he never used the word *religion*—by finding analogous institutions surviving at another world's end.

As a statesman of intercolonial science Spencer promoted anthropology as a science. When he edited the *Handbook of Melbourne* (1890), while acting as general secretary to the second AAAS Congress, he asked Fison to contribute a chapter on the Aborigines. Spencer wrote a major anthropological section for the 1914 British Association meeting handbook. He remained an executive officer of the AAAS from its inaugural congress until he became congress president in 1921. He valued its New Zealand connection and maintained links with scientists there during several visits. That the British Association met in Australia was due chiefly to the efforts of David, Masson, and Spencer from 1909. Both Masson and Spencer attended meetings in London at different times in order to arrange schedules.

Because Spencer wanted visiting association anthropologists to experience Aboriginal field conditions, he arranged various excursions for them, but these were curtailed by the outbreak of the war. He did stage an exhibition of 10,000 stone artifacts in his museum, however, with the vain intention of demonstrating to European typologists that their taxonomic theories were erroneous.[48] One legacy of the war was Malinowski's enforced sojourn in Papua. In Spencer's biography, I have indicated that Spencer's support was crucial in ensuring the success of that celebrated field initiation. It is possible that without Spencer, Malinowski would have been expelled from Papua. Spencer interceded with the Commonwealth government and the Papuan administrator and, in wartime London, Spencer raised funds from Mond, Malinowski's sponsor, for his return field visit.

Spencer's own fieldwork benefited in 1901 from the petition signed by the British Liberal and academic establishment which Frazer arranged for presentation to the Victorian and South Australian governments. Frazer wanted to try again in 1908, to ensure that Spencer and Gillen visited the Kimberley region. When Spencer declined, Frazer and Marett turned elsewhere and raised support for the expedition that transformed A. R. Brown into Radcliffe-Brown.[49]

In an attempt to lobby for a chair of anthropology in Sydney, around 1923 Spencer and Masson (by this time Malinowksi's father-in-law) sought the intervention of Frazer and Haddon in Australian affairs. Both men had figured two years earlier in the establishment of the Australian National Research Council (ANRC), when they wrote to the government to emphasize the importance of research. It was the ANRC which now sought overseas submissions to back up its anthropological case to the reluctant and parsimonious government. After years of negotiation, Spencer was the effective ANRC spokesman at the time the chair was finally established in 1926.

Australia had proved an evolutionary quarry for well over half a century by that time, and Australian "facts" bolstered conflicting social models ranging from those of L. H. Morgan and Tylor to Andrew Lang and Durkheim. They influenced theories such as totemism, kinship systems, marriage institutions, and origins of prehistoric art. It is ironic that Spencer's university has never admitted anthropology as a discipline, despite the simultaneous presence on campus of Spencer, of Howitt, who served on the Council, and of Fison at Queens College. It is little known that in 1919 Spencer volunteered to act as an honorary reader in anthropology in order to initiate the teaching of social anthropology. As it happened, Spencer was appointed a representative of the ANRC on the Advisory Committee on Anthropology at Sydney, which was presided over by Australia's first full-time academic anthropologist, Radcliffe-Brown.

What, then, have been the status and achievement of anthropology? In colonial times local anthropologists showed deference to overseas social evolutionist gurus, and their research themes borrowed much from overseas models. Yet they retained a healthy independence combined with a growing professionalism, often proving more rigorous than their mentors. Australian anthropology had a brief and telescoped history in comparison with natural or biological science. It is doubtful, therefore, whether it can really be adapted into Basalla's three-stage model for the spread of science. Neither is it wholly conformable with the natural science format discerned by Moyal: "classificatory and descriptive and confined to taxonomic fields."[50] Australian clients were encouraged to provide only facts and descriptions by their patrons. While the resulting kinship tables or classificatory data provided those patrons with taxonomic information which bolstered their arcane systems, some Australian ethnographers did not abdicate their right to formulate their own theories.

Even so, anthropology failed to gain the respectability of an academic teaching discipline until almost four decades after it had been accepted as an independent AAAS section. It was also unfortunate in its first professor, who took a most restricted definition of his discipline and who influenced his students profoundly during his brief tenure.[51] Radcliffe-Brown's own department offered no prehistory before 1962; partly because the Depression and the War intervened, no other anthropology or linguistic department was established until the 1950s; all research or teaching in physical anthropology was performed by anatomists in their spare time; the first teaching chair in prehistory was taken up in 1971 and the first full-time lecturer in physical anthropology was appointed even more recently; not a single state enacted legislation to protect Aboriginal relics before 1965.

Misguided social Darwinists certainly stimulated Australian anthropological research and rated Aboriginal society as scientifically significant. Baldwin Spencer informed visitors to the National Museum of Victoria that Aboriginal society "may be regarded as a relic of the early childhood of mankind left stranded."[52] Only in recent years have anthropologists and prehistorians recognized it for what it is—one of the most precious and individual cultures, significant not only to science but also to humanity.

1. D. J. Mulvaney, "The Australian Aborigines: Opinion and Fieldwork," *Historical Studies in Australia and New Zealand* 8 (1958): 131–51, 297–314; Mulvaney, "Fact, Fancy, and Aboriginal Australian Ethnic Origins," *Mankind* 6 (1966): 299–305; Mulvaney, *The Prehistory of Australia* (Ringwood: Pelican, 1975).

2. George Basalla, "The Spread of Western Science," *Science* 156 (5 May 1967): 611–22.

3. Cf. Nathan Reingold, "Definitions and Speculations: The Professionalization of Science in America in the Nineteenth Century," in *The Pursuit of Knowledge in the Early American Republic*, ed. A. Oleson and S. C. Brown (Baltimore: Johns Hopkins University Press, 1976), p. 52.

4. J. Urry, "Notes and Queries on Anthropology and the Development of Field Methods in British Anthropology, 1870–1920," *Proceedings of the Royal Anthropological Institute*, 1972.

5. British Association for the Advancement of Science, 22d Report (Belfast, 1852), pp. 243–44.

6. Sir James G. Frazer, *Questions on the Customs, Beliefs, and Languages of Savages* (Cambridge: Cambridge University Press, 1907), p. 5.

7. R. R. Marett and T. K. Penniman, *Spencer's Scientific Correspondence* (Oxford: Oxford University Press, 1932), p. 23.

8. Ibid., p. 126.

9. *The Colonial Intelligencer; or, Aborigines' Friend* (London, 1847–48), p. 69.

10. T. F. Bride, *Letters from Victorian Pioneers* (Melbourne: Government Printer, 1898).

11. "Report of the Select Committee of the Legislative Council on the Aborigines," in *Votes and Proceedings of the Legislative Council of Victoria*, 1858–59, vol. 5, pp. 25–72.

12. Ibid., p. 468.

13. J. C. Greene, "Science, Learning, and Utility: Patterns of Organization in the Early American Republic," in *Pursuit of Knowledge*, ed. Oleson and Brown, pp. 1, 19.

14. A. P. Elkin, "Anthropology in Australia: One Chapter," *Mankind* 5 (1958): 225–42.

15. Oxley, *Journals of Two Expeditions* . . . (London, 1820); Sturt, *Two Expeditions into the Interior of Southern Australia*, 2 vols. (London, 1833); Sturt, *Narrative of an Expedition into Central Australia* (London, 1849); Mitchell, *Three Expeditions into the Interior of Eastern Australia*, 2 vols. (London, 1839); Mitchell, *Journal of an Expedition into the Interior of Tropical Australia* (London, 1848); Leichhardt, *Journal of an Overland Expedition in Australia* (London, 1847); Eyre, *Journal of Expedition* . . . , 2 vols. (London, 1845); Grey, *Journals of Two Expeditions of Discovery*, 2 vols. (London, 1841); and G. F. Angas, *South Australia Illustrated* (London, 1846).

16. Grey, *Journals of Two Expeditions*, vol. 2, p. 228.

17. Reingold, "Definitions and Speculations," p. 38.

18. Smyth, *The Aborigines of Victoria*, 2 vols. (Melbourne: Government Printer, 1876).

19. Curr, *The Australian Race*, 4 vols. (Melbourne: Government Printer, 1886).

20. Ibid., vol 1, p. xi.

21. D. J. Mulvaney, "The Ascent of Aboriginal Man: Howitt as Anthropologist," in *Come Wind, Come Weather*, ed. W. H. Walker (Melbourne: Melbourne University Press, 1971), pp. 299, 305, 323.

22. 10 August 1873, Howitt Papers, La Trobe Library, Melbourne.

23. Taplin, *The Narrinyeri* . . . (Adelaide: Government Printer, 1874), and *Folklore, Manners, Customs, and Languages of the South Australian Aborigines* (Adelaide: Government Printer, 1879); T. Worsnop, *The Prehistoric Arts, Manufactures, Works, Weapons, etc. of the Aborigines of Australia* (Adelaide: Government Printer, 1897).

24. E. Salter, *Daisy Bates* (Sydney: Angus and Robertson, 1971), pp. 152, 176.

25. Elkin, "Anthropology in Australia," p. 226.

26. R. R. Marett, *Anthropology* (London, 1911), p. 2.

27. Mulvaney, "The Australian Aborigines."

28. J. J. Wild, "Outline of Anthropology," *Proceedings of the Australasian Association for the Advancement of Science* (Sydney, 1889), pp. 443, 445.

29. D. J. Mulvaney, "The Australian Aborigines"; Mulvaney, "Fact, Fancy"; Mulvaney, "The Anthropologist as Tribal Elder," *Mankind* 7 (1970): 205–17; Mulvaney, "Ascent of Aboriginal Man"; Mulvaney, *Prehistory of Australia*; "Gum Leaves on the Golden Bough: Australia's Palaeolithic Survivals Discovered," in *Antiquity and Man*, ed. J. D. Evans (London: Thames and Hudson, 1981); D. J. Mulvaney and J. H. Calaby, *So Much That Is New: Baldwin Spencer, 1860–1929* (Melbourne: Melbourne University Press, 1985).

30. Howitt papers, La Trobe Library; quoted in part in *Come Wind, Come Weather*, ed. Walker, p. 212.

31. 10 August 1873, Howitt papers, La Trobe Library.

32. Cf. Walker, *Come Wind*, pp. 296–97, 300.

33. L. H. Morgan to L. Fison, 5 February 1872, typewritten copy, La Trobe Library.

34. Smyth, *Aborigines*, vol. 2, pp. 300–309, 323–32.

35. B. J. Stern, "Selections from the Letters of L. Fison and A. W. Howitt to L. H. Morgan," *American Anthropologist* 32 (1930): 262.

36. Ibid., p. 447.

37. Walker, *Come Wind*, pp. 229, 300.

38. Taplin, *Folklore*, p. 1.

39. Mulvaney, "Australian Aborigines," pp. 36–47; Mulvaney, "Anthropologist as Tribal Elder"; Mulvaney, "Gum Leaves."

40. Much of the documentation for what is said in the remaining pages of this essay is contained in the biography of Spencer, *So Much That Is New*, by Mulvaney and Calaby.

41. See, for example, *Nature* 61 (1900): 277; 63 (1901) : 592, 616; 64 (1901): 132.

42. *The Australian Critic* 1 (1890): 185.

43. W. B. Spencer, "The Necessity for an Immediate and Co-ordinated Investigation into the Land and Fresh-Water Fauna of Australia and Tasmania," *Victoria Naturalist* 37 (1920): 120–22.

44. Commonwealth Archives A3 NT 13/11442; R. T. M. Pescott, *Collections of a Century* (Melbourne: National Museum, 1954).

45. W. B. Spencer, "J. J. Fletcher," *Proceedings of the Linnean Society of New South Wales* 52 (1927): xxxiv.

46. Royal Commission on the University of Melbourne, *Victoria: Papers Presented to Parliament*, vol. 2 (1904).

47. Marett and Penniman, *Spencer's Scientific Correspondence*.

48. A. S. Kenyon and D. J. Mahony, "Stone Implements of the Australian Aborigine," *Guide* (Melbourne: National Museum, 1914).

49. Marett and Penniman, *Spencer's Scientific Correspondence*, p. 112.

50. Ann Mozley Moyal, *Scientists in Nineteenth Century Australia* (Sydney: Cassell Australia, 1976), p. 3.

51. Mulvaney, *Prehistory of Australia*, p. 121.

52. Baldwin Spencer, *Guide to the Australian Ethnographical Collection in the National Museum of Victoria* (Melbourne: Government Printer, 1901), p. 12.

The Zoological Exploration of the Australian Region and Its Impact on Biological Theory

KATHLEEN G. DUGAN

As Europeans began to explore the more remote areas of the world, European scientists received with great excitement reports of new plants and animals. They were eager to learn more, but their researches were frequently hampered by the geographic isolation of the new lands and the scientific ignorance of many observers and collectors. The accuracy and completeness of European knowledge about the alien environment depended on the establishment of an effective system for the collection of data. It was therefore in the interest of the European scientific community to promote an interest in science among colonists and to support the growth of colonial scientific institutions. Colonial scientists and collectors who provided the raw data expected recognition of their contribution and were interested in the theoretical interpretation placed by the European scientific community on the new data. An examination of the growth of knowledge about the zoology of the Australian region during the nineteenth century illustrates the difficulties involved in establishing cooperation between two distinct scientific communities separated by educational, philosophical, and social differences.

During the early stages of colonial science, scientists concentrated on learning about the natural environment and about the plants, animals, geology, and geography of the new land. This kind of information was of particular importance to nineteenth-century biologists who were confronted with problems of adaptation, geographic distribution, fossil history, classification, and comparative anatomy. Biologists were seeking a theoretical framework that would explain all these satisfactorily. The answer was eventually provided by the Darwinian theory of evolution, but this theoretical synthesis was achieved only after many vigorous debates, in which evidence from distant lands figured significantly.

The peculiar plants and animals of the Australian region provided many surprises. Since the area was explored comparatively late, European biological theories developed largely without reference to its flora and fauna. The new discoveries could not easily be explained by existing theories, so biologists were forced to look for new explanations. In particular, the pouched mammals (marsupials) and the egg-laying mammals (monotremes) posed special problems of reproduction, classification, and past and present geographic distribution. Those problems, which were of great theoretical importance, aroused much interest and debate among European scientists, with evolutionists and antievolutionists taking opposing sides.

The inability of European scientists to observe these animals in their natural environment frustrated their efforts to learn more about them. The distance of Australia from Europe and time involved in travel ensured that few recognized scientists would visit the continent. For this reason, scientists were eager to develop a system for the collection of information that would ensure a steady flow of observations and specimens into the laboratories and museums of Europe. European scientists solicited the assistance of colonists who were interested in natural history.

The Australian collectors and scientists who made the observations and contributed the specimens were unable to participate in the theoretical debates that were sparked by their evidence. Many lacked adequate scientific training to understand the issues involved. Those with the requisite education lacked up-to-date information. They could not take part in the meetings and conversations that provided a forum for the exchange of ideas among the European scientific community. The absence of good libraries with adequate scientific collections, moreover, and the slow trickle of current scientific journals and books into Australia ensured that the colonists remained largely outside even the more formal systems of information exchange. Further, the social and intellectual climate within the colony sometimes made European scientific theories unacceptable to colonial scientists. The Australian scientific community, for example, remained largely anti-Darwinist long after Darwinian evolution had achieved general acceptance among European scientists.[1]

Because of these differences, European scientists were not interested in the theoretical opinions of Australian scientists. They wanted only the raw data, which they then interpreted within the framework of European theoretical debates. They did not regard the Australians as their scientific colleagues or intellectual peers. Because they doubted the scientific competence of the Australian observers they were likely to reject any empirical evidence that violated European expectations unless it could be verified independently by a member of the European scientific community.

The marsupials of the Australian region provided important evidence concerning both geographic distribution and fossil history. The Darwinian theory of evolution provided a much more satisfactory explanation of these new facts than competing, antievolutionary theories. The information-gathering network within colonial science was thus able to contribute important data to current theoretical debates. The case of the egg-laying mammals, however, shows that the information-collection system was inadequate for the identification and demonstration of unsuspected flaws in European biological theories. I shall examine here these three different cases, paying particular attention to the collection of the empirical evidence, its effect on European scientific theories, and the interaction between the European and Australian scientific communities.

In 1830 the first important discovery of vertebrate fossils from Australia was made in the Wellington Caves of New South Wales. These fossils, the first evidence of past mammalian geographic distribution on a continent known for its biological peculiarities, had a significant effect on European biological thinking.[2] Clergyman John Dunmore Lang examined the fossils with the help of Aborigines and interpreted them as evidence of a worldwide flood.[3] Lang offered the Australian fossils as support for the catastrophist theory of English geologist William Buckland, who argued that cave fossils were the remains of animals destroyed by a divinely caused calamity. These species were succeeded, Buckland claimed, by a new, unrelated set of animals miraculously created to replace them.[4] Lang believed that the fossils offered scientific proof of the biblical account of Noah's flood and evidence of the benevolent foresight of the Creator.

For if this territory were over-run with such beasts of prey as the antediluvian inhabitants of the cave at Wellington Valley, it would not have been so eligible a place for the residence of man as it actually is. The tiger or hyaena would have been a much more formidable enemy of the Bathurst settler than the despicable native dog, though indeed they would certainly have afforded a much nobler game to the gentlemen of the Bathurst Hunt.[5]

Lang's views were published in Europe, but his interpretation was entirely ignored. European theorists reversed Lang's argument, using the Wellington fossils to attack Buckland's theory and cast doubt on special creationism.

European scientists were impressed with the fact that the fossils represented extinct species that are closely related to living Australian species. This suggested that the laws of geographic distribution (which currently isolate marsupials in Australia) operated in the recent geological past as well. When Darwin returned from the *Beagle* voyage, Richard Owen identified his Argentinian fossils as extinct species closely related to living South American species. This evidence, combined with the Wellington Caves fossils, led Darwin to formulate the "law of the succession of types" stating that "existing animals have a close relationship in form with extinct species."[6]

To support his catastrophist theory, Buckland emphasized the differences between extinct and living species. The law of succession, by contrast, stressed the similarities. The law established an organic continuity between past and present that strongly suggested an evolutionary connection. The special creationists could not adequately explain why two separate, independent creations should bear such a close relation to each other. If, however, the theory that new species evolve from pre-existing ones is accepted, then living species must be closely related to recently extinct species. The law of succession had a significant effect on Darwin's thinking. It was this law, combined with his observations on geographic distribution in the Galápagos, that led him to doubt the stability of species and begin serious research on evolution.

The Wellington Caves fossils provided important empirical evidence to suggest and support an evolutionary theory. But the Australians who discovered this evidence and communicated it to the European scientific community adopted a very different theoretical interpretation consistent with their conservative theological beliefs. As will be seen, the inability of the colonial scientific community to participate in the European theoretical debates and their reluctance to accept the theories developed in Europe hindered scientific progress in both places.

Australia is now famous as the "land of living fossils." The similarity between the living plants and animals of Australia and the fossil species of the European Mesozoic, first noticed in 1838, was regarded as evidence of the primitive nature of the Australian environment. At first the correlation attracted little attention.

The subject became the focus of intense debate in 1844 when Robert Chambers used the correlation to support his own evolutionary theory.[7] He argued that the plants and animals of the European Mesozoic resembled the living flora and fauna of Australia because both were at the same early stage of evolutionary development. Chambers wished to show that species had gradually evolved by a natural process governed by God-given natural law. He used the Australian evidence to attack the theories of the antievolutionists, who attributed the origin of species to the special, miraculous exertions of a divine Creator. He eventually concluded that,

Development is a matter of time and in the case of [the Galápagos Islands and Australia] the full time has not yet elapsed. It is therefore exactly what we might expect, upon the natural hypothesis that, in these regions, animals [sic] life should have yet hardly reached the mammalian stage, the point which was attained in our elder and greater province about the time of the [Mesozoic]. On the other hand, no rational cause for this imperfect zoological show can be presented in consonance with the plan of special exertions.[8]

Chambers believed that development was only a function of time. Therefore the primitive nature of the Australian plants and animals proved that Australia was a young continent, only recently emerged from the ocean.

Chambers's theories provoked bitter attacks from special creationists who objected to his amateurish, inaccurate science and regarded his atheistic and materialistic tendencies as fundamentally dangerous to Christianity and society. They argued that the similarities between modern Australia and the European Mesozoic were attributable to similar environments rather than age. According to the principle of adaptation, one should expect to find similar species in similar environments. Adaptation, they believed, provided positive evidence of God's beneficent design. In a further effort to discredit Chambers, special creationists argued that Australia was not younger than other continents but was, in fact, much older. Special creationists claimed that evolutionists could not explain the absence of organic change on this ancient continent.[9] Australian scientists did not actively participate in this debate. Lacking the necessary information about the fossil history of Europe, they were not in a position to notice or comment upon the similarities. To the extent that they expressed any opinions on the subject, they agreed with the theories of special creationists.[10]

When Darwin formulated a new theory of evolution by natural selection, special creationists attacked it with the same arguments they had used against Chambers. How could two different sets of plants and animals, so widely separated in space and time, be physically related to each other? They claimed that the discontinuities in fossil history and geographic distribution made a gradual process of evolutionary development impossible in this case. The Darwinists, moreover, failed to explain why such an ancient continent has such primitive, undeveloped flora and fauna.

To answer these criticisms, Darwinists quickly reinterpreted the Australian evidence. For those naturalists who believed in the stability of species, the principle of adaptation was the only possible explanation for the observed similarities between the plants and animals of the European Mesozoic and those of modern Australia. This principle was also applied to explain the peculiarities of present distribution. Why are marsupials almost exclusively confined to the Australian region and why are no large placentals to be found there? Antievolutionists argued that there must be some special feature of the Australian environment that made it uniquely suited to marsupials. The adaptation explanation collapsed under Darwinist criticism because theorists failed to identify any environmental feature characteristic of the entire Australian region that was to be found nowhere else.

Richard Owen argued that marsupials were especially well adapted to the Australian deserts. Mothers could carry their young in a pouch on their long searches for water.[11] Darwin regarded this explanation as a "gigantic hallucination," pointing out that marsupials are found in the rain forests of Brazil. New Guinea, moreover, with a very wet climate, is also populated extensively by marsupials.[12] Owen's theory also failed to account for the similarity to European fossils because the lush vegetation of the European Mesozoic indicated abundant rainfall.[13] Most damning to the adaptation argument was the fact that when placentals such as cats, rabbits, and goats are introduced into Australia they thrive at the expense of the native marsupials. The failure of marsupials to compete successfully with introduced placentals proved, Darwinists argued, that they were not perfectly adapted to their environment.[14]

Against special creationist explanations of the Australian evidence, Darwinists developed their own explanation that they based on isolation and evolution by natural selection.[15] At the time of the Mesozoic, "primitive" plants and animals had worldwide distribution. Australia and Europe must therefore have been linked by a land bridge. With the disappearance of this land bridge at a comparatively early period in the earth's history, Australia became isolated.[16] In other parts of the world, placentals evolved in response to competitive pressures and environmental change. Unable to compete with the superior placentals, marsupials died out. But in Australia there had been reduced competition and little environmental change, so the "primitive" flora and fauna were little changed.[17]

The Darwinists' explanation was clearly superior to that of their special creationist opponents. No longer could the ancient, unchanged character of the Australian flora and fauna be used as an argument against evolution. The Darwinists, moreover, could explain the competitive success of placentals introduced into Australia. Darwinian theory explained why dissimilar environments in Australia were inhabited by similar animals and why similar environments in Australia and other continents were populated by dissimilar animals. Adaptation failed to explain these facts. The Darwinist explanation soon became widely accepted within the European scientific community.

Colonial scientists were, however, unwilling to accept this interpretation. Australian scientific activity was strongly influenced by religious belief. Among the few colonists who had the education and leisure to engage in scientific research, a significant number were clergymen working within the tradition of natural theology. They studied creation the better to understand the Creator. A general lack of good scientific training combined with the difficulty of obtaining up-to-date information from Europe may have contributed to their scientific conservatism. Colonial scientists continued to cling to theories that had become outdated in Europe. The Australian scientific community strongly objected when Australian evidence was used to support a theory that they found scientifically and morally repugnant. Henry Barkly, president of the Royal Society of Victoria and governor of that colony, called upon Australian scientists to join in the holy war against Darwinism and refute "errors so pernicious to the very existence of Christianity."[18]

Frederick McCoy, a leading member of the Australian scientific community and chief representative of institutionalized science in Victoria, rose to the challenge. McCoy came to Australia in 1854 as professor of natural science at Melbourne University and later became the first director of the National Museum of Victoria. McCoy had been working on Australian paleontology for some time but did not enter into the debate about the age of the Australian flora and fauna until threatened by the Darwinist arguments. McCoy wanted to develop a coherent explanation of the Australian evidence that would combat evolutionary theory. He argued that Australia had not, as Darwinists claimed, existed for a long period with little geological change. Rather, the continent had been almost entirely submerged, as had most of the world, during the early Tertiary, and all the plants and animals had been destroyed. The flora and fauna that now populate Australia represent a new creation, entirely unrelated to Mesozoic fossil species.[19] As an alternative to Darwinian evolution, McCoy proposed a theory of discontinuous progression with a succession of catastrophic disasters followed by miraculous creations. This theory had once been popular in Europe and had been defended by Adam Sedgwick, McCoy's teacher at Cambridge.

McCoy's theories were published in Europe but were virtually ignored there.[20] Australian scientists, however, found them convincing. Catholic priest Julian Tenison-Woods, a prominent geologist in the Australian scientific community, used McCoy's theories to combat Darwinism. Colonial theorists had no influence in the European scientific debate.[21] Despite their efforts, the Australian evidence was given a Darwinist interpretation.

The observations and specimens provided from Australia added to a growing body of knowledge about worldwide geographic distribution. The adaptation argument collapsed under the weight of all this evidence. Australia provided just one example, although the uniqueness of the Australian flora and fauna made it a particularly graphic example. The similarity of the Australian plants and animals to those of the European Mesozoic was particularly important because it pointed to relations between extinct and living species that must be explained as a result of past geological changes. Darwinists were able to explain the Australian distribution as a logical consequence of evolution by natural selection, demonstrating the superior explanatory power of the new theory.

In all these debates, the Australian evidence was important only insofar as it could be used as evidence for or against a particular theory. When the theoretical argument shifted, the significance of the observation changed. It was therefore extremely difficult for Australian naturalists to participate in the debate. Although they were in a better position to collect evidence than their European counterparts, their isolation from the centers of scientific activity made it impossible for them to evaluate its significance in the light of current theoretical problems.

These same factors made it nearly impossible for the Australians to identify and demonstrate significant errors in European theoretical beliefs even when directly confronted with evidence, as the case of the egg-laying mammals shows. The monotremes (platypus and echidna) are confined entirely to Australia and New Guinea, so Europeans had had no previous acquaintance with mammals that laid eggs. European taxonomy was based on the assumption that all milk-giving animals gave birth to live young; all warm-blooded, egg-laying animals were classified as birds; and all egg-laying quadrupeds were placed among the reptiles. The monotremes, combining characteristics from all three groups, violated a well-established regularity of nature. The egg-laying mammals forced a reordering of traditional taxonomic categories. These transitional creatures, linking distinct taxonomic groups, offered evidence in favor of evolution. More than eighty-five years after the first platypus specimen arrived in Europe, European scientists finally accepted that the monotremes laid eggs. The delay reveals the weaknesses of the system of collecting information in colonial science and the resistance of the European scientific community to evidence that violated their theoretical preconceptions.[22]

The first dried platypus specimen to arrive in Europe, in 1797, created great excitement because the animal appeared to combine general mammalian features with a duck's bill. Skeptical biologists suspected a hoax, but once its authenticity had been established, the creature raised no serious theoretical problems.[23] Its peculiar anatomy could easily be explained by the principle of adaptation. Its anomalous "beak" was well suited for feeding in the mud of riverbeds. Naturalists assumed that the platypus was a normal, viviparous, milk-giving mammal.[24]

These assumptions were challenged when the arrival of the first spirit specimens enabled Europeans to examine its internal anatomy. Platypus reproductive structures resembled those of a lizard, and anatomists could find no evidence of mammary glands. This led to astonished speculations that the creature laid eggs.[25] This argument was particularly attractive to evolutionists. The sharp, well-defined divisions between the various vertebrate classes constituted a serious objection to theories of transmutation. Evolutionists seized on the monotremes as transitional forms, linking the lower vertebrates to the mammals.[26]

As long as anatomists failed to identify the mammary glands, European biologists were ready to agree that monotreme reproduction was abnormal.[27] Sporadic reports from Australia—by Aborigines, colonists, and visiting scientists—confirmed that monotremes laid eggs.[28] European biologists did not challenge these claims. Between 1801 and 1823 nearly everyone agreed that monotremes were anomalous creatures differing significantly from normal mammals.[29]

The discovery of platypus mammary glands in 1824 began a decade of vigorous debate concerning the true nature of monotremes.[30] Biologists were convinced that milk-giving animals always produced live young.[31] If monotremes gave milk, they reasoned, then they must reproduce just like other mammals. Theorists split into two camps. Led by evolutionist Étienne Geoffroy Saint-Hilaire, those who wished to emphasize the anomalous nature of the monotremes claimed that the animals laid eggs and did not give milk.[32] The opposition, led by antievolutionist Richard Owen, argued that they were normal mammals, giving milk and producing live young.[33] Unable to believe in animals that both gave milk and laid eggs, both sides tried to force the evidence to fit their preconceived categories. The European debate was based primarily on evidence from pre-served specimens because this was the only evidence that could be examined in Europe. Owen and his followers pointed triumphantly to the mammary glands.[34] Geoffroy and his supporters argued that these glands could not possibly give milk and so must serve some other useful function. Perhaps they were scent glands or lubricating glands.[35] Maybe they produced a kind of mucus to nourish the young platypuses, or possibly they secreted carbonate of soda to form the eggshell.[36] Geoffroy argued that anatomical dissection re-vealed eggs inside the body of the female platypus, thus conclusively demonstrating that the animals were oviparous. Owen had to admit that the internal reproductive anatomy was unusual. He agreed that eggs were formed, but he believed that they were incubated and hatched inside the mother's body; the young were born live. The debate could not be resolved with the evidence available in Europe.

The central empirical problems were, Do these animals give milk? Do they lay eggs? These questions could only be answered by field observations in Australia. For more than fifty years European scientists failed to find the correct solution, not because Australian colonists were unable to provide the correct answers but rather be-cause European scientists were unwilling to believe them.[37] Lauder-dale Maule, a military officer in New South Wales, decided to inves-tigate the truth of the "generally accepted belief"—accepted in Australia, that is—that the platypus laid eggs and gave milk. He lo-cated a platypus nest, observed milk flowing from the mammaries of the mother, and found remnants of eggshell in the debris in the nest.[38] Owen used Maule's account to support his claims that mono-tremes gave milk, but dismissed the eggshells, suggesting that they were merely excrement coated in urine salts.[39] Conversely Geoffroy was quite ready to believe in Maule's eggs but dismissed all colonial observations of milk on the grounds that there was no proof that the substance really was milk.[40]

Geoffroy eagerly seized on the Australian reports of platypus eggs, but Owen and his supporters rejected these as unsubstantiated, sec-ond- and third-hand information. When he heard that platypus eggs had arrived in England, Geoffroy believed that his case had been conclusively proven. He was acutely embarrassed when it was demonstrated that the eggs, later identified as turtle eggs, were too large to pass through the pelvis of the platypus.[41]

Geoffroy had no direct contacts with Australian colonists, so he had to rely on random bits and pieces of information that other European scientists saw fit to publish. He could not request specific information nor could he evaluate the reliability of the observers. Owen, on the other hand, could take full advantage of the colonial relationship between Australia and England. He was provided with a steady flow of information from a number of Australian sources and could direct specific requests to individuals and institutions.[42] He could control the publication and initial interpretation of the evidence.

Owen's main informant was a Sydney physician, George Bennett, one of the most respected members of the Australian scientific community. The intellectual and emotional relationship between Bennett and Owen, as shown in their correspondence, reveals something of the personal side of colonial science.[43]

Bennett and Owen became friends while they were both young men at the Royal College of Surgeons in London. Owen remained in England while Bennett traveled and finally settled in Sydney. Bennett's medical training enabled him to understand the major theoretical debates of European science and made him a careful, reliable observer. Bennett would have preferred a career in the study of Australian natural history, but the lack of financial support for colonial science forced him to devote much time and attention to his medical practice.

In 1829, on a sea voyage in the New Hebrides, Bennett came across a living specimen of the pearly nautilus. Recognizing its rarity—none had been described since 1602—and its importance, Bennett sent it to Owen, who was at that time a little-known young scientist. Owen's description of that specimen won him international recognition and election as a Fellow of the Royal Society. For the next fifty years Bennett provided Owen with vast numbers of specimens of Australian animals. These provided the raw material for a large portion of Owen's scientific work and helped make his reputation as one of the world's greatest comparative anatomists.

Bennett made a great financial and emotional commitment to this exchange of information. He sent Owen countless letters, observations, scientific papers, and specimens. He financed his own research expeditions, and when he was too old for the vigorous demands of field research in remote areas, he sent his son, George Frederick Owen Bennett, to continue to collect the specimens Owen requested. In return, Bennett received some letters—about one tenth as many as he wrote—and some copies of Owen's published papers. He was continually frustrated by his inability to obtain up-to-date scientific information from Europe. He complained that his ignorance of European descriptions of the Australian specimens hindered his own scientific work. In recognition of his contributions, Bennett received a number of international honors, such as Fellow of the Linnean Society of London, Fellow of the Royal College of Surgeons of England, and Fellow of the Zoological Society of London, which secured for him a respected position within the Australian scientific community. Within the European scientific community Bennett was recognized for providing valuable observations and specimens; Owen received the credit for establishing their theoretical significance.

Clearly the exchange of information was not evenly balanced, yet it was Bennett who felt obligated to Owen. Recognizing his own scientific inferiority, Bennett gratefully acknowledged his intellectual debt to his mentor. Bennett, adopting Owen's theoretical convictions, argued that monotremes must give milk and give birth to live young. These preconceptions controlled the direction of his research and prevented him from discovering the true nature of platypus reproduction.

One group of people who obviously had the necessary knowledge to resolve the dispute were the Aborigines. Other colonists had cited Aboriginal reports as evidence that the platypus laid eggs. Bennett decided to question the Aborigines himself "to find out the cause of error." When he was told that the animals laid eggs, he assumed that either his informant misunderstood the question or he himself had misunderstood the answer. He repeated his questions until he eventually got an answer that he interpreted to mean that the platypus produced live young. He attributed the confusion to the "fact," alleged by Richard Owen, that eggs form and hatch within the mother's body. Bennett used Aboriginal accounts in support of his claim that monotremes gave milk, but, with respect to reproduction, concluded "that no dependence can be placed on native accounts, but that naturalists must seek for information in their own investigations."[44] The system of colonial science left scientists completely unable to collect biological information from the people best qualified to provide it. Although language differences may have been a contributing factor, the principal reason for this failure was unwillingness to accept information from people outside the scientific community.

Bennett recognized that three important points needed to be demonstrated: monotremes gave milk, eggs were formed inside the mother's body, and the young were born live. He quickly found the necessary evidence to prove the first two claims. He wanted European recognition for these discoveries so he sent his specimens and observations to Owen, requesting that he arrange for publication.

Owen published Bennett's account as requested. At the same time Owen identified milk in the stomach of a young platypus that Bennett had sent him.[45] Owen, rather than Bennett, was credited with finally proving that the monotremes gave milk. Scientists who had previously rejected all Australian reports now accepted Owen's conclusions because they were based on anatomical evidence available in Europe and because no one would question the scientific judgment of an observer of Owen's stature.

In 1834 Geoffroy conceded defeat, admitting that the platypus produces milk and gives birth to live young.[46] Owen's demonstration ended the controversy because scientists continued to assume a necessary connection between lactation and live birth. For the next fifty years, reports of platypus eggs were either ignored or explained away. In 1864, for example, Owen received an account from a Victorian physician claiming that a recently captured platypus had laid two eggs. Owen rejected this evidence on the grounds that the eggs had not been preserved so he was unable to examine their contents. He suggested that perhaps the trauma of capture had caused the animal to abort spontaneously.[47] European scientists were clearly unwilling to accept colonial evidence that challenged their conclusions.

Meanwhile, George Bennett spent years searching for the last piece of evidence to prove Owen's theory: a female with a hatched fetus inside her body. Since Bennett had limited time and money, he often shot the animals while they were feeding instead of investing the time and labor necessary to dig them out of the nest. Using this procedure, he could never hope to find platypus eggs. Bennett failed to find the answer because he was looking for the wrong thing in the wrong place.

European biologists had made a serious mistake and did not know it. Australian scientists, more easily convinced by European authority than by the empirical evidence available in Australia, were led into error. Yielding to the scientific judgment of Owen and Bennett, most Australian scientists agreed that monotremes gave birth to live young.[48] Others weren't sure or considered the question unresolved. Meanwhile, Aborigines and other Australians not educated within the European scientific tradition continued to believe in platypus eggs.

This situation continued until 1884, when two different researchers simultaneously and independently discovered echidna eggs. Wilhelm Haacke, director of the South Australian Museum, acquired a pair of echidna in order to study their early development. Convinced that their young were born live, he was quite astonished to find an egg in the female's pouch. He hastened to announce his discovery and display the egg at the next meeting of the Royal Society of South Australia.[49] Adelaide newspapers took little notice, commenting only briefly in their regular coverage of the scientific meetings that Haacke's egg "proves to a certain extent" the theory that monotremes laid eggs.

During that same week William H. Caldwell, working in the Queensland bush, also discovered an echidna egg. Caldwell, a British scientist funded by Cambridge University, had come to Australia to study the reproduction of transitional animals, specifically monotremes, marsupials, and Ceratodus (lungfish). Caldwell sought and received the cooperation of colonial governments and colonial scientific institutions. He chose to go to Queensland because the large number of Aborigines there supplied him with the cheap labor necessary for a mass collecting effort. Caldwell used Aborigines as a convenient source of labor but clearly regarded them as inferior people with no valuable scientific information to offer. He remarked,

A skilful black, when he was hungry, generally brought in one female Echidna, together with several males, every day.... The blacks were paid half-a-crown for every female, but the price of

flour, tea, and sugar, which I sold to them, rose with the supply of Echidna. *The half-crowns were, therefore, always just sufficient enough to keep the lazy blacks hungry.*[50]

That such blatant racism was acceptable within the European scientific community suggests another reason that European scientists failed to make use of Aboriginal knowledge about Australian natural history.

Like other European scientists, Caldwell was firmly convinced that the monotremes gave birth to live young. He was quite surprised to discover an echidna egg. Eager to publicize his discoveries among his European colleagues, Caldwell notified Professor Liversidge in Sydney requesting that he cable members of the British Association for the Advancement of Science, which was then meeting in Montreal. As a matter of courtesy, Caldwell also privately informed Bennett. Caldwell's telegram, "Monotremes oviparous, ovum meroblastic," was read to the Montreal meeting on September 2, 1884, the same date that Haacke announced his discovery to the Royal Society of South Australia. The reaction to Caldwell's announcement, "the most important scientific news that had been communicated to the present meeting," was sensational. The discovery made headlines on three continents. No one doubted Caldwell's claim.

Caldwell did not feel that it was necessary to inform the Australian scientific community of his discovery, so they learned of it only indirectly. The news was cabled from Sydney to Montreal to London and then back to Sydney. In the course of these many transmissions, a serious error was made and the Australian press announced the astonishing discovery that monotremes were viviparous. Since scientists had long believed that monotremes gave birth to live young, such news would have astonished no one. It was not the discovery itself that was considered newsworthy but rather the reaction of the European scientific community to the evidence from Australia. Haacke reminded the Adelaide press that he had made the same discovery, but his remarks were ignored outside Adelaide. The Australian scientific community, eager to learn more, got very little information from Caldwell. He agreed to describe his discoveries to the Royal Society of New South Wales but categorically refused to discuss their theoretical importance.[51] He cut short his Australian research and returned to England to write up his results. Such behavior had the effect of barring Australians from even learning about the European theoretical debate.

Haacke wanted credit for his discovery but he was almost entirely ignored, even within Australia. Caldwell was honored because he was able to make his discovery known to a much wider scientific community and because his scientific credentials commanded respect. Responding to the widespread publicity, a number of colonists protested that they had always known that monotremes laid eggs. The *Sydney Morning Herald* dismissed their claims on the grounds that knowledge without proof was not enough. Scientific proof required that the evidence be "examined and reported on by scientists in whom the world has faith, then all the scientific world will stand convinced and will believe where they have not seen." Australian colonists, lacking international and even local credibility, could never provide such proof.

In a few months Caldwell had solved the problem Bennett had been investigating for a lifetime. Bennett was naturally disappointed, complaining to Owen, "Who would have thought that an animal with so large a milk gland should actually demean itself by laying small white eggs?"[52] Caldwell was not looking for monotreme eggs. He found them because he conducted a mass collecting campaign over an extended period with the assistance of dozens of Aborigines. Lacking institutional support, Bennett never had the necessary time and money for such research.

Caldwell's discovery forced taxonomists to revise their definition of a mammal to include creatures that lay eggs. During the 1830s the controversy concerning monotreme reproduction was part of a larger debate between the evolutionists and antievolutionists. A half century later, the European scientific community no longer seriously doubted the truth of evolution. Egg-laying mammals easily fitted into the theoretical framework of Darwinian evolution and were used as evidence in debates about alternative paths of evolutionary development.

Conclusion

The history of Australian zoology during the nineteenth century reveals some of the strengths and weaknesses of colonial science. Data from the periphery, transferred to Europe by colonial collectors and observers, was interpreted within a theoretical framework provided by European professionals. Colonial evidence was particularly important because European theories developed without knowledge of it. When existing theories contradicted the evidence or failed to explain it, European theorists were compelled to look for new explanations. In the ensuing debates, new theories were judged on the basis of the extent to which they explained the colonial evidence.

Distance forced European scientists to rely on colonial observers and collectors for information. Evidence from the colonies was subject to more theoretical manipulation than evidence that was easily available in Europe because European scientists were unwilling to trust the scientific judgment of their informants. The system of information collection was most likely to break down when the evidence violated theoretical beliefs that were generally accepted within the European scientific community.

Colonial scientists were unable to make a significant theoretical contribution. They lacked the necessary knowledge because of their isolation from the centers of scientific discussion. Theories adopted in Europe were unacceptable within Australia because of differences in philosophical, religious, or political perspectives. The colonial evidence was important only when it forced a change in prevailing theories. A colonial scientist did not have enough credibility within the European scientific community to force such a change. Colonial scientists followed the intellectual leadership of their European mentors. They allowed European scientists to judge what discoveries were true, what evidence was important, and who should be given credit. Under these circumstances, colonial scientists were unable to identify and demonstrate errors in European scientific beliefs.

Most of the research for this essay was done while the author was Research Scholar in Science and Humanities at the National Museum of Victoria and was supported by grants from the Myer Foundation and the Ian Potter Foundation.

1. Ann Mozley, "Evolution and the Climate of Opinion in Australia, 1840–76," *Victorian Studies* 10 (1967): 411–30.

2. For a more detailed discussion see Kathleen G. Dugan, "Darwin and *Diprotodon:* The Wellington Caves Fossils and the Law of Succession," Linnean Society of New South Wales, *Proceedings* 104 (1979): 265–72.

3. Although colonists generally tended to regard the Aborigines as ignorant savages, they were sometimes able to take advantage of their superior knowledge of the

natural environment. Lang noted that although it is presumptuous of one who is not acquainted with comparative anatomy or osteology to identify fossils, "the Aborigines are very good authority on this point in the absence of such men as Professor Jameson, or Professor Buckland, or Baron Cuvier." "L" [John Dunmore Lang], "Letter to the Editor," *Sydney Gazette*, 25 May 1830.

4. William Buckland, *Reliquiae Diluvianae* (London: John Murray, 1824), esp. pp. 183–84.

5. John Dunmore Lang, "Account of the Discovery of Bone Caves in Wellington Valley," *Edinburgh New Philosophical Journal* 10 (1831): 364–68.

6. Camille Limoges, *La Sélection naturelle* (Paris: Presses Universitaires de France, 1970), pp. 17–18; Charles Darwin, *Journal of Researches into the Geology and Natural History of the Various Countries Visited by H.M.S. Beagle, from 1832 to 1836* (London: Henry Colburn, 1839), pp. 209–10.

7. [Robert Chambers,] *Vestiges of the Natural History of Creation*, reprint of 1st ed. (New York: Humanities Press, 1969), p. 258.

8. [Robert Chambers,] *Vestiges of the Natural History of Creation*, 10th ed. (London: John Churchill, 1853), pp. 117–18.

9. Adam Sedgwick, *A Discourse on the Studies of the University of Cambridge*, 5th ed. (London: John W. Parker; Cambridge: John Deighton, 1850), pp. 263–64.

10. William Sharp Macleay, [Letter to the Editor,] *Sydney Morning Herald*, 2 December 1847; W. B. Clarke, "Mr. Turner's Diprotodon," *Sydney Morning Herald*, 6 December 1847; Ludwig Becker, "Über das Alter der lebenden Thier-und-Pflanzen-Welt in *Australien*," *Neues Jahrbuch für Mineralogie, Geognosie, und Petrefaktenkunde* (1858) p. 536–39.

11. Richard Owen, *On the Classification and Geographical Distribution of the Mammalia* (London: John W. Parker and Son, 1859), pp. 28–29.

12. Charles Darwin to Charles Lyell, 23 [September 1860] quoted in P. Thomas Carroll, *Annotated Calendar of the Letters of Charles Darwin in the Library of the American Philosophical Society* (Wilmington, Delaware: Scholarly Resources, 1976), no. 227, p. 85.

13. Andrew Murray, *The Geographical Distribution of Mammals* (London: Day and Son, 1866), p. 285.

14. Searles V. Wood, "On the Form and Distribution of the Land-Tracts during the Secondary and Tertiary Periods Respectively; and on the Effects upon Animal Life Which Great Changes in Geographical Configuration Have Probably Produced," *Philosophical Magazine*, 4th ser., 23 (1862): 391–92.

15. Ibid., 393.

16. Franz Unger, *Neu-Holland in Europa* (Vienna: Wilhelm Braumüller, 1861), p. 20.

17. Thomas Henry Huxley, "Anniversary Address of the President," Geological Society of London, *Quarterly Journal* 26 (1870): lix–lxi.

18. Henry Barkly, "Anniversary Address of the President, 8 April 1861," Royal Society of Victoria, *Transactions* 6 (1861–64): xxiv–xxvi.

19. Frederick McCoy, "Note on the Ancient and Recent Natural History of Victoria," *Catalogue of the Victorian Exhibition, 1861* (Melbourne: John Ferres Government Printer, 1861), pp. 159–74. McCoy elaborated on these arguments in a series of newspaper articles published under the pseudonym Microzoon and entitled "Why Is Australia Odd?" *Australasian*, 6 August 1870, 17 September 1870, 24 September 1870.

20. Frederick McCoy, "Note on the Ancient and Recent Natural History of Victoria," *Annals of Natural History*, 3d ser., 9 (1862): 138–45.

21. Julian Tenison-Woods, "The History of Australian Tertiary Geology," Royal Society of Tasmania, *Papers and Proceedings* (1876): 76–79.

22. The theoretical debate was focused on reproduction as a criterion for classification. The issue was quite complicated, with different naturalists arguing for oviparous, ovoviviparous, and viviparous reproduction. Only the broad outlines are summarized here. For further information, see Kathleen G. Dugan, "Marsupials and Monotremes in Pre-Darwinian Theory" (Ph.D. dissertation, University of Kansas, 1980).

23. George Shaw, *The Naturalist's Miscellany* (London: Nodder & Co., [1799?]), vol. 10, plates 385–86.

24. Johann Friedrich Blumenbach, "Über das Schnabelthier (Ornithorhynchus paradoxus) ein neuendecktes Geschlecht von Säugthieren des funften Welttheils," *Magazin für den Neuesten Zustand der Naturkunde* 2 (1800): 211–14.

25. "Zur Anatomie des Schnabelthiers," *Magazin für den neuesten Zustand der Naturkunde* 3 (1801): 724–25.

26. Jean Baptiste de Lamarck, *Zoological Philosophy* (New York and London: Hafner Publishing Company, 1963), p. 74.

27. Everard Home, "A Description of the Anatomy of the *Ornithorhynchus paradoxus*," Royal Society of London, *Philosophical Transactions* 92 (1802): 69–82; Friedrich Tiedemann, *Zoologie* (Heidelberg: Landshut, 1809), vol. 1, p. 589; Etienne Geoffroy Saint-Hilaire, "Note où l'on etablit que les monotrêmes sont ovipares, et qu'ils doivent former une cinquième classe dans l'embranchement des animaux vertébrés," Société Philomathique de Paris, *Bulletin*, 1822: 95; John Fleming, *The Philosophy of Zoology* (Edinburgh: Archibald Constable & Co.; London: Hurst, Robinson & Co., 1822), vol. 2, p. 215; Jan van der Hoeven, "Mémoire sur le Genre Ornithorhinque," Deutsche Akademie der Naturforscher, *Nova Acta Leopoldina* 11 (1823): 366–68; Everard Home, *Lectures on Comparative Anatomy* (London: G. and W. Nicol, 1823), vol. 3, pp. 364–65.

28. George Caley, letter to Joseph Banks, 12 March 1804, in Banks Papers—Brabourne Collection, vol. 8, Caley, p. 141, in Mitchell Library, Sydney (A79-1); John Jamison, ["Observations on *Ornithorhynchus paradoxus*,"] Linnean Society of London, *Transactions* 12 (1818): 584; Thaddeus Bellingshausen, *The Voyage of Captain Bellingshausen to the Antarctic Seas, 1819–21* (London: Hakluyt Society, 1945), vol. 2, p. 353; Patrick Hill, ["Observations on *Ornithorhynchus*,"] Linnean Society of London, *Transactions* 13 (1822): 623–24.

29. Henri Ducrotay de Blainville was a notable exception. See *Dissertation sur la place que la famille des ornithorhynques et des echidnés doit occuper dans les séries naturelles* (Paris: Lebeque, 1812), pp. 92–101.

30. Johann Friedrich Meckel, "Die Säugthiernatur des Ornithorynchus," *Notizen aus dem Gebiete der Natur- und Heilkunde* 6 (1824): 144.

31. Meckel and Baer were exceptions, arguing that animals that give milk need not necessarily give birth to live young. See Johann Friedrich Meckel, *Ornithorhynchi paradoxi Descriptio Anatomica* (Leipzig: Gerhard Fleischer, 1826), p. 58; Karl Ernst von Baer, "Noch eine Bemerkung über die Zweifel, welche man gegen die Milchdrüse des *Ornithorhynchus* erhoben hat, und Betrachtung über das Eierlegen und Lebendiggebaren," *Archiv für Anatomie und Physiologie*, 1827: 571–575. Richard Owen agreed with them in theory. See "On the ova of the *Ornithorhynchus paradoxus*," Royal Society of London, *Philosophical Transactions* 124 (1834): 556. Nevertheless, in practice he used the evidence for lactation to support his arguments that the platypus gives birth to live young.

32. Etienne Geoffroy Saint-Hilaire, "Sur un Appareil glanduleux récemment dé-

couvert en Allemagne dans l'ornithorhynque situé sur les flancs de la region abdominale et faussement considéré comme une glande mammaire," *Annales des sciences naturelles* 9 (1826): 458–60; "Sur les Glandes abdominales des ornithorhynques, faussement présumées mammaires, lesquelles secrètent, non du lait, mais du mucus, première nourriture des petits nouvellement eclos," *Gazette médicale de Paris*, 2d ser., 1 (1833): 157.

33. Henri Ducrotay de Blainville, "Sur les Mamelles de l'Ornithorhynque femelle, et sur l'ergot du mâle," Société Philomathique de Paris, *Nouveau Bulletin* (1826): 138; "Sur la Génération de l'Ornithorhynque," Société Philomathique de Paris, *Bulletin*, s.4 (1833): 48; Richard Owen, "On the Mammary Gland of the *Echidna hystrix*, Cuv.," Zoological Society of London, *Proceedings* 2 (1832): 180.

34. Richard Owen, "On the Mammary Glands of the *Ornithorhynchus paradoxus*," Royal Society of London, *Philosophical Transactions* 122 (1832): 517–38; "Response to Etienne Geoffroy Saint-Hilaire, Memoir on the Abdominal Glands of the Ornithorhynchus," Zoological Society of London, *Proceedings* 1 (1833): 30–31; "Response to 'New Observations on the Nature of the Abdominal Glands of *Ornithorhynchus*' by E. Geoffroy Saint-Hilaire," Zoological Society of London, *Proceedings* 1 (1833): 95–96.

35. Etienne Geoffroy Saint-Hilaire, "Sur un Appareil glanduleux," pp. 458–60; "Considérations sur des oeufs d'*Ornithorinque*, formant de nouveau documens pour la question de la classification des monotrêmes," *Annales des sciences naturelles* 18 (1829): 160.

36. Etienne Geoffroy Saint-Hilaire, "Sur les Glandes abdominales," p. 156; "New Observations on the Nature of the Abdominal Glands of *Ornithorhynchus*," Zoological Society of London, *Proceedings* 1 (1833): 92; "Mémoire sur les glandes mammellaires pour établir que les Cétacés n'allaitent point comme à l'ordinaire leurs petits, et qu'ils pourraient s'en tenir à les nourrir de mucus hydraté!" *Annales des sciences naturelles, (Zool.)*, 2d ser., 1 (1834): 176; Etienne Geoffroy Saint-Hilaire, ["Reflections on Dr. Weatherhead's Communication Respecting the *Ornithorhynchus*,"] Zoological Society of London, *Proceedings* 1 (1833): 15.

37. Sociologist Ron Westrum noted the general unwillingness of the scientific community to accept evidence provided by laymen who lack accepted scientific credentials. This unwillingness helps to maintain high professional standards within science, but it sometimes leads scientists to reject important evidence that they cannot verify personally. See articles by Westrum, "Science and Social Intelligence about Anomalies: the Case of Meteorites," *Social Studies of Science* 8 (1978): 461–93; "Knowledge about Sea-Serpents," in *On the Margins of Science: the Social Construction of Rejected Knowledge*, ed. Roy Wallis, Sociological Review Monograph no. 27 (Keele, England: University of Keele, 1979); and "Social Intelligence about Anomalies: The Case of Unidentified Flying Objects," *Social Studies of Science* 7 (1977): 271–302.

38. Lauderdale Maule, ["Habits and Oeconomy of the *Ornithorhynchi*,"] Zoological Society of London, Committee of Science and Correspondence, *Proceedings* 2 (1832): 145.

39. Richard Owen, "On the Mammary Glands of the *Ornithorhynchus*," p. 534.

40. Geoffroy Saint-Hilaire, "Analyse d'un mémoire intitulé: Découverte de glandes monotrémique chez le rat d'eau et dissertation sur l'essence, les rapports, et le mode de formation de ce nouveau système d'appareils glanduleux," *Institut* 1 (1833): 28.

41. Geoffroy, "Sur les Glandes abdominales," p. 79; "Considérations sur des oeufs," pp. 162–63; Harry Burrell, *The Platypus* (Sydney: Angus & Robertson, 1927), 38.

42. See Ann Mozley Moyal, "Sir Richard Owen and His Influence on Australian

Zoological and Palaeontological Science," Australian Academy of Science, *Records* 3 (1976): 41–56.

43. V. M. Coppleson, "The Life and Times of Dr. George Bennett," Post-Graduate Committee in Medicine, University of Sydney, *Bulletin* 2 (1955): 207–64.

44. George Bennett, "Notes on the Natural History and Habits of the *Ornithorhynchus paradoxus* Blum.," Zoological Society of London, *Transactions* 1 (1835): 240.

45. Richard Owen, "On the Young of the *Ornithorhynchus paradoxus*, Blum.," ibid., 226.

46. Etienne Geoffroy Saint-Hilaire, "Nouvelle Révélation d'oviparité dans les monotrêmes," *Institut* 2 (1834): 339–40.

47. Richard Owen, "On the Marsupial Pouches, Mammary Glands, and Mammary Foetus in the *Echidna hystrix*," Royal Society of London, *Philosophical Transactions* 159 (1865): 684.

48. Ibid., 673.

49. Wilhelm Haacke, "Meine Entdeckung des Eierlegens der *Echidna hystrix*," *Zoologischer Anzeiger* 7 (1884): 647–53.

50. William H. Caldwell, "The Embryology of Monotremata and Marsupialia— Part II," Royal Society of London, *Philosophical Transactions* 148 (1887): 465.

51. William H. Caldwell, "On the Development of the Monotremes and *Ceratodus*," Royal Society of New South Wales, *Journal* 18 (1884): 118.

52. Quoted in Coppleson, "Life and Times of George Bennett," p. 259.

Environment, Economy, and Australian Biology, 1890–1939

C. B. SCHEDVIN

On first inspection the history of Australian science appears to suffer from lack of symmetry. It has a classical past and a modern era, but lacks a middle age and an early modern period. The classical age falls between Cook's exploration of the east coast of Australia in 1770 and Charles Darwin's visit on HMS *Beagle* in 1836. In this period of exploration and discovery Australia was a naturalist's wonderland;[1] the botanical and zoological specimens that were taken for study in Europe helped to transform contemporary understanding of natural process. The modern era begins abruptly in the 1930s, and throughout the following twenty or thirty years Australian scientists and their institutions were able to make significant contributions to "western" science in such diverse fields as medical immunology and radio astronomy.

My intention is not to suggest that the century between the age of discovery and the push for fundamental knowledge should be characterized as the dark age of Australian science, a period of systematic retreat from intellectual endeavor. Ann Moyal, Michael Hoare and others have shown that scientific associations were established at the time and some flourished throughout long periods.[2] There were intermittent bursts of geographical and geophysical exploration. Increasingly an attempt was made to collect and understand the natural fauna and flora with the establishment of museums and botanical gardens. By the 1880s some university departments were attempting to break out of the tyranny of the lecture theater and demonstration room. Even the colonial and state governments were beginning, at the turn of the century, to appoint biologists who were able to devote a proportion of their time to applied research. My purpose in playing down the significance of scientific endeavor in the "dark ages" is twofold. First, the small scientific community was not sufficiently strong to generate a broadening and deepening of scientific endeavor from within its own ranks. In other words, we should be wary of the linear model of scientific progress, with the view that the growth of science was inherently continuous from the small associations of the nineteenth century to the great institutions and outstanding achievements of the twentieth century. The history of Australian science is littered with examples of frantic endeavor followed by collapse, unbounded optimism followed by pessimistic indifference, and a lack of public trust in long-term intellectual endeavor.

The second reason follows from the first. The prehistory of Australian science was greatly influenced by external pressures. As indicated, material factors were predominant. Closely related was growing pressure on the natural environment, which affected economic performance. Also relevant were the political framework within which economic and scientific decisions were taken and the growth in biological understanding from the end of the nineteenth century—particularly in the wake of Louis Pasteur and Robert Koch.

Accordingly I interpret the spread of Western biological science to Australia between the 1890s and 1930s within the context of economic and environmental change. Broadly, science was accepted as legitimate and worthy of public support because its offspring—applied science and technology—were progressively incorporated into the productive system. This coincided with fundamental changes in the pattern of economic development, from the nineteenth-century model of exploitation of resources with a low technological base and toward the twentieth-century model of intensive exploitation of existing resources by means of more sophisticated technology. Case studies will be used to illustrate the theme, but a few words are necessary about the nineteenth-century background.

The phase I have called the classical age of Australian science, of course, was only nominally Australian. It was part of European imperial expansion and shared many of the exploitive characteristics of that process. Specimens of Australian natural history were collected and exported; little of permanence was left behind after the departure of the naturalists.

With the waning of European interest in nature's wonderland, there was little in the local scene to sustain intellectual endeavor. Indeed, almost everything was against the development of science: isolation from Europe, the small size of the population, and an emerging egalitarian and anti-intellectual tradition more intense than in North America. Further, the Australian colonial economies were based on the exploitation of readily available natural resources and the export of a narrow range of staple commodities such as wool, gold, fishery products, and, toward the end of the nineteenth century, grain. Low population density in relation to available resources meant that economic expansion could be achieved by incorporating more land and minerals into the system. The main innovations in the pastoral industry, for example, involved the provision of permanent fencing and permanent water for stock depastured at a distance from river frontages;[3] innovation included selective breeding of merino sheep, using methods pioneered by Robert Bakewell, by taking advantage of a gene pool extended by the importation of stud stock from Germany, France, and North America.[4] In the case of agriculture the industry was reliant on seed varieties imported from Europe. Innovation was restricted to the use of soluble phosphates to correct for mineral deficiency, particularly on the light soils of South Australia, and to the development of special machinery to clear and cultivate native species of Eucalyptus. Throughout most of the nineteenth century agriculturalists were protected from the worst European diseases because of distance and the dry climate, and this was an additional reason for the low level of technological self-consciousness.

Change in the level of technological awareness is evident as early as the 1880s, but is much more pronounced after the economic collapse and long period of drought in the 1890s and early 1900s. The period under discussion corresponds with Basalla's third phase in the spread of Western science—the struggle for an independent scientific tradition and the creation of self-supporting institutions—and thus is of particular interest.

A complex series of economic changes was at the heart of the increasing interest in technology and ultimately in science. The 1890s marked the end of an era in the sense that economic expansion could no longer proceed by outward movement of the frontier from the relatively fertile southeastern crescent. Almost all available land suitable for livestock and wool production had been occupied. Future increases in rural production required more investment, particularly in the form of railroad construction, and more sophisticated husbandry techniques that would allow mixed production of wool, meat, and grain on properties which in the nineteenth century had specialized in one commodity or another.

The improvement in efficiency, however, was not sufficiently great to compensate for the closure of the frontier. Despite the heavy investment in transport and the encouragement given to new industries such as exports of beef and butter based on refrigerated shipping, rural production was no longer able to sustain the level of increase in material welfare and rate of population growth that was such a marked feature of the nineteenth century. By the 1920s this trend was sufficiently clear to cause a crisis in development policy. Should the country continue to rely on its traditional role as an exporter of food and raw materials to the United Kingdom? If so, immigration and population growth would need to be severely curtailed. In this vision Australia would remain a small, homogeneous, high-income society heavily dependent on the metropolitan economy. The alternative was to embark on industrialization, encourage population growth, and accept high tariff protection and a lower rate of increase in average material welfare. The goal was a larger and more independent society; ultimately increased size might lead to higher welfare than under the free-trade alternative.

During the depression the die was cast in favor of industrialization, although more by necessity than choice. During the 1920s, however, the imperial ideal predominated—the vision of a self-sufficient British Empire of which Australia would be a leading part as an exporter of commodities and an importer of surplus British labor and capital. At this stage those who warned about the limited capacity of the land to absorb more families, in the tradition of the yeoman farmer, were in a distinct minority. To this end Australian scientific resources were to be mobilized and linked with British science in pursuit of the economic integration of the empire. Government support for science on a significant scale by formation of the Council for Scientific and Industrial Research (CSIR) was provided in this context. It is significant that a second government agency was established at the same time and for an ancillary purpose. The Development and Migration Commission was created to investigate projects which would lead to the settlement of British migrants on the land in the yeoman tradition.[5]

Pressure on the rural economy was caused also by a complex series of environmental changes. These changes can be divided into two broad categories: the accumulated effect of intensive grazing on native pastures and of constant cropping on light soils relatively low in plant nutrients and organic material; and the cumulative effect of introduced pests and diseases—notably the rabbit, sheep blowfly, internal and external parasites of animals, and a range of noxious weeds.

Australia's native grasslands had evolved under conditions of periodic drought, low soil fertility, and light grazing by marsupials. Kangaroos, wallabies, and other marsupials had evolved to traverse long distances; they foraged lightly, passing quickly from place to place. Under these conditions tall grasses such as the deep-rooted perennial *Phalaris tuberosa* flourished. Marsupials fed on the tips of plants and did not damage root systems.[6]

The introduction of intensive-grazing European animals, particularly the various strains of European merinos, rapidly denuded the sparse native pastures. Particularly in the 1870s and 1880s, with the rapid increase in sheep members and the occupation of the semiarid western districts of New South Wales and Queensland, the condition of the natural grasslands deteriorated to such a degree as to promote large-scale erosion and soil drift. During the long end-of-century drought the rate of deterioration accelerated.[7]

Overgrazing by sheep was by no means the only cause. Excessive clearing of land for plowing, overcropping, and the explosion of the rabbit population contributed significantly. One consequence of more intensive animal production was that soil and pasture, already low in mineral nutrients, were further depleted of phosphate, potassium, nitrogen, and the minor elements. Of course, the introduction of the merinos was not wholly detrimental; eventually Australian pastures were transformed when native grasses and shrubs were replaced by nitrogen-fixing species such as clover, lucerne, and *Medicago* varieties. The merino played a part in inadvertently introducing seeds of leguminous grasses—as well as weeds—on the backs of live sheep, and they were also instrumental because burr seed clings tenaciously to wool. But introduced species were not able to transform Australian grasslands until minerals—mainly phosphate with trace elements—had been added artificially, and this did not occur on any significant scale until after World War II. For most of the first half of the twentieth century grasslands continued to deteriorate as a result of the combined effects of intensive grazing, exponential growth of the rabbit population, and expansion of wheat-growing.

The economic effect of deteriorating grasslands was accentuated by the cumulative effect of animal and plant pests and diseases, most of them imported. Again the incidence of these pests and diseases appears to have been much more pronounced after the turn of the century than before, which suggests a broad-based series of ecological changes which are not yet understood adequately. The effect of the rabbit has already been mentioned. In economic terms this was the most serious pest because by the interwar years it had grown to plague proportions, in most pastoral districts sharply cutting the carrying capacity of sheep runs. Before the 1890s the sheep blowfly, *Lucilia cuprina*, was almost unknown as a cause of fly strike in sheep but became a serious pest in the half dozen years before World War I. Entomologists suggest that this particular species of blowfly was imported in the form of pupae from South Africa or India and may have evolved from a carrion feeder to a species that completes its life cycle on live animals.[8] Whatever its pattern of evolution, the insect had adapted brilliantly to the dry inland; by the 1930s it was causing losses in the form of lower production and higher costs equivalent to around 6 percent of the value of wool ouput. Control measures included constant inspection of flocks, removal of surplus wool from around the crutch, jetting with chemicals, and the surgical removal of skin folds from the crutch area (the Mules operation, still controversial in the 1930s).

A broadly similar pattern is evident in internal parasites of sheep, particularly liver fluke (*Fasciola hepatica*); external parasites such as cattle tick and buffalo fly; virus infections such as bovine pleuropneumonia; and a wide range of noxious weeds, notably the cactus species *Opuntia* (prickly pear) which had dominated more than 20 million hectares of good grazing country in south-central Queensland and northern New South Wales. *Opuntia* had been introduced in the 1830s in the Anglocentric expectation that it might provide good hedging. By world standards Australia was still a clean environment. Foot and mouth disease was almost unknown; outbreaks of anthrax and rinderpest were sporadic; rabies was present in Tasmania in the 1860s and 1870s but was eliminated promptly. As A. E. Pierce has remarked, "It is possible that the acute devastating diseases with a high mortality, such as rinderpest, and haemorrhagic septicaemia, with a high mortality and no evidence of the carrier state, were self-eliminating from susceptible species during the voyage to New Zealand or Australia."[9] What was important in contemporary perception was that the nineteenth-century advantage of a natural quarantine was being diminished: in the period between the wars Australian pastoralists were faced almost for the first time with the challenge of scientific husbandry in a time of rising labor costs and much slower increases in physical productivity.

This outline of environmental and economic history is intended to suggest a materialist interpretation of Australian biological science, indeed of Australian science in general. Economic considerations were uppermost in influencing—not in determining—the pattern and extent of scientific endeavor: applied science was required to win its spurs in technology before "higher" levels of intellectual activity could achieve support with any degree of consistency.

In the long struggle to graft the ethics of science onto a materialistic culture, four stages can be distinguished.

1. Appointment to Australian chairs of chemistry, natural philosophy, and biology in the 1880s of British scientists with the ambition to make room for research in the relentless round of teaching. David Orme Masson in the chair of chemistry at Melbourne is the clearest example. J. T. Wilson's encouragement of physiological studies of marsupials in the Sydney medical school is another.

2. The handful of research-oriented schools and departments achieved some success, but they were seriously affected by the long economic depression of the 1890s. In particular, Melbourne University failed to regain the optimistic spirit of the 1880s until well after World War II, and most departments had great difficulty in maintaining momentum in research. The feature of this stage, however, is the training of Australian graduates for subsequent postgraduate

work in Europe, a trend that gathered some momentum shortly before and after World War I. Two winners of an 1851 Exhibition illustrate the pattern. W. L. Waterhouse, a Sydney graduate in agriculture and subsequently a pioneer in plant breeding, received postgraduate training at Imperial College, London, an experience cut short by World War I. Waterhouse was one of the first to hold a research chair—at Sydney—in an Australian university. Although from the field of chemical physics rather than biology, David Rivett is another example. Rivett, a Melbourne graduate in chemistry, undertook graduate work at Oxford and at the Nobel Institute, Stockholm, then succeeded Masson in the Melbourne chair of chemistry in 1924.

3. During the 1920s the growing ambition of science departments to undertake research was seriously curtailed by the rapid growth in student numbers and the attempt by state governments to cut the level of expenditure after the inflationary years of World War I. Melbourne was most severely affected because of the Gladstonian parsimony of a succession of governments controlled or influenced by the Country Party, but Sydney and Adelaide also suffered. Thus, at a time of steady increase in the number of well-trained scientists and growing opportunities for applied work, the universities were in no position to respond in a positive manner. The exception is the University of Adelaide, which established the Waite Agricultural Research Institute in 1924. The inability of the universities to respond flexibly was one reason for the creation of the Commonwealth research organization CSIR to mediate between science and industry.

4. After the worst years of the 1930s support for science grew at an unprecedented rate, mainly in the form of additional finance for specific CSIR projects with a strong utilitarian flavor. Historians of science have highlighted the protean influence of war on scientific endeavor. The effect of economic depression should be more fully understood. The attitude toward agricultural and biological science—and also toward physics—was transformed in the early 1930s. The collapse of income and prices made agricultural producers much more technologically conscious and willing to support applied science in the hope of improving economic efficiency. The Australian Wool Board, for example, was created in 1936 to support research—and publicity—in an endeavor to counter competition from synthetic fiber. It is significant that the CSIR received the bulk

of the Wool Board's funds; except for the Waite Institute the universities were not beneficiaries.

The increasing utilitarian orientation of biology is reflected in the changing institutional structure of Australian science. During the second half of the nineteenth century the universities, museums, botanical gardens, and specialist societies were the sole custodians of the scientific ethos. During the 1890s most colonies established departments of agriculture in response to the growth of biological knowledge and the severity of the economic collapse. One effect of the depression was a shift in the pattern of production from large-scale wool growing to small- and medium-scale agriculture, dairying, viticulture, and orcharding. The departments were not scientific institutions in the customary sense. They were established for administrative purposes and to disseminate existing knowledge; they were oriented toward small producers such as wheat growers and dairy farmers, who were in no position individually to acquire knowledge of new techniques and whose industry organizations were weak and fragmented.[10] Significantly the departments did not provide extension services for pastoralists who preferred independence: state agricultural services were designed for the rural yeomanry and smallholders, not for the upper strata.

Nevertheless, the departments were engaged in some applied science, notably wheat breeding. They undertook experiments in the cold storage of fruit and meat, encouraged some taxonomic work of plant diseases such as rust, appointed a few entomologists to tackle the blowfly problem, undertook some investigations in the area of veterinary pathology, and initiated bacteriological studies in connection with the production of milk and cheese. With the support of the newly established university departments of agriculture and veterinary science, the state departments were seen as fulfilling their responsibilities satisfactorily before World War I. After the war, however, they were less and less able to respond to the economic pressure and the growth in biological knowledge. A frequent complaint was that the small number of agriculturalists and veterinarians were obliged to divide their time between administration, extension, and the laboratory; as a consequence research activity was intermittent and lacked consistent direction. Scientists were subject to short-term political pressures and obliged to respond to any disaster that cropped up, such as an outbreak of brown rot in apples or contagious mamitis in dairy cattle. It is easy to exaggerate the poverty of departmental science: excellent descriptive and applied work was completed, particularly in plant pathology. But the lack of positive research planning was a serious handicap during the 1920s.

The perceived inadequacies of departments of agriculture and universities, both creations of state governments, encouraged the national government to support a third institutional tier. The Commonwealth established an Advisory Council of Science and Industry in 1916 in the wake of formation of the British Department of Scientific and Industrial Research (DSIR) the preceding year. In 1920 the Advisory Council was given continuing status in the form of an Institute of Science and Industry, but the institute languished for lack of funds.[11] As mentioned earlier, the reorganization of the institute on a more satisfactory basis in 1926 by the creation of the CSIR was undertaken within a specific economic and political context. At least in conception the CSIR was intended to be of central importance in Prime Minister Stanley Melbourne Bruce's vision of British imperial economic and cultural partnership. The council would act as a funnel for the transfer of British biological science to the antipodes. Rural export efficiency would be enhanced, which would allow Australia to absorb more British capital and promote the settlement of more immigrants on the land in the yeoman tradition.[12]

These ideals, of course, were to crumble under the pressure of economic collapse, but during the late 1920s and early 1930s they were a powerful influence on Australian scientific development, largely through the agency of the Empire Marketing Board. The board, through the CSIR and the Waite Institute, was instrumental in encouraging research for the pastoral industry, a much neglected field. It helped finance the early stage of dryland agrostological research at Adelaide on the basis of collaboration between the CSIR and the Waite. An insectary at Canberra for entomology was financed partly by the board in an attempt to find a natural predator of the sheep blowfly. Imperial enthusiasts envisaged a chain of tropical research stations in Africa, Ceylon, Australia, and the West Indies in the expectation that the British Empire could become largely self-sufficient in tropical products. A large all-purpose laboratory was intended for Queensland. In the event a much more modest laboratory was established in Townsville by the CSIR, with assistance from the Empire Marketing Board, for the development of a vaccine against bovine pleuropneumonia, a disease that had become virulent in tropical Australia. Funds were also provided on a cooperative basis to assist the botanical program in Canberra.[13]

Even more significant than these specific funding proposals was the reliance on overseas scientific experts to launch the major research programs. British authorities were asked to prepare guidelines for the development of research strategies. Sir Arnold Theiler, previously of the Veterinary Research Institute at Onderstepoort, South Africa, for example, spent six months in 1928 planning the CSIR's animal health program. A similar task was undertaken for forest products by A. J. Gibson of the Indian Forestry Service. The purpose was to obtain useful advice, seek recognition within the British scientific community, and possibly attract the visiting expert to head the new research division. On the surface the CSIR appears to have accepted its status as a client of metropolitan science.

In the event, this was not the beginning of a strong center-periphery relationship, at least not in biology. The reports by visiting experts were ineffective because they were not tempered by sufficient knowledge of Australian conditions. The visitors could not be attracted to lead the new research divisions. Further, as noted earlier, the severity of the depression led to the abolition of the Empire Marketing Board and disenchantment with the concept of economic empire. Probably more important, however, was the independent spirit of scientific inquiry, which, given the opportunity, will reject any authoritarian tendency. During the interwar period Australian scientists were able to gain some protection from day-to-day political pressure. The CSIR offered most protection, but some university departments were also able to make room for concentrated research. The growth in biological knowledge offered an increasing range of opportunities for applied investigations in the Australian environment, and scientists were not slow to struggle for recognition as equals with their British peers—preferably by election to the Royal Society.

The interaction between economic retardation, environmental stress, and center-periphery linkage will be elaborated in three case studies drawn from the early scientific history of the CSIR. The

studies are drawn from diverse fields of biology: nutritional biochemistry, veterinary medicine, and entomology. In different ways
they illustrate how science at the periphery struggled to assert its
independence and its own normative structure.

Scientific Nationalism on the Grasslands

Thornburn Brailsford Robertson was born in Edinburgh in 1884. His
parents migrated to South Australia when he was a child, and he
was educated at the University of Adelaide, taking first class honors
in science in 1905. Most of the next fourteen years was spent in the
United States, where he obtained a Ph.D. in physiology and subsequently taught at the universities of California and Toronto. Robertson's dedication to research was a product of his experience in North
America, where he obtained a considerable reputation in both physiology and biochemistry. In 1919 he returned to Adelaide in the
chair of physiology, where he planned to continue his fundamental
studies of human and animal growth. But the constraints of an Australian laboratory in the 1920s were oppressive. Robertson was intent on determining the effect of nucleic acid and thyroid on the
growth of animals; mice were to be used for the experiments. An
Animal Products Research Foundation was established which was
able to provide a modest $1,000 a year to help finance the experi-
ments, but much more space and laboratory assistance were needed
to mount an effective program.

Nutritional physiology was not intended to be one of the foundation programs of the CSIR, but Robertson was able to make the best
of a lucky break. The South Australian representative on the Council was the director of the Waite Institute, A. E. V. Richardson, who
was abroad at the time of the first meeting in 1926. Robertson replaced Richardson and persuaded David Rivett, the new chief executive, that his research program could be redirected toward the nutrition of sheep and that such studies could be of fundamental
importance in the long run. It would be misleading to suggest, however, that Robertson's persuasiveness was the crucial factor. David
Rivett, who believed strongly in the heuristic value of uncommitted
research and conjecture, was casting around for distinguished scientists and a challenging program that could serve as a scientific

flagship, a way of acknowledging the norms of science in a new organization that would inevitably be preoccupied with application, technology, and even extension. Acceptance of Robertson's overtures was an act of faith in science itself.

Robertson built his hypothesis boldly on several questionable sets of assumptions. He assumed, first, that a limiting factor in animal production and wool growth was the amount of plant protein available for ingestion. He assumed further that the semiarid pastures that were common in South Australia had been severely affected by overstocking and that the way to increase wool production was to enrich these pastures by some means. In the circumstances of the 1920s it is understandable that he did not ask the economic question: For a given amount of research expenditure, what was the probability of increasing national wool production by concentrating on problems of the semiarid zone against the more fertile grasslands of the southeastern crescent?

A second set of assumptions concerned the relation of protein and essential mineral intake to the growth of wool fiber. Robertson was an adherent of the external hypothesis, the view that there exists a strong positive relation between intake and growth (and the quality of the fiber). In particular, he anticipated that the amino acid composition of semiarid pastures might be deficient in cystine, which appeared to be required for the growth of wool fiber, and that any deficiency in cystine might be overcome by the artificial addition of protein to the diet of sheep in these zones. The alternative and more accepted contention among biochemists was that amino acid production was a more complex and "internal" process and that the rate and quality of wool growth was unlikely to respond significantly to any *specific* regimen of supplementary feeding. An additional though subordinate Robertsonian hypothesis was that minerals such as magnesium, iodine, fluoride, zinc, and copper might be critical to the healthy growth of tissues.[14] This was a more productive line of enquiry. Important minerals such as phosphorus, calcium, and iron were known to be vital physiologically, but the function of minor elements had not yet been established.

Almost immediately Robertson's program was subjected to stringent criticism from the United Kingdom on both scientific and economic grounds. The scientific critique was that Robertson would fail to extract *all* the protein from cellulose and thus be unable to explain fully the composition of wool fiber, an essential step in his research design. He was advised to try the more empirical approach of testing various feed additives against a control flock of sheep. The economic criticism, offered in 1928 by J. B. Orr, director of the Rowett Institute, Aberdeen, was that much more positive results would be achieved by applying existing knowledge rather than wagering on fundamental research. The CSIR was advised, for example, that more would be achieved by improving pastures in higher rainfall areas by known techniques, such as superphosphate topdressing, than by attempting to discover nutritional deficiencies by direct means.[15] Both sets of advice were sound enough on their own terms, but they contained a clear additional message: the "parent" scientific institutions in the United Kingdom were in the best position to advance knowledge; Australian scientists should concentrate on the application of established knowledge to local circumstances.

Robertson combined the stubbornness of a Scot with the disciplined drive of a Puritan. The paternalism of his compatriots was a spur to even more heroic feats of endurance in the generation of experimental evidence. He thundered to Rivett:

If we are to be kept in a state of economic and intellectual parasitism upon Great Britain, then the Empire will not grow stronger, but progressively weaker. . . . [We will become] an emasculated horse-racing provincial community incapable of being of any use to [ourselves] or to the Empire. . . . I feel with you that we must make a strong stand against any kind of paternal dictation which would reduce us to the status of lab-boys of British scientific administrators. We have our own judgements, weak or strong as time will show, and for better or worse we must stand by them.[16]

Within a year of writing these words Robertson died, a victim of pneumonia and overwork, at the age of 45.

The program was continued by Hedley Marston, one of Robertson's students who helped to establish the CSIR's animal nutrition laboratory. Marston was a brilliant biochemist whose sense of personal insecurity was protected by the display of vanity and a tendency to dominate his subordinates. After Robertson's death the original hypothesis became sacred, as did the primacy of pure knowledge over practical wisdom. The grip of the founding theory was gradually loosened by Charles Martin, who had retired shortly before as director of the Lister Institute, London, and who wished to repay a debt to his youth and to Australia. Martin had spent a formative period in the 1880s with J. T. Wilson at the University of Sydney working on the physiology of marsupials; subsequently he was professor of physiology at Melbourne University before returning to England in 1903 to head the Lister.

Martin was too shrewd a judge of men and of science to propose any fundamental alteration of the Robertson-Marston program. Instead, he sought to build bridges between Marston's physiological approach and adjacent disciplines such as agrostology, soil chemistry, and biochemistry. He also encouraged the acceptance of applied or field problems, particularly in the desperate economic circumstances of 1931–32. In his second annual report he wrote "however complete knowledge of animal nutrition may become, it alone will not solve [practical problems in sheep-rearing and wool production]."[17] With remarkable clarity he anticipated the accepted wisdom of the 1960s: fundamental gains in the understanding of nutrition were most likely to be achieved by focusing on the interaction of soil, plant, and animal systems. Accordingly an agrostologist was appointed to the research group, and closer links were established with the division of soils, which at that time was housed at the Waite Institute.

The field problem that was eventually to vindicate Martin's ecological approach had already received some attention by the division of animal nutrition before Martin's arrival in 1931. The condition was described as coast disease, because it was common along the calcareous littoral that stretched along the coast from Cape Otway in Victoria to the west of South Australia; although known by a different name a similar syndrome was found in Western Australia. In South Australia the disease affected sheep, primarily, although it soon became clear that all ruminants were vulnerable. Since the nineteenth century, graziers had complained about a wasting condition on coasty country that produced anemia and often death, particularly in the spring, even though supplies of fodder were adequate. When stock were moved to "sound" ironstone country they would soon recover, even though the general condition of these pastures was poorer.

Although a variety of veterinary and other explanations were bandied about, mineral deficiency—probably iron—was a fairly obvious deduction. The Waite Institute had reached this conclusion when it was involved briefly before 1930; so also had two Western Australians, J. F. Filmer and E. J. Underwood, who were investigating the cause of Denmark disease of cattle, which produced symptoms similar to coast disease.[18] A relation between the syndromes, however, was not at first suspected.

Involvement of the CSIR commenced in 1930, but three years later Martin regarded the progress of the investigation as unsatisfactory. The iron-deficiency theory had been discarded because animals had not responded to the administration of pure iron oxide, a conclusion also reached by the Western Australians at about the same time in 1933. Because Filmer and Underwood had found that the disease responded moderately to treatment with limonite, they proceeded to fractionate this substance to identify the impurity or impurities that might be active, a sound if unspectacular empirical approach. The Adelaide strategy was to identify promising trace elements from a long potential list, tactics that relied on good luck for early success. In the event the South Australians made the lucky break. One of the division's chemists, R. G. Thomas, made one of those deductive leaps that occasionally bring great reward. Thomas had been associated indirectly with the investigation for some years and was familiar with the symptoms of wasting and anemia. In chemical abstracts he came by chance across a paper which indicated that an excess of cobalt could lead to overproduction of red blood cells in rats. While he would have known that the highly calcareous soils were deficient in cobalt, obviously the connection was highly speculative. But it was soon shown to be correct: by August 1934 penned sheep had responded dramatically to small doses of cobalt nitrate. The Western Australians confirmed the finding early in 1935, and it was soon generalized to the ruminants as a group.[19]

The discovery was of great significance for Australian biologists in a number of different respects. First—and most important from an economic and environmental point of view—the discovery consolidated the research program on the function of trace elements in soil-plant-animal systems. During the next dozen years or so copper, molybdenum, selenium, zinc, and the other minor elements were found to be vital to plant and animal growth. The economic consequence was that the area available for animal and crop production was extended by several million hectares, notably in South Australia. Environmental quality was enhanced in the long term because a more varied and denser pasture sward reversed the trend toward sand drift, erosion, and depletion of grasslands. In the clearest possible way science had passed the pragmatic and economic tests, although the full ramifications were not evident until after World War II.

Second, the dramatic success provided the materials from which legends are made. Hedley Marston, as leader of the research team, was able to claim the efficacy of the Robertsonian fundamental approach. In fact, cobalt was a godsend because it allowed the unproductive work on cystine to be phased out without loss of face, and Marston was able to slip into the trace element work while still insisting on the superior heuristic status of fundamental research. In fact, much of the credit was due to Charles Martin's encouragement of research related to a field problem and interdisciplinary team work.[20]

The Pragmatics of Veterinary Medicine

The process of transmission of the science and technology of veterinary immunology is comparatively uncomplicated and conforms closely to the pattern outlined earlier. The techniques of preparation of antiserum and immunology for animals were known in Australia from the latter part of the nineteenth century as understanding grew of bacteriology and virus infection. J. A. Gilruth, foundation professor of veterinary science at Melbourne University, experimented with a number of preparations, but such work drew more from art than from science.

Harold Woodruff succeeded Gilruth in Melbourne in 1913 and brought a more sophisticated knowledge of bacteriology. Woodruff had learned his trade at the Royal Veterinary College and University College Hospital, London, where he held a chair in veterinary hygiene and subsequently in veterinary medicine. In Melbourne he struggled to combine teaching with research. Despite the economic importance of the livestock industries, veterinary science attracted only a handful of students. Employment prospects were grim because animal health was a minor concern of the state departments of agriculture and graziers did not employ veterinarians to service commercial livestock. Only a few could gain a livelihood from work with thoroughbred horses and small animals.

Nevertheless, Woodruff was able to establish a small veterinary research institute in large part by undertaking routine testing for the Victoria department of agriculture. He attracted a few graduate students in his special field of bacteriology and immunology, including H. E. Albiston, lecturer from 1923, and A. W. Turner, who commenced postgraduate work in 1923. Turner undertook the investigation of so-called black disease of sheep, which was the cause of irregular though often heavy losses in northern Victoria and southern New South Wales. He succeeded in implicating a species of bacteria (B. oedematiens) and was successful in preparing antiserum. Then in 1926 he took bacterial specimens with him to the Pasteur Institute, Paris, for intensive study. The experience converted a promising graduate student into a skillful professional. On his return he joined the CSIR because of the imminent closure of the Melbourne veterinary school as a result of a sharp decline in the numbers of students, not the first or last instance of the new Commonwealth organization taking advantage of the unstable condition of university research. Immunology required good detective work to identify strains of the relevant baccilli and cookery skills in the preparation of antiserum. In 1928 Turner obtained the full cooperation of graziers, which he did not have before his departure, an indication of the growing economic consequences of a variety of animal diseases. Within twelve months Turner had gathered samples of the baccillus and prepared a vaccine that reduced mortality by 60 percent in the field trials.[21] In several senses this was an unspectacular result. Vaccination was of moderate effectiveness on its own. It needed to be combined with control of internal parasites, particularly liver fluke; it was comparatively expensive to administer because of the labor cost involved in additional sheep handling. But it was the most effective animal vaccine that had been prepared in Australia, and it provided clear evidence of the economic potential of veterinary medicine. Turner's work was a classic example of applied science in Australian context and the importance of economic linkage.

The example was repeated a few years later with the preparation of an even more effective vaccine against bovine pleuropneumonia, an infection that had been introduced in Victoria during the 1850s but that had become enzootic in northern Australia, where there was minimum control of the movement of stock. Most of the work was at a research station established near Townsville, northern Queensland, as one of the Empire Marketing Board's projects. By 1936–37 commercial quantities of pleuro vaccine V.F.5 were being produced and found to be highly effective. Henceforth the main difficulty was herd management: the prevention of reinfection by carriers who had escaped vaccination and detection. Nevertheless, veterinary medicine had consolidated its utilitarian image as a leading example of applied science.

The Vicissitudes of Entomology

In veterinary medicine Australian science was clearly dependent on theoretical understanding, which was available in its most concentrated form in Paris. In entomology there were no obvious centers of concentrated knowledge. Entomology is an ancient discipline. Aristotle taught it to his pupils in the peristyles of Athens. The Chinese were pioneers in many fields, and were probably the first— around A.D. 300—to use natural enemies to control insect pests. They also developed an elaborate classification of insects, which began in the first century B.C. and which contributed powerfully to traditional pharmacology and medicine.[22] But perhaps because of the heterogeneity of species and the conservative character of taxonomy, the discipline remained in the descriptive stage for an extended period. Fundamental studies of evolution and phylogeny, anatomy, and physiology did not begin in a concerted manner until the early years of the twentieth century; the study of insect ecology was largely a product of the period between the wars.[23]

The economic potential of applied entomology came under notice in 1889, when the lady beetle genus *Vedalia* was used to control the cotton-cushion scale *Icerya purchasis* in the citrus groves of southern California. *I. purchasis* had been identified as of Australian origin and appeared to be controlled by the lady beetle. The ecological principle was simple and established the classical method of biological control. The geographical movement of species often involved the separation of pests and their natural enemies. The reunion of host and predator was the basis of the classical method.[24]

Biological control was soon found to be no panacea, but a further demonstration of the method helped to launch Australian economic entomology at the end of the 1920s. As is well known, the occasion was the collapse of the prickly pear pest, *Opuntia*, under the on-slaught of a natural predator, *Cactoblastus cactorum*, which had been discovered in central America and reared in Texas for export to Australia. This was one of the country's most spectacular scientific events, perhaps the first occasion when the possibilities of applied science in the Australian environment were widely appreciated. By 1920 more than 25 million hectares of good grazing and arable land in southern Queensland and northern New South Wales had been occupied by *Opuntia* to a degree that sharply reduced its economic value, and the pest was spreading at the rate of a million hectares a year. The cactus could be controlled by mechanical and chemical means, but the cost of around A$45 per hectare often exceeded the commercial value of the land when it was used for grazing and offered no defense against reinfestation.

C. *cactorum* was first released in 1926 and within eighteen months had demonstrated its capacity to control the main *Opuntia* species. Between 1929 and 1932, however, the insect expanded at such a spectacular rate as to cause the literal collapse of the plant across millions of hectares by the combination of caterpillar infestation and bacterial and fungal rot.[25] After collapse the remaining pear could be cleared by hand, and resettlement was well under way in the mid 1930s. The *scientific* cost of clearing was about 4.5 cents per hectare for land valued at from $30 to $40 per hectare, a cost–benefit ratio that would gladden the heart of the most niggardly economist.[26]

The small entomological fraternity was understandably jubilant, and even David Rivett—a skeptical chemical physicist—was impressed. Biological control seemed to possess almost unlimited scope. The list of pests for control was a long one—sheep blowfly, buffalo fly, codlin moth, grasshopper, blackberry, weeds such as St. John's Wort, *Xanthium* or burr varieties, skeleton weed, and so on. Sheep blowfly was the most serious pest economically and became the new focus of attention. Entomologists, as well as other members of the scientific community, however, did not at first fully appreciate the unique circumstances of prickly pear control. These were that predators were comparatively abundant in the natural environment of *Opuntia* and that the lack of any similarity between the plant and the native flora removed the possibility of an indiscriminate attack. Such circumstances were combined only rarely.

The CSIR's division of economic entomology was established in 1928, with biological control of the sheep blowfly and the more important pasture weeds as its central mission. The chief was Robin J. Tillyard, F.R.S., a brilliant systematist of Australian flora and fauna and a recent convert to biological control.[27] Tillyard had had one of those predictably unpredictable entomological careers: Mathematics Tripos (Cantab.), read Oriental languages and theology as preparation for education, schoolmaster at Sydney Grammar School, member of the Linnean Society of New South Wales with a passion for the study of *Odonata*, Linnean Macleay Fellow at the University of Sydney, special investigator of the relation between insect populations and the depletion of freshwater fisheries in New Zealand, head of biology at the Cawthron Institute, New Zealand, chief of the division of economic entomology, CSIR.[28] Tillyard possessed unlimited capacity to observe, record, and understand the world of insects, but he was not well equipped to manage colleagues. He alternated soaring self-confidence, even grandiosity, with deep, destructive depression. In research science such a temperament can be an asset; in the management of scientific programs it is often disastrous. Tillyard's case was no exception.

The biological control program had one hand tied behind its back from the outset because of the intrusion of imperial interests. The Empire Marketing Board shared in the general entomological enthusiasm and was eager to finance the building of insectaries. Farnham Royal was keen to establish itself as the parent organization: the British institute would help search for, examine, and breed promising predators for subsequent testing and release in Australia. This was to be the ideal of imperial scientific collaboration with the mother country undertaking the core science and the dominion the field trials.

As noted, the sheep blowfly (*L. cuprina*) was the main target of the CSIR biological control program. The search for a predator was undertaken without firm knowledge of the evolution or life cycle of this highly durable and adaptable menace. The search included the south of France, where a natural predator of European blowfly species was judged to possess distinct promise. The predator, *Alysia manducator*, was reared and tested at Montpellier and was introduced, quarantined, and acclimatized briefly in 1928 before release around Canberra in 1929 and 1930. Tillyard spoke optimistically and often about the chances of success. But according to Ian Mackerras, leader of economic entomology's blowfly section, *Alysia* was never sighted in the field.[29] Tillyard, the inveterate scientific punter, had wagered too heavily on a rank outsider; the classical principle of biological control had not been followed. Not nearly enough was known about *L. cuprina*—indeed the species of *Lucilia* had not yet been correctly identified—insufficient time had been given to breeding and acclimatization, and in any case a European import was most unlikely to be able to dominate an evolutionary superfly such as *L. cuprina*.

In 1931 Tillyard's mood was near rock bottom. He announced, without first warning David Rivett, that the attempt to control sheep blowfly biologically had been abandoned. Other projects were faring just as badly. There seemed to be no chance of success in finding an effective predator for buffalo fly, and prospects for weed control were gloomy although not abandoned. In the midst of the great economic depression entomology was without a coherent program and was threatened with closure as the most expendable of the CSIR's research divisions. Tillyard was seriously ill in 1933 and retired early the following year. Rarely has a research program collapsed so ignominiously after having been launched with such a fanfare only a few years before.

It would be misleading, however, to suggest that the *Alysia manducator* episode and other disappointments forced the division to undertake fundamental studies of insect behavior. Research of this kind had been under way from the outset. A small team with the inglorious title of Blowfly Section, for example, had been investigating the ecology of carrion decomposition, and by the mid 1930s detailed knowledge was available of the succession of insects responsible for interring the corpse.[30] Such information was of some utilitarian value. Previously departments of agriculture recommended the burying or burning of dead sheep, on the assumption that the exposed carcass was a breeding ground for blowflies. Through ecological and taxonomic studies it was possible to show conclusively that *L. cuprina* was the one significant species that completed its life cycle on the live animal, so removal of the exposed carcass did not influence the size of the *L. cuprina* population. Experiments also made it clear that trapping, even when tens of thousands were caught in a limited area or period, made no difference to the total population or rate of strike. The more facts that were discovered the more monumental appeared the task of control. While the collapse of biological control was not responsible for the initiation of fundamental insect ecology, failure was a strong spur to basic understanding. Indeed, it has remained so since World War II, when *L. cuprina* has been the subject of fundamental population and genetic studies. The latter involve genetic manipulation. Since 1965 attempts have been made to modify genotypes in the field population to render the insect unable to survive winter temperatures and by attempting "genetic death" by rearrangement of chromosomes to produce sterile males, blind females, and so forth.[31] The research continues.

In other respects, too, fundamental work was stimulated by the great complexity of the biological phenomena under investigation and the difficulty of achieving applied results. A. J. Nicholson, Tillyard's successor, for example, pioneered the theory of a regulatory or feedback mechanism that was density-dependent in its operation as an explanation of insect population dynamics.[32] It is important to note that the support given to entomology throughout a long period during which its work did not lead to utilitarian application was an indication of the maturing of Australian science. It is true that the division was almost abandoned during the financial crisis of the CSIR in 1933, but this was more the result of Rivett's irritation with Tillyard than of research failure. The CSIR and an increasing number of university departments were able to give protection to scientists seeking knowledge for its own sake involved in basic research for long periods without utilitarian results. In the case of entomology, local scientists were able to contribute to understanding in the complex field of insect ecology. In the context of the CSIR as a whole, there is no doubt that the pragmatic success of the veterinarians and nutritionists strengthened the organization and allowed cross-subsidies to be paid.

In conclusion the following points can be made without elaboration. The progressive linkage of Australian science and the rural economy between 1890 and 1939 enabled scientists to achieve some independence from the short-term interests of producers and bureaucrats, as well as from metropolitan European science. By the 1930s, however, the linkage was more complex. Until the end of the 1920s scientists and technologists were alternately too closely associated with the bureaucratic or productive function, as in the case of agriculture department employees, or too distant, as were the universities. The CSIR as a statutory corporation devoted to the linkage of science and industry enabled a number of the more positive characteristics of the university and the government department to be combined. The veterinarians were able, for example, to perform in some respects like a good departmental team, but there was also an opportunity for the entomologists to pursue fundamental understanding for long periods without successful application. The linkage, therefore, was at a healthy distance.

Undoubtedly the depression accelerated the process, particularly in the pastoral industry. The change in attitude before and after the crisis is striking. From 1934 and 1935, when recovery from the worst years was sufficiently advanced, industry funds were made available to the CSIR in increasing quantity for specific research projects—a rare event in the financial history of Australian science. And in a number of fields Australian biologists were largely independent of their European counterparts. It can be argued, of course, that biology was backward as a branch of knowledge and its phenomena subject to all sorts of regional variations that conferred advantages on local empirical enquiry. The point has some validity, but to accept it without reservation would be to acknowledge physics as the paradigm case.

A somewhat different version of this essay appeared in *Historical Studies* no. 82 (April 1984): 11–28.

1. Ann Mozley Moyal, ed., *Scientists in Nineteenth Century Australia: A Documentary History* (Sydney: Cassell, 1976).

2. Michael E. Hoare, "Science and Scientific Associations in Eastern Australia, 1820–1890" (Ph.D. dissertation, Australian National University, 1974).

3. N. G. Butlin, *Investment in Australian Economic Development, 1861–1900* (Cambridge: Cambridge University Press, 1964).

4. H. E. Fels estimates that selective breeding was the most important source of improvement in productivity in the wool-growing industry of Western Australia, and his estimates would apply to the country as a whole. See H. E. Fels, "Sheep Improvement," in *Agriculture in Western Australia*, ed. G. H. Burvill (Nedlands: University of Western Australia Press, 1979).

5. A concise discussion of the problem of the 1920s is to be found in W. A. Sinclair, *The Process of Economic Development in Australia* (Melbourne: Longmans Cheshire, 1976).

6. There are many technical accounts of natural grasslands ecology. A popular account is given by H. C. Trumble in *Blades of Grass* (Melbourne: Georgian House, 1946); see especially chap. 4.

7. Francis Ratcliffe, *Flying Fox and Drifting Sand* (Sydney: Angus & Robertson, 1948); F. N. Ratcliff, "Soil Drift in the Arid Pastoral Areas of South Australia," CSIR Pamphlet no. 64 (Melbourne, 1936).

8. K. R. Norris, "The Ecology of Sheep Blowflies in Australia," in *Monographiae Biologiciae*, vol. 8, *Biogeographie in Australia*, chap. 32 (1959); M. J. Whitten and others, "The Genetics of the Australian Sheep Blowfly, *Lucilia cuprina*," in *Handbook of Genetics*, vol. 3, ed. R. C. King (New York: Plenum Press, 1973).

9. A. E. Pierce, "An Historical Review of Animal Movements, Exotic Disease, and Quarantine in New Zealand and Australia," *New Zealand Veterinary Journal* 23 (7) (1975): 125–36.

10. B. D. Graham, *The Formation of the Australian Country Parties* (Canberra: Australian National University Press, 1966).

11. Sir George Currie and John Graham, *The Origins of CSIRO: Science and the Commonwealth Government, 1901–1926* (Melbourne, 1966).

12. W. K. Hancock, *Problems of Economic Policy, 1918–1939*, vol. 2, pt. 1, of *Survey of British Commonwealth Affairs* (London: Oxford University Press, 1942).

13. CSIRO Archives.

14. T. Brailsford Robertson, "Animal Nutrition Problems," *CSIR Journal* 1 (1927–28).

15. J. B. Orr, "Memorandum on Research on Nutrition in Australia," CSIR Pamphlet no. 10 (1929).

16. Robertson to Rivett, 15 April 1929, CSIRO Archives, series 3, file PC/147/2.

17. Division of Animal Nutrition, report for the year 1931–32, pp. 1–2, CSIRO Archives.

18. Denmark disease was named after the Denmark district on the southeast coast of Western Australia, near Albany.

19. E. J. Underwood, *Trace Elements in Human and Animal Nutrition*, 3d ed. (New York: Academic Press, 1971); H. J. Lee, "Trace Elements in Animal Production," in *Trace Elements in Soil-Plant-Animal Systems*, ed. D. J. D. Nicholas and Adrian R. Egan (New York: Academic Press, 1975).

20. A distressing footnote to the cobalt episode is that priority for discovery has been the subject of an underground dispute between the Western Australians and the South Australians. Writing after the event Eric Underwood told the "cobalt story" in such a way as to imply that most of the credit was to Filmer and himself, an interpretation that has led to indignation in South Australia; see Underwood, *Trace Elements*. There is little doubt that cobalt deficiency was diagnosed several months earlier in Adelaide, and Underwood acknowledged as much in an article, "Enzootic Marasmus: The Determination of the Biologically Potent Element (Cobalt) in Limonite," written jointly with Filmer, in *Australian Veterinary Journal* 11 (3) (June 1935): 84–92. Apart from the understandable disappointment at having been beaten by a short half head, what seems to have rankled was Marston's unscrupulous self-seeking in promoting the discovery in his own name—the team effort took second place—and the studious neglect of any mention of the Western Australian work. Although Marston was not closely associated with the work, he announced the discovery in 1935 at ANZAAS in

Melbourne and subsequently rushed a note into the *CSIR Journal* to preempt the
Western Australians. My interpretation is that there was no serious dispute about
priority but that Underwood felt bound to make known the close second achieved by
the Western Australians.

21. A. W. Turner, *Black Disease (Infectious Neerotic Hepatitis) of Sheep*, CSIR
Bulletin no. 46 (Melbourne, 1930).

22. J. Needham, *Science and Civilization in China, I. Introductory Orientations*
(Cambridge: Cambridge University Press, 1954).

23. Chapters on evolution and phylogeny, anatomy and morphology, and physiology in *History of Entomology*, ed. R. F. Smith, T. E. Mittler, and C. N. Smith (Palo
Alto, Calif.: Annual Reviews, 1973).

24. K. S. Hagen and J. M. Franz, "A History of Biological Control," ibid.

25. An illustration of the remarkable reproductive power of the insect can be given.
In February 1926 some 81,000 eggs were distributed by hand near Emerald in central
Queensland. Two and a half years later five men were able, in one week, to gather 22
million cocoons in the same area. Field entomologists estimated that to destroy one
hectare required 22 million larvae. Most of the 25 million infested hectares had been
cleared by 1940.

26. A. P. Dodd, "The Biological Control of Prickly Pear," in *Biogeography and Ecology in Australia*, ed. J. A. Keast, R. L. Crocker, and C. S. Christian (The Hague: Junk,
1959).

27. Tillyard's main publications were *The Biology of Dragonflies* (1977) and *The
Insects of Australia and New Zealand* (1925).

28. J. W. Evans, *The Life and Work of Robin John Tillyard, 1881–1937*, Macrossan
Lectures 1962 (Brisbane, 1963).

29. Personal interview, 2 July 1976.

30. R. J. Tillyard and H. R. Seddon, eds., "The Sheep Blowfly Problem in Australia,"
Joint Blowfly Committee Report no. 1, CSIR Pamphlet no. 37 (Melbourne, 1933);
M. E. Fuller, *The Insect Inhabitants of Carrion: A Study in Animal Ecology*, CSIR
Bulletin no. 82 (Melbourne, 1934).

31. G. G. Foster and M. J. Whitten, "The Development of Genetic Methods of Controlling the Australian Sheep Blowfly, *Lucilia cuprina*," in *The Use of Genetics in
Insect Control*, ed. R. Pal and M. J. Whitten (Amsterdam: Elsevier, 1974).

32. I. M. Mackerras, "Alexander John Nicholson," *Records of the Australian Academy of Science* 2 (November 1970).

The United States Experience

Graduate School and Doctoral Degree: European Models and American Realities

NATHAN REINGOLD

By achieving independence in the eighteenth century and before experiencing the Industrial Revolution, the United States of America necessarily had a different "colonial" relationship to Great Britain and to Western Europe from those of other former colonies. Although political independence provided an opportunity to follow different cultural patterns, the United States had no other models to emulate, to modify, or to reject. Political independence did not entail a divorce from Western civilization. Throughout most of its history, the cultural inferiority and the dependence of the United States was assumed on both sides of the Atlantic. In this case, *colonialism*, in the absence of political hegemony, is related to *provincialism*.

After World War II, the high status of the sciences in the United States was widely recognized, posing a problem of historical causation. One nationalistic interpretation located the source of the new status in the graduate school. Doctorates and universities existed in Europe, particularly in Germany, but the graduate school was distinctive to the United States.[1] So too then was the academic multiprofessorial department, and some observers identified it as the locus of United States achievements.[2] I prefer to discuss the graduate school but without in any way demeaning the importance of the department.

In what follows, I am interested in the period to 1920, because the basic pattern of higher education existed by then, but I will make a few forays into later events. My hypothesis is deceptively simple.

Because of an American misunderstanding of the German situation and the way this misunderstanding fitted into purely indigenous factors, from 1900 to 1930 (roughly), the better universities developed a policy that produced a doctoral program not only different from the German but superior in certain respects. Although my research is concentrated on the physical and biological sciences, I suspect that the situation was similar in the humanities and the social sciences.

From my standpoint, *colonialism* simply means being different according to some Western European yardstick. I refuse to accept any necessary ascription of inferiority. To understand what was different in the rise of the graduate school requires a comparative consideration of the relations of "higher" and "lower" education; the institutional and ideological roles of vocationalism and of the applications of knowledge; the way concepts of general knowledge and moral upbringing influenced perceptions of research and its inevitable fellow traveler, specialization; and last—but by no means least—the way these were sometimes related to questions of class, ethnicity, and gender. In a brief essay, I can only suggest some of these relations.

By any measure, the German university of the nineteenth and early twentieth centuries was a significant contributor to Western civilization.[3] Andrew D. White, the first president of Cornell University, was exaggerating a bit at the end of the last century when he declared, "Germany was looked upon in the United States as a kind of second mother-country," because of the German university.[4] White and other university reformers did look to Germany for models and a yardstick to measure their accomplishments. They accepted much of the German educational ideology and sometimes the sentimentality implicit in such musical works as the *Academic Festival Overture* and *The Student Prince*, even as they used the German experiences selectively for their own purposes.

Originating in the trauma of defeat and resultant reforms of the Napoleonic era, especially in Prussia, the academic ideology had a decided nationalistic bias. Nearly a century later, Friedrich Paulsen could describe the reforms as "a cause which did not [only] concern the German population but the whole of mankind, for the conviction [was] that the distinctive character of the German people was indispensable to mankind."[5] To the German educational reformers, the aim of education was a process of self-cultivation (*Bildung*), the unfolding of personality by a steeping in general knowledge based on fundamental philosophic principles.[6] The yardstick for this education was provided by classical antiquity, to quote Paulsen: "Lastly, since the end of the eighteenth century, the Germans have yielded themselves to the influence of the Hellenic spirit with greater fervour than any other nation."[7] Despite warm words for the sciences, Paulsen could easily conclude, "Intellectual pride finds a much more congenial soil in the domain of aesthetic criticism, of linguistic and literary studies than of natural science and technology."[8] Interestingly, the attitudes explicit and implicit in Bildung were echoed by some German scientists.

Although the programs of the Prussian university reformers were only in effect briefly, they persisted as an ideology, becoming particularly strong during the Hohenzollern Empire. Within the faculties of the university—law, medicine, theology, and philosophy—philosophy was the site of the ideology and the special pride of the Germans. Originally an inferior body that prepared students for the professional faculties, the philosophy faculty and its professors were what most Americans meant when they talked of emulating the German universities. Another product of the reform, the classical gymnasium, was recognized as an essential element in the success of the German university and also entered into U.S. discussions of university reform.[9]

Within each faculty, professors supposedly had an absolute right to teach what they wished (*Lehrfreiheit*), while students were free to enroll anywhere with the *Abitur* from a gymnasium and to take any course (*Lernfreiheit*). Paulsen and others were proud that teachers were expected to be researchers: "In Germany the scientific investigators are also the instructors of academic youth. . . . The important thing is not the student's preparation for a practical calling, but his introduction into scientific knowledge and research."[10]

Not all American academics believed that Greeks populated the universities in the Rhine Valley and eastward. In 1852, responding to a German-influenced reform movement, Henry Vethake of the University of Pennsylvania noted that in Germany few were in the faculty of philosophy and that most of those were preparing for teaching, diplomacy, and the civil service.[11] Whatever the actual numbers, Paulsen and others recognized that many more students were at the universities for *Brotstudium* than for *Bildung* in the strict sense. As the lower degrees (bachelor's and master's) were in decay, the universities offered only two degrees, the doctorate and the *venia legendi*. Most students left without taking the former, having prepared themselves, for example, for one of the state examinations for a position in government service. What did *Lernfreiheit* mean for those many in *Brotstudium* who needed preparation for particular examinations? To an American, it is odd that those serious, hard-working young men were looked down on by the professoriate.[12]

The Ph.D. was required of gymnasium teachers and for further advancement in the academic world. Of course, the degree was taken by some for prestige and for various personal reasons. It was awarded after at least three years' standing beyond the gymnasium, an oral examination, and "a smaller dissertation." The larger dissertation—perhaps not the right term—was part of the habilitation, a process by which one proved worthy of becoming a peer of the university professoriate, receiving the *venia legendi*. The two stages exactly paralleled the Prussian civil service procedure. Despite words uttered against vocationalism, the universities clearly served the utilitarian needs of the state. Analogously, in the English system and in parts of the U.S. collegiate world in past eras, preparing for the ministry was not considered vocational in the pejorative sense that some applied to engineering.[13]

In the philosophical faculties, there were full (*ordentliche*) professors in whom power resided. In this period the associate (*ausserordentliche*) professors and the lowest rank (*Privatdozenten*) largely depended on the good will of the ordentliche professors. Both were unsalaried throughout most of this period. A peculiar feature was that the Dozents could give private lectures, often on the most advanced topics, while the Ordenarius presented the general course in the field. As the professors' salaries depended on enrollment, giving a course required for a state examination had obvious economic benefits. As late as the 1920s and 1930s, some U.S. academicians praised the arrangement by which young Ph.D.s could give advanced courses rather than a heavy schedule of elementary classes.

Forgotten in such praise was the absence of salaries during most of the period 1800–1920. The entire system was decidedly skewed to upper socioeconomic strata and served to produce and to maintain a *Bildungsbürgertum*—what Fritz Ringer denoted as the German Mandarins, borrowing a term from Max Weber—an intellectual elite in governing circles and universities serving as standard-bearers of a national cultural ideology. Although these Mandarins have peculiarly German traits, to North American eyes they clearly have relatives in Great Britain and France.

Before turning across the Atlantic to the American republic, a brief consideration of the gymnasium, the university institute, and the *Technische Hochschule*. The first was overwhelmingly classical. Despite glowing praise by Germans and others, its work in mathematics and natural sciences was modest until late in the last century. The closest recent analog is the English grammar school on the arts side after the eleven-plus examination of the post–World War II years. That is, both are examples of premature specialization. As the Abitur was essential for university entrance and, consequently, for admission to many positions, accounts of German education are filled with the struggle to gain acceptance for other secondary schools that emphasized science and modern languages.[14]

The university institutes were directly funded by the *Kultusministerium*, not out of the regular university budgets. They were research entities, not part of the formal instructional process. Pending future research in all academic fields, not only the sciences, it is an open question to me how great a share the institutes actually contributed in preparation for the doctorate and the venia legendi. They were clearly not like the United States departments, which, in general, controlled both instruction and research. In United States academic circles today, the unity of teaching and research is a widely accepted cliché and loosely credited to a German precedent, which may not exist in the American sense.[15]

The institutes acted as a device for separating new, specialized knowledge from the mainstream of formal education. The separate institutes helped keep the generality of knowledge from being overwhelmed by specialization, particularly in the sciences. Despite every noble phrase about philosophic unity, specialization was on the increase in the German university in the nineteenth century. Discovering particulars was acceptable, but many German academics apparently regarded specificity as akin to and leading to vocationalism; realia were déclassé.[15]

Of all the vocations, engineering most troubled the Mandarins, some later writers say because of the French example and also in reaction to the cameralism of the eighteenth century. The Technische Hochschulen developed independent of the university, attaining the right to award degrees only at the end of the last century. Somehow, engineering as an applied profession violated some Germans' class sensibilities and cultural ideology.[16] Even academic chemists such as Liebig, strongly involved with applications, carefully and with apparent sincerity proclaimed their allegiance to purity, to the generality of knowledge. In the German university, as in the larger society, word and deed often did not match.

In nineteenth-century America, the problem was different: reality and aspiration were believed to be incompatible. During the early decades the four-year residential colleges provided education to a small portion of the population. A few of the colleges were associated with professional schools of law and medicine. Throughout the century and afterward conventional wisdom equated the first two college years with the work in the gymnasium, the grammar school, and the lycée and the last two years, hopefully, with the European university level. Elevating the entire college course to university level often appeared to be beyond attainment.

Colleges are still important in American education. Even into this century, they retained their early purposes: moral inculcation; development of mental faculties (originally from the classics and mathematics); and provision of the knowledge regarded by the society as necessary for both cultivation and practical needs. The pattern was quite like what Nicholas Hans has disclosed for an eighteenth-century England in which preparation for adult life combined elements of both liberal education and transmission of specific vocational knowledge. In both eighteenth-century Britain and in early nineteenth-century America, the assumption and reality were that hardly anybody would attend a real university. For that reason, the colleges and Hans's schools had to provide what each society required in the way of both cultivation and preparation for selected occupations.[17]

Movements for upgrading the colleges in the United States usually involved adding courses and professors to accommodate new definitions of cultivation and emerging vocations—from law, medicine, and theology, for example, to science and engineering. From Stanley Guralnick's work, we know of the significant improvements within the colleges in the sciences during the years 1830–60.[18] In many institutions there was a desire to expand the higher courses and the professional schools further in emulation of Europe, particularly Germany.

The phenomenon of lower-level educational entities raising themselves, some even to true university status, is a persistent aspect of United States life. During the pre–Civil War era a number of academies—the analogs of Hans's "private schools"—aspired to and attained collegiate status; a few even aimed higher. After World War II, some normal schools, junior colleges, and even community colleges "bettered" themselves, a few to university level. The lack of any national regulations or standards lured academic entrepreneurs into higher education. The coming of the graduate school, 1847–1920, was simply an instance of a general continuing trend.

The motivation of some of the U.S. professoriate for higher courses and for research opportunities is usually stressed in the historical literature. Perhaps as important, in most instances, was the relation to undergraduate education. The resultant historical development presents the following typology:[19]

a. Institutions in which the undergraduate college, later commonly designated as of "arts and science," was the funnel through which students passed before entrance into the graduate and professional schools. Only the college provided education to the bachelor's level. Harvard and Yale are the best examples, but their paths to that position were quite different. Princeton is an interesting variant. Yale's is by far the most important single U.S. influence on university development, particularly graduate education.[20] Three notable, influential new universities of the 1860–1900 era, Cornell, Johns Hopkins, and Chicago, were founded by Yale graduates—Andrew D. White, Daniel Coit Gilman, and William Rainey Harper, respectively. The Yale experiences entered into the actions of all three men significantly. Both White and Gilman studied in Germany; Harper, a Yale Ph.D., did not.

b. Cornell is the prototypical U.S. university, and consequently, Andrew D. White is perhaps the most significant of the university builders in the United States. Cornell was a private university, but one that received some of the Morrill land grant funds. While most of White's policies had earlier precedents, his example influenced the adoption of the pattern now typical of state universities and most others. Cornell was notable in its acceptance of women and the early place accorded alumni and faculty among the governing trustees. White favored the equality of fields in the university— whether pure or applied—and the pattern of "breadth and depth" now common in undergraduate education. Unlike Harvard and Yale, Cornell and the state universities did not exalt *the* undergraduate college of arts and science but allowed coequal schools to issue bachelor's degrees. Only in graduate education did Cornell—and the state universities—lag as an innovator and as a model. Although Cornell's first Ph.D. was awarded in 1872, to Henry Turner Eddy, who became a notable engineer and engineering educator, leadership in graduate studies was elsewhere.[21] In 1910 a writer reported that 3,471 doctorates had been awarded during the preceding dozen years. A sample of fourteen leading universities produced four fifths of that total, but nine of the private universities in the sample, including Cornell, accounted for 2,634 and the five state universities for only 89. Only after 1910 did the state universities expand graduate output, a process still going on.[22] The usual explanation for the lag of the state universities is their emphasis on vocationalism, reflecting the viewpoints of state legislators. This explanation rings false, not only because of the expansion of graduate education in many of the state universities after 1910, but also because of the pre-1910 pattern, in which efforts were made in many institutions to give an education beyond the narrow confines of vocational preparation.

c. In a third category of institution are those founded with deliberate slants toward the university level, whether graduate or postgraduate, such as Johns Hopkins, Clark, and Chicago.[23] In all three unsuccessful efforts were made to bar or to diminish undergraduate education. Only a few, rare examples exist today in the United States—the Rockefeller University, for example, formed after World War II by adding a formal educational function, Ph.D. and postgraduate, to an existing research entity, the Rockefeller Institute.

To return to Yale, in 1847 it formed its Department of Philosophy and the Arts, which included the School of Applied Chemistry; in 1852 the School of Engineering was added. The department was the first in the United States to give earned master's degrees, and in 1861 it awarded the first U.S. doctorate to a physicist, A. W. Wright.

In 1863 J. Willard Gibbs received his, not a bad start for an educational tradition.

The two schools were undergraduate entities, which merged to form the Yale Scientific School in 1854, renamed the Sheffield Scientific School in 1861 in honor of a wealthy donor. Sheffield, not Yale College, received Connecticut's land grant allotment under the Morrill Act. Not only was Sheffield in competition with Yale College on the undergraduate level, but it and its parent department embodied a quite different educational view, stressing the primacy, not of the college, but of the philosophical faculty. It provided an undergraduate education in which the sciences, modern languages, and what we now call the social sciences had pride of place. Sheffield even organized a "Select Course" designed as preparatory to nontechnical pursuits for future businessmen, offering a three-year B.S. Sheffield pioneered into newer fields. The study of English language and literature at Yale originated in Sheffield, not the college. To James Dwight Dana and others around 1870, the department and its Sheffield School provided an opportunity to form a real university. In retrospect Dana's view overlaps to some extent the idea of a polytechnic university, influencing W. B. Rogers in his contemporary founding of the Massachusetts Institute of Technology. The selection of Noah Porter as president of Yale killed Dana's initiative. A good part of Yale's subsequent history is the attempts of Yale College's adherents to develop graduate education while stifling Sheffield, which continued to exist in an attenuated state as late as the post–World War II era.[24]

An analogous policy existed at Harvard under its great president, Charles W. Eliot. Harvard's Lawrence Scientific School never attained the strength or other characteristics of Sheffield. Like the Yale administrators, Eliot favored the college and stifled any possibility of a rival undergraduate body by placing the science teaching labs under the control of the college in the interests of economy.[25] Yale followed suit. Eliot, a chemist, and the Yale leaders had no intention of dropping or diminishing the sciences. Perhaps on the same principle as the Oxford Examination Statute of 1850, which required the Literae Humaniores before entrance to the schools of Natural Science, Law, and Medicine, Eliot and his peers at Yale wanted the American equivalent as a requirement for entrance into graduate and professional work.[26] Eliot successfully attenuated undergraduate engineering at Harvard during the last century; at Yale that goal was reached only in 1962.

Both universities initially favored a slow growth of graduate education during the period after the Civil War but followed different strategies. In New Haven, fending off Sheffield to maintain the primacy of Yale College produced a policy of regarding the graduate school as something added on, not intrinsically related to the college. Arthur Twining Hadley, Porter's successor, thought that pure research belonged in research institutes while the college fostered good citizenship, not expertise. Training teachers, not advancing research, was the aim of the graduate work.[27]

Eliot is a fine example of an American patrician liberal of the nineteenth century. He too saw the college as training for leadership in the society. Clearly influenced by *Lernfreiheit*, Eliot developed a complete elective system for undergraduates, a form of laissez faire applied to the formation of intellect and character. The graduate program was to develop naturally out of the elective system under the assumption of a continuity of identity of undergraduate and graduate education, perhaps by analogy with the German model. Not only did Eliot favor the college, but he initially doubted the viability of full graduate work in the United States.[28]

In 1876 a former professor at Sheffield shattered the assumptions that prevailed at Cambridge, Massachusetts, and at New Haven.[29] Daniel Coit Gilman and his trustees deliberately launched Johns Hopkins University as a graduate-level institution. A Berlin Ph.D., he called his creation the "philosophical faculty" and reluctantly had a nominal undergraduate enrollment, which later grew under community pressure. The small, high-quality faculty avoided formalities of structure and procedure. Throughout its golden age under Gilman, Johns Hopkins was a great success. In the years 1878–89, while Yale produced 101 doctorates and Harvard 55, Hopkins granted 151 Ph.D.s.[30] As so often remarked, a great part of the intellectual history of the United States during the subsequent quarter of a century is the result of the labors of these men. One even became president of Princeton and then of the United States.

On the English model, Hopkins established fellowships as a necessity for a successful graduate program. The German precedent produced a short-lived version of the *Privatdozent* system. But Hopkins had an unmistakable American flavor. As Hugh Hawkins puts it, the European aspects were subtly Americanized by their earlier stay at Sheffield. Like Hadley, Gilman did not see Hopkins as a research institution but as a place for teaching research. Most of its graduates—184 out of 212 by 1891—joined college and university faculties, spreading the word that real universities could and should exist in the United States.[31]

Gilman went on to establish the Medical School of Johns Hopkins, of great influence in subsequent developments in the United States. Even before Gilman's retirement, in 1901, Johns Hopkins was in trouble and went into something of a decline. The endowment proved inadequate to the needs. Expanding the undergraduate program did not lead to the acquiring of new funds, only to new problems and a loss of the singularity of the original conception.[32] More important, other universities with greater resources expanded their graduate programs.

By far the most spectacular new academic presence was the University of Chicago, which opened in 1891. It was well endowed by Rockefeller money, but its greatest asset was its president, William Rainey Harper, a specialist in Semitic languages. He was a Wunderkind, a Yale Ph.D. at eighteen, given to grand conceptions, elaborate administrative schemes, and great ability in day-by-day operations. Like Eliot and like Nicholas Murray Butler of Columbia, Harper had the same skills visible in such men as Rockefeller. Conventional wisdom correctly describes Chicago as a great university from the moment its doors opened. The success challenged the conventional wisdom at Harvard and Yale about the necessity and desirability of universities developing from or around *the* college.

Chicago was very American in some ways; in others it clearly reflected Europe. Of the original faculty, most had earned U.S. degrees, with Yale in the lead—sixteen, of whom nine were Ph.D.s. Even at the doctoral level there were fourteen German degree holders to twenty-one or twenty-two from the United States. Only five of the faculty were Hopkins graduates. Chicago rejected the extreme elective system of Eliot and chose the American norm of breadth and depth—that is, a pattern of distribution in all areas plus a concentration in one, the whole designed to provide both general cultural and specialized skills.

From its inception the graduate school was described as non-professional for pure research on the German model. There was a sharp distinction made between the first two undergraduate years, junior or Academic College, and the last two, the senior or University College. Harper tried unsuccessfully to banish the former from his campus. Harper had two graduate schools—sciences, arts and literature—reflecting an effort to have broad subject-area faculties rather than departments as the unit of governance. As in the similar developments in Columbia University, the German influence is clear, but in the long run the departments rather than faculties became the sites of actions and power. Another possible German influence is the complete absence of engineering, unique among major United States universities.[33]

Hopkins and Chicago, like Cornell, clearly influenced the then modest state universities. More important for the development of graduate education before 1920 was the effect on the older private establishments, such as Harvard, Yale, and Princeton. Eliot had advised the Hopkins trustees not to go into graduate education. Spurred by Gilman's achievement, Eliot deployed Harvard's more extensive resources into improving the faculty so as to expand the graduate degree program. That success was linked in Eliot's mind to the elective system, the one justifying the other. Samuel Eliot Morison was probably correct in his typical Harvard mixture of complacency and self-mockery when describing his university as the "premier American Ph.D. mill" by 1900.[34]

In contrast, the Yale situation is best described as modesty and anguish. The faculty improved in quality, and the graduate program expanded accordingly. But the Yale College–Sheffield split hindered a full commitment to a university program. Among the Yale College faculty, the alumni, and the trustees (the Yale Corporation), strong allegiance to the ideal of the college and hostility to Sheffield with its Select Course and its science orientation persisted.

Although also dating from before independence, Princeton, earlier the College of New Jersey, was smaller than Harvard and Yale and notably different in lacking professional schools in law and medicine.[35] But for the successful ambitions of some faculty, it might have remained a four-year liberal arts college, as did Amherst College in Massachusetts. In 1868 James McCosh became president of Princeton. He was a Scot who was both a Presbyterian clergyman and a philosopher in the tradition of common-sense realism. Mc-

Cosh instituted graduate study to the master's level in 1877–81 and established fellowships, citing the precedent of Edinburgh University. The early program consisted largely—and curiously—of physics and philosophy. Despite lukewarm sentiment among trustees and alumni, the doctorate was authorized in 1887.

Earlier, in 1872, the School of Science was formed, requiring Latin for admission. Two years later the School of Engineering opened. (It is a peculiar anomaly that, to this day, only Princeton of the Ivy League trinity has a continuous history of undergraduate professional engineering.) By the end of the century Princeton began resembling the Yale College–Sheffield situation in at least one respect: The lesser classical requirements for admission to Sheffield and to the School of Science at Princeton—coupled with the attractiveness of the Select Course and the more "modern" subjects—produced a rising enrollment in both to the dismay of the adherents of the traditional college. Having successfully attenuated the Lawrence Scientific School and having relaxed the rigidities of entrance and bachelor degree requirements, Eliot looked complacently on the problem of his colleagues to the south.

In actuality all three institutions were in trouble by 1900. The crux of their problems was the social and intellectual status of the undergraduate college. Those problems and their resolution had significant implications for graduate education. Earlier, McCosh was distressed by signs that Princeton was becoming a fashionable school for the wealthy with a rising stress on sports, eating clubs, and other undergraduate social activities. He wanted to attract talented, poor youths to the ministry, as Princeton had in earlier days. Graduate study, with its fellowships, was a way of restoring a more serious tone to Princeton.[36]

Yale graduates had a group memory of a golden age of the college as a republic of learning and morality with a simplicity of life style for both faculty and students—the early Roman republic infused with Congregational piety. Whether true or not, by the latter years of the nineteenth century Yale had become fashionable, sports-conscious, and filled with examples of undergraduate frivolity. Despite Eliot's genuine desire to have Harvard truly open to talent from whatever social source, Harvard came in time to present a picture at variance with his rather austere ideals. It had a "Gold Coast" and a student body a significant element of which was unconcerned with patrician ideals of leadership, let alone vocational and intellectual motives. At both Harvard and Yale around 1900, the reality of undergraduate education was a student body largely skilled in finding easy courses in order to graduate with a minimum of disruption to a preferred anti-intellectual style of student culture. A similar pattern was in evidence at Princeton. An ambiguous situation, if not a clash, existed between the emerging culture of the graduate school and the culture of the undergraduate college—specifically, the pattern of behavior of those undergraduates whose goals were not the same as those of the minority going on to the Ph.D. and the like-minded entering the traditional professional schools.

When Eliot's forty-year tenure ended in 1909, he was succeeded by A. Lawrence Lowell, who taught government at Harvard. Lowell was primarily concerned with the undergraduate college and did away with such Eliot reforms as the complete elective system. He worked successfully to raise the caliber of Harvard undergraduate education by instituting honors work, tutorials, and examinations— all clearly showing English influences. Like the Cambridge reformers discussed by Rothblatt in *The Revolution of the Dons*, Lowell sought a sense of community, leading to the establishment of a common freshman year, followed by later years in residential houses.[37]

Lowell's preference was decidedly for the humanities and the social sciences, but he was not necessarily hostile to the natural sciences. The sciences were included in his vision of community if performed in the proper spirit, an American version of old Göttingen and old Oxbridge. His role in the McKay bequest eventually led to an applied science program at Harvard. At Harvard and elsewhere there were complaints about the specialization and vocationalism of the Ph.D. No less a personage than William James wrote against the "Ph.D. Octopus."[38] (Although the degree was commonly associated with the natural sciences, from 1873 to 1928 only 25 percent of the Harvard doctorates were awarded in that sector.) Harvard cheerfully demolished the Union, used by graduate students, to make way for its new scheme of undergraduate housing.[39] At the same time, the departments pushed graduate education, which became a principal concern of James B. Conant, who succeeded Lowell in 1933. Where Lowell stood was manifest by the founding that year, at his instigation, of the Harvard Society of Fellows, a collegial body giving three-year appointments to those not more than twenty-five years of age who would take no courses nor study for any degree. It was old Oxbridge transmuted to Cambridge by the Charles River.

At Princeton the classicist dean of the Graduate School, Andrew F. West, had a parallel vision of a residential graduate college that would provide a community for graduates and postgraduates at Princeton.[40] At the same time Woodrow Wilson embarked on an upgrading of the undergraduate program, stressing tutorials and a sense of intellectual community. When West won the struggle over the graduate college, Wilson entered politics. Veysey is probably correct in stressing the basic similarity of the two men, but Wilson's position was closer to Lowell's than was West's. Despite his speciality, it was West, the champion of graduate education, who received and refused an offer to head M.I.T., which was by then improbable for either a Wilson or a Lowell.

At Yale the faculty of the College introduced honors work and moved to eliminate the easy courses. Resistance to change in both Yale College and Sheffield prevented many major reforms before World War I. In 1907, the historian George Burton Adams proposed a radical solution. A proponent of Yale College, Adams feared that the popularity of the Select Course and the pressure to relax the traditional entrance requirements would lead to its demise. Not at all hostile to science or to other newer fields, Adams wanted Yale to have three undergraduate colleges: the time-honored Yale College, a scientific and technological school, and a school for the social sciences and modern languages. It was a solution in the spirit of Cornell and the state universities in that all fields were treated equally, and undergraduates had more than one path to the bachelor's degree. Perhaps it failed for that reason.[41]

In 1919, taking advantage of faculty absence in war work, the Yale alumni, with aid from elements in the administration, pushed through a "great reorganization" despite a later near revolt from the faculty. Sheffield lost its separate undergraduate program. The college still had its prideful place. A common freshman year was followed by a community experience in residential colleges. Entrance and graduation requirements were reformed. Sheffield still existed in the form of the science and engineering departments at the undergraduate level and at the graduate level in certain arcane senses. Freed of the worst aspects of the old struggle, Yale's graduate program expanded.[42]

Sheffield's last dean, Charles H. Warren, according to Furniss's short history of the Yale Graduate School, was "wary of the purposes of high university authority, suspecting that policies injurious to the sciences might be incubating in secret"—this in the university of Silliman, Dana, and Gibbs! It was only after World War II that Sheffield became completely a legal fiction, its undergraduate program merging completely with the college while the graduate school became the holder of the historical tradition.[43]

Dean Furniss recognized the historical reality that much of Yale's graduate training and research traditions came from Sheffield, not the college, even a field such as English language and literature. And his brief account discloses a different spirit from that evident in Pierson's history of the college. Furniss notes that the great reorganization was supposed to overcome the reluctance of Yale's traditional families to send their sons into science. At first without comment, Furniss called attention to the emphasis on "inspirational teaching" and the claim "that advancement to the highest faculty rank would be awarded on teaching excellence alone." Although a College graduate, Furniss represented a different world view. He rejected the idea of "broad cultivation of the mind." He saw the doctorate as a professional degree, leading in most instances to employment in higher education. Believing in research as the function of the true university and of its graduate school, Furniss writes: "These young people [the graduate students], putting first things first, seek to qualify for remunerative employment. Having obtained this primary objective they are prepared to espouse lofty non-utile aims. The Professor, fortunately, has a secure grip on his meal ticket and is in a position to rise above mundane things in his avowal of professional purposes." The graduate school was a vocational institution, and "training in research was [its] inescapable duty." Long before Furniss wrote, in the era during which graduate instruction became established, his views represented the reality of higher education, not the increasing rhetoric of the exponents of the liberal culture ideal, as represented by Lowell and Wilson.[44]

That reality reinforced a policy based on a misapprehension about the German doctorate. Many United States academics assumed it was *the* prerequisite for teaching at the university level. That the German doctorate served as a prerequisite for gymnasium teaching also reinforced the high United States opinion of that institution

which stressed its overlap with the college. The point about its relation to habilitation and the *venia legendi* usually vanished from
academic arguments. Nor was it common for American university
reformers to note the basic structural distinction between the single
Ordinarius, the Professor, and the multiprofessorial departments
becoming common in United States colleges and universities from
1890 to 1920. If a "doctorate" was to be required for teaching beyond
the high school, the United States required a graduate system that
would produce more neophyte professors and by a faster procedure
than habilitation but clearly differentiated from both the liberal arts
bachelor and the increasingly numerous products of the normal
schools.

Originally, doctoral requirements in the United States extended
the time beyond the B.A. for advanced study modestly, perhaps with
an eye to the realities of resources and to the German tradition. In
1887 Princeton asked for two years of study beyond the bachelor's
degree—a major and a cognate study—plus a thesis. If the two last
years of undergraduate work are considered to be university level,
the formal requirements were not much different from those in Germany. By 1899 a minimum of two years of graduate work was fairly
standard, with the proviso—obviously German-inspired—of not
more than one year in residence. The National Association of State
Universities in 1908 specified three years of graduate study. By 1912,
the prestigious Association of American Universities noted "never
less than 2 years." By 1916 the association declared: "The amount
and character of the work should be such that the degree rarely
could be attained in less than 3 years following the attainment of a
bachelor's degree or equivalent." That is, continuing the previous
thrust, the doctorate now required a minimum of five years of university work, while only three were required in Germany.[45] Ultimately this structural change, aided by other trends, helped transform once colonial or provincial universities into the peers of the
leading European schools.

What is being hypothesized here is not a claim that professors
were superior, that students were more intelligent, that specific
courses were more detailed or deeper, that theses were better or
more original, or even that administrators were more enlightened
and farsighted. It eludes me how anyone could properly make such
comparisons between the United States and Germany, let alone all

Western Europe. Any such claims are made in disregard of the obvious point that educational systems fit national traditions, social structures, and intellectual aspirations. The more sweeping the comparisons, the more futile the effort.

The claim here is more modest and arises out of a consideration of circumstances within the United States. The longer period specified for the doctorate made possible a formal requirement for a more comprehensive educational preparation, particularly in the leading schools. This produced a growing number of individuals trained, on the average, in a broader range of specialized topics, reflecting the increasing size and intellectual diversity of the academic departments. These individuals provided a human resource, a surplus of trained talent, for the further expansion of higher education, industrial research, the government science bureaus, and the independent research institutions during the years 1900 to 1940.

Particularly striking in retrospect is the way the revival of liberal culture figured in this outcome. The seekers after community and a common intellectual tradition often looked to Oxbridge for inspiration—or, rather, to an idealized vision of Oxbridge. The introduction of tutorials is an obvious example, as are the expressions of hostility to specialization and vocationalism, often accompanied by affirmations of intellectual purity. But the reformers of schools such as Harvard, Yale, and Princeton—and their imitators elsewhere—could only turn for yardsticks to the graduate school; attempts to foster other modes of education have had only limited success to this day. When another Yale Wunderkind, Robert M. Hutchins, came to Chicago before World War II, his Great Books educational strategy had its effect, but a limited one, even in his own institution.

To illustrate the improvements in Yale College before World War I, Pierson proudly listed honors theses, as if to show that near graduate-level work was achieved. Improvers of the undergraduate college often ended up supporting improvements in graduate programs in order to get better teachers. Adherents of the research-oriented graduate school and disciplinary department could readily assent to improvement of undergraduate education in order to get better-prepared doctoral candidates. No one wanted a college substantially involved in remedying the deficiencies of the high school, a nearly universal complaint throughout the last century.

Both groups could agree that abstract knowledge, now defined as "pure," was the necessary preliminary to both culture and utility; many Americans, in fact, had trouble differentiating the two. It was an old belief, present even in antebellum America, but not exactly British or German, by the turn of this century. Perhaps settlers in what was formerly British North America south of Canada carried European mirrors across the ocean in which to look at themselves. In the westward translation the mirrors may in time have acquired a distinct distortion. In the United States, liberal culture never had the same degree of antiscience bias present in Germany and Great Britain. Like the undergraduate college, the graduate school was also of "arts and science."

1. Laurence R. Veysey, *Emergence of the American University* (Chicago: University of Chicago Press, 1965) Richard J. Storr, *The Beginnings of Graduate Education in America* (Chicago: University of Chicago Press, 1953); W. C. Ryan, *Studies in Early Graduate Education* (New York: Carnegie Foundation for the Advancement of Teaching, 1939); W. C. John, *Graduate Study in Universities and Colleges in the United States* (Washington, D.C.: Government Printing Office, 1935).

2. The best-known statement on the importance of the multiprofessor department is Joseph Ben-David, "The Universities and the Growth of Research in Germany and the United States," *Minerva* 7 (1968–69): 1–35.

3. Perhaps the best place to begin is with the writings of Friedrich Paulsen, *The German Universities and University Study* (New York: Macmillan, 1906) and *German Education, Past and Present* (London: Scribner's, 1908). Among recent historical treatments, I am indebted to Charles McClelland, *State, Society and University in Germany, 1700–1914* (Cambridge: Cambridge University Press, 1980), and Fritz K. Ringer, *The Decline of the Mandarins: The German Academic Community, 1890–1933* (Cambridge: Cambridge University Press, 1969).

4. Paulsen, *German Universities*, p. 9.

5. Paulsen, *German Education*, p. 184.

6. See also, W. H. Buford, *The German Tradition of Self-Cultivation: "Bildung" from Humboldt to Thomas Mann* (Cambridge: Cambridge University Press, 1975).

7. Paulsen, *German Education*, p. 179.

8. Ibid., pp. 294–95.

9. Ibid., pp. 188–93. During the years before World War I, the "classic" German university was changing, and Paulsen saw this as part of the influence of the coming of the masses, who had the option of free immigration to the new world. He also saw the Anglo-American college as a possible model for an institution between the gymnasium and the university. Ibid., pp. 179, 204, 297; *German Universities*, pp. 278–79. McClelland, in *State, Society, and University*, p. 137, notes the American example as having influenced discussions of technology.

10. Paulsen, *German Universities*, p. 3.

11. Storr, *Beginnings of Graduate Education*, p. 79.

12. McClelland, *State, Society, and University*, pp. 104, 201.

13. Ibid., pp. 104, 166–67. This position was, in part, a reaction against the French example.

14. Ibid., p. 110.

15. Ibid., pp. 285–86; Paulsen, *German Education*, p. 137f.

16. McClelland, *State, Society, and University*, p. 104.

17. Nicholas Hans, *New Trends in Education in the Eighteenth Century* (London: Routledge & Kegan Paul, 1951).

18. See his *Science and the Ante-Bellum American College* (Philadelphia: American Philosophical Society, 1975) and "The American Scientist in Higher Education," in *The Sciences in the American Context: New Perspectives*, ed. Nathan Reingold (Washington, D.C.: Smithsonian Institution Press, 1979), pp. 99–141. For an account of German influence in the humanities, see Carl Diehl, *Americans and German Scholarship, 1770–1870* (New Haven: Yale University Press, 1978).

19. This typology is indebted to, but differs from, those given by Storr, *Beginnings of Graduate Education*, pp. 131–32, and by George W. Pierson, *Yale College, an Educational History, 1871–1921* (New Haven: Yale University Press, 1952) pp. 44–45, basically because I am not that much impressed with the essentiality of the factors that differentiate Harvard and Yale.

20. In addition to Pierson, *Yale College*, see R. H. Chittenden's *History of the Sheffield Scientific School* (New Haven: Yale University Press, 1928) 2 vols; E. S. Furniss, *The Graduate School of Yale* (New Haven: Yale University Press, 1965); and D. C. Gilman, *The Relations of Yale to Letters and Science* (Baltimore: Johns Hopkins University Press, 1901).

21. See White's *Autobiography* (New York: Century, 1905) and Morris Bishop, *A History of Cornell* (Ithaca, N.Y.: Cornell University Press, 1962). I am not arguing for primacy, let alone downgrading other influential universities. What developed at Cornell—not wholly White's doing—is what most U.S. universities tended toward then—and even now; consider the belated admission of women to Princeton and Yale undergraduate life.

22. E. E. Slosson, *Great American Universities* (New York: Macmillan, 1910), p. 317.

23. Hugh Hawkins, *Pioneer: A History of the Johns Hopkins University, 1874–1889* (Ithaca, N.Y.: Cornell University Press, 1960); Richard J. Storr, *Harper's University: The Beginnings* (Chicago: University of Chicago Press, 1966).

24. Pierson, *Yale College*, pp. 55–59, 377f.

25. Hugh Hawkins, *Between Harvard and America: the Educational Leadership of Charles W. Eliot* (New York: Oxford University Press, 1972). Still useful is S. E. Morison, *The Development of Harvard University since the Inauguration of President Eliot, 1869–1929* (Cambridge: Harvard University Press, 1930).

26. Michael Sanderson. *The Universities in the Nineteenth Century* (London: Routledge & Kegan Paul, 1975), p. 75. This and his *Universities and British Industry* (London: Routledge & Kegan Paul, 1973) are extremely valuable for comparative studies of higher education and necessary correctives for the limited vision of Sheldon Rothblatt's, *The Revolution of the Dons: Cambridge and Society in Victorian England* (New York: Basic Books, 1968).

27. Pierson, *Yale College*, pp. 126–27.

28. Hawkins, *Pioneer;* p. 13; Robert A. McCaughey, "The Transformation of American Academic Life: Harvard University, 1821–1892," *Perspectives in American History* 8 (1974): 239–332.

29. Veysey, *Emergence of the American University*, p. 160.

30. Hawkins, *Pioneer*, p. 122.

31. Ibid., pp. 37, 64, 291.

32. Ibid., pp. 239–41.

33. Storr, *Harper's University*, passim.

34. S. E. Morison, *Three Centuries of Harvard, 1636–1936* (Cambridge: Harvard University Press, 1936), p. 335.

35. T. J. Wertenbaker, *Princeton, 1746–1896* (Princeton, N.J.: Princeton University Press, 1946), and Willard Thorp and others, *The Princeton Graduate School: A History* (Princeton, N.J.: Princeton University Press, 1978), are the best general sources.

36. A new study of McCosh, J. David Hoeveler, Jr., *James McCosh and the Scottish Intellectual Tradition* (Princeton, N.J.: Princeton University Press, 1981), came out too late for use in the writing of this essay.

37. For Lowell, see the excellent *DAB* entry by Hugh Hawkins.

38. William James, "The Ph.D. Octopus," *Harvard Monthly* 36 (1906): 1–9.

39. The percentage is from Morison, *Development of Harvard University*, p. 458; see also Morison, *Three Centuries*, pp. 335, 370–71, 479. Even in Germany the percentages for science in the Ph.D. population were overestimated by those against vocationalism and specialization, who blamed "science" for undesirable events and trends. Nor was Harvard unique; witness the statistics for the University of Pennsylvania in 1908–9 given in Slosson, *Great American Universities*, p. 366. Whether to stress reality or express his own inclination, Gilman at Hopkins took pains to repudiate charges of overstressing science; see Hawkins, *Pioneer*, p. 50. A modern historian, disenchanted with the consequences of science, found only a small percentage of the early Ph.D.s avoiding the stain; see Veysey, *Emergence of the American University*, p. 173. Perhaps Gilman repudiated to placate those in his day who had positions analogous to Veysey's.

40. For West, see the excellent *DAB* entry by Laurence R. Veysey.

41. Pierson, *Yale College*, pp. 377f., 480.

42. Furniss, *Graduate School of Yale*, passim; and Pierson, *Yale College*, passim.

43. Furniss, *Graduate School of Yale*, pp. 83–89.

44. Ibid., pp. 20–30, 90–91, and passim. The history of Yale is anything but typical of the U.S. situation. Despite some points that reflect the peculiarities of that history, Furniss gives what was both the reality of graduate education and the accepted wisdom, whatever the rhetoric used on formal occasions. Of particular interest are his last two chapters.

45. Thorp and others, *Princeton Graduate School*, p. 19; Wertenbaker, *Princeton*, p. 379; John, *Graduate Study*, pp. 28, 35, 46f.; Storr, *Beginnings of Graduate Education*, pp. 158–59.

European Origins of the American Engineering Style of the Nineteenth Century

EDWIN T. LAYTON, JR.

In 1837 a British engineer, David Stevenson, visited the United States and returned home to write an account of what he had observed. His *Sketch of the Civil Engineering of North America* is one of the best accounts there is of the emergence of a distinctive engineering tradition in America, one that differed from Old World practices in many important respects. Unlike many other visitors he did not interpret *difference* as *inferiority*. He saw that there were good reasons for the differences he saw, even the occasional coarseness, the sometimes unfinished appearance, and the seemingly temporary nature of many of the works that he observed.[1] What Stevenson saw was not a crude imitation of old-world technology, but the appearance of an engineering style adapted to the needs of the New World. That Stevenson was able to sympathize with this style, even to admire it, was partly attributable to his engineering insight, which led him to see the reasons for the differences that he observed. But a second factor may have been that the style he observed drew heavily upon a common Anglo-Saxon heritage. He may have grasped intuitively that there were some profound similarities in the ways that British and American engineers formulated problems and went about solving them.

National differences in technology have been a source of endless fascination, particularly in the older, less critical historical works. The clear intent of these older works has been nationalistic and ideological. Uncritical celebrations of heroic inventors purport to show national priority and superiority in technology, a superiority usually traced to the unique qualities of the particular culture. Remnants of this cultural nationalism are still to be found in the modern multi-volume histories of technology. These, however, are archaic survivals from the past. Within the last generation a professional discipline of history of technology has emerged. There has been a strong tendency to reject cultural nationalism and even more the naive and uncritical scholarship that underlay much of this old literature. The development of technology, like that of science, is best seen as part of universal history: a history that transcends national boundaries and petty nationalistic considerations. Major inventions are no longer considered unique achievements of one or two "heroic" individuals, but rather as complex systems that evolve through the efforts of many people, usually drawn from more than one country.[2]

Recently, however, there has been a revival of interest in national differences and styles in technology. The animus in this scholarship is neither parochial nor uncritical. Rather, national, or better, cultural, differences make it possible for us to address some fundamental problems concerning the nature of technology, its development and diffusion, and its relation to culture. A useful starting point might be the well-known fact that the introduction of new technologies often has a disruptive effect, particularly on primitive societies. The effects of the snowmobile on the culture of the Lapps is a good case in point.[3] This raises the question, Why is not technological diffusion equally destructive to larger, more complex, more advanced societies? The answer appears to be that they are more adaptable. In general there are two kinds of adaptation to technology, social and technical. A society can adapt to offset the harmful effects of a new technology. Thus, the harsh inequalities of the factory system have been buffered by the rise of trade unions, self-help, and government action. But technologies can be changed also, in order to make them serve the needs of society better or to minimize harmful effects. Both forms of adaptation are important, and they usually go together. But I shall be concerned primarily with the second: changing the technology to fit it to the needs and values of a culture. This is one of the primary causes of national differences or styles in technology. Thus, Americans modified the technologies imported from the Old World, not through any mysterious force of

Yankee ingenuity or any inherent superiority, but rather as a cultural necessity: to adapt the technology to their needs.

A few briefly sketched examples may clarify the sort of technological adaptations that were made in America. The steam engine was introduced into America, along with the textile factory system, the railroad, and the machine tool industry, from England during the latter eighteenth and early nineteenth centuries. In each instance they were modified in significant ways, but after a delay. This can be seen as a two-stage process. In the first, British mechanics carried the new technology to the New World and installed it in essentially its original form. This was the case, for example, with the Watt steam engine. But this engine did not fit American needs. There were few shaft mines and little need for steam engines as mine pumps. Nor was the Watt engine well adapted to American industrial needs; water power was abundant and cheap. But America did have a pressing need for steam transportation. The Watt engine was pressed into the service of transportation from the start: Robert Fulton's steamboat, the *Clermont,* was driven by an engine imported from England. But the atmospheric engine with its clumsy condenser was ill adapted to transportation; an engine with more power in relation to weight was needed. Oliver Evans designed an original high-pressure engine independent of the parallel work of Richard Trevithik in England. But the high-pressure, noncondensing engine became the norm in America, not only for steamboats and railroads but for stationary power as well, despite the considerable hazards from boiler explosions. Thus, British and American steam engine technologies tended to diverge through the 19th century.[4]

A similar case can be shown for the railroad. The first American railroads were direct imports from Britain, and locomotives, rails, and rolling stock continued to be imported for some time. But the British railroad was not adapted to American conditions. Capital was scarce, so expensive cuts and fills were not possible, and rail lines had to conform to the topography. British locomotives had only limited turning and climbing ability, both of which were necessary under American conditions. Thus through a series of innovations railroading was adapted to American conditions: John Jervis developed the pivoted forward truck for locomotives, to allow them to negotiate sharp turns; Mattias Baldwin doubled the steam pressure, increasing tractive power; others developed lever systems that helped keep railroad cars on the rails despite rough, uneven road-beds. The effect of these and other changes was to adapt the railroad to American geographical and economic conditions.[5]

In the case of textile manufacturing a two-stage process can also be seen. In 1789 an English mechanic, Samuel Slater, brought over the technology of the water frame and set up the first modern spinning mill at Pawtucket, Rhode Island. These first mills were very similar to those in England. But here too adaptive pressures worked for change. Slater arrived at a time when there was a surplus of labor in parts of New England, and a factory system based on cheap labor was possible. But the opening of the West produced a mass exodus that had left many abandoned villages in New England by the 1820s and 1830s.[6]

A group of Boston capitalists developed a novel approach, first at Waltham, then at the new city of Lowell, where they could tap the power of an entire river, the Merrimack. Their innovations rested upon economies of scale inherent in large, integrated plants: raw cotton went in at one end and finished textiles came out the other. Weaving, spinning, and all other processes were combined under one roof. These large plants were energy-intensive; they needed about 800 to 1,000 horsepower each. But waterpower was abundant in America. The Boston capitalists led by Lowell recruited a new source of labor: farm girls who worked for a few years to amass dow-ries and who lived in strictly supervised dormitories. A number of technical changes contributed to the success of these early ventures, including ring spinning, the extensive use of belting, and highly efficient hydraulic turbines based on Fourneyron's French prototype.[7]

In contrast to the case with textiles, the machine-tool and metal-cutting industries were able to adapt to scarcity of labor and high wages. Here, too, the initial technology was imported from Britain, but it was soon adapted to the lower skills and greater cost of American labor. The story is too complex even to outline here. But fundamentally Americans built labor-saving machinery in great variety and learned to integrate special purpose machines efficiently through improved systems of management.[8] It is possible, however, to speak of two outputs: mass production, which reached a sort of culmination with Henry Ford, and new management systems, the most famous being Frederick W. Taylor's "scientific management."

These can be regarded as innovations, though both were complex systems involving many separate innovations, and both managerial methods and production systems were dynamic, evolving systems, not static entities. Thus, both are difficult to characterize in simple terms.[9]

The patterns of diffusion that I have sketched briefly appear to be quite widespread. A similar pattern is to be found in the diffusion of American innovations to the rest of the world. Mass production and scientific management were among the most important American innovations diffused in the twentieth century. But in no instance did Europeans install carbon copies of these American systems. Individual elements did diffuse, such as time-and-motion studies and the assembly line. But automobile manufacturing methods and management systems developed independently in Europe. That is, these innovations were adapted to the particular cultures that absorbed them.[10] The same could be said of Japanese technology. After a period of imitative direct diffusion, the Japanese, after the Second World War, adapted and improved upon imported Western technologies with results known to all.

In most studies diffusion is assumed to be direct. That is, an artifact or process is simply transported to a new location without change. The pattern I have described is perhaps better understood by the concept of "stimulus diffusion" proposed by the anthropologist A. L. Kroeber in 1940. He used the terms *stimulus diffusion* and *idea diffusion* synonymously, defining them as the situation whereby

a system or pattern as such encounters no resistance to its spread, but there are difficulties in regard to the transmission of the concrete content of the system. In this case it is the idea of the complex or system which is accepted, but it remains for the receiving culture to develop a new content.[11]

Stimulus diffusion or idea diffusion appears to be the norm, not the exception. All borrowing involves some invention and all invention some borrowing. This is to do no more than restate an anthropological truism, that cultures resist change by assimilating alien intrusions. It is only to be expected that technological systems which produce cultural change or disruption will be assimilated through mechanisms of social and technological adaptation. But this expectation raises a number of questions. What is the nature of the "difficulties" that cause change? How "new" is the new content of the system?

In presenting the diffusion of important technologies from Europe and their modification in America, I have followed the conventional wisdom in stressing factor endowments as economic determinants of technical change. There is no doubt about their importance, but a good deal of evidence suggests that they do not explain everything.

In my own studies of the history of engineering in America I have found that ideas and values served as cultural filters. The cultural elements that modulated engineering thought and practice can be referred to as part of a style in engineering. For American engineers in the nineteenth century this was expressed, on one level, by a strong philosophical bias against the analytical and deductive style of engineering developed in France and on the continent of Europe. American engineers followed a deep-rooted Anglo-Saxon philosophical bias in favor of an empirical, inductive style. In this respect, then, British and American engineers had much in common. But American engineers carried these shared traits much further.

An extreme example of the American engineers' rejection of deductive mathematical theory can be found in Benjamin F. Isherwood's *Experimental Researches in Steam Engineering*, published in 1863. Isherwood was engineer-in-chief of the United States Navy during the Civil War. Apart from his wartime services, Isherwood was an important advocate of higher professional standards. The function of the navy in mechanical engineering was analogous to that of the army in civil engineering: the navy served as a training school for mechanical engineers. It sponsored a good deal of early scientific research in engineering. Many of the first professors of mechanical engineering at American universities were former naval engineers. Thus, Isherwood carried a good deal of influence.[12] In his book Isherwood denounced the use of all deductive methods and mathematical theory in engineering, thus setting a style that many others were to follow.[13]

Isherwood's philosophy is explicitly drawn from British sources, principally Hume, Bacon, and Scottish common-sense philosophy. His objection to mathematical theory was that it was, of necessity, highly idealized; that is, it postulated such things as perfect elasticity, frictionless machines, and the like. He argued that these were metaphysical abstractions without real existence and that such theories ignored the complex factors actually at work. He held that "we must not seek for physical truth in metaphysical abstractions." Instead Isherwood argued for a Baconian empiricism. To him "the true method of constructing a sound theory on any subject in physical science, is to commence by ascertaining the value of every quantity by direct experiment." The data should be reduced to tabular form. Then "from these tables we can form general laws by gradually rising from particulars to generals." Citing Hume and Scottish common-sense philosophy, Isherwood denied that science had or should have any metaphysical content. The only reality was the laws, and these laws were "merely generalized facts."[14]

The rejection of mathematical theory by American engineers presents something of a paradox. There is ample evidence that they used it. And this was particularly true of the ablest of them, the very ones who so vehemently protested against mathematical theory. There is, in fact, much evidence that American engineering was deeply influenced by French engineering theory. In 1933 Ralph Shaw did a study of engineering books available in America in 1830. Briefly, Shaw found that of the books available at that date 252 were British, 223 were American, and 188 were French;[15] that is, the number of French engineering books was only slightly smaller than the number of British books. If allowance is made for the fact that many of the British and American books were translations of French books or based upon French works, then the French influence in numerical terms was at least equal to the British influence.

This finding is clearly anomalous. As Brooke Hindle and others have pointed out, the British influence on American technology was much greater than that of France, if innovations are considered.[16] That is not to say that the French influence was insignificant. The idea of interchangeable parts in the manufacture of firearms was French, as was the hydraulic turbine. But when one turns from innovations to engineering theory the picture is quite different. The first American engineering school was the U.S. Military Academy at West Point. The school was remodeled after the French engineering schools by Sylvanus Thayer after 1816. He sent Dennis Hart Mahan, a promising graduate who taught engineering to cadets from 1832 to 1871, to study at French engineering schools. For the early years many of the texts were French. The system then in use encouraged officers to enter civilian practice, and West Point engineers constituted the elite of American civil engineering for more than a generation. Mahan did a textbook that sold more than 15,000 copies. Through his teaching, his texts, and his translations, Mahan was a primary source for the transmission of French and continental engineering theory to America.[17]

Mahan was by no means unique. The second important engineering school in America was Rensselaer Polytechnic Institute, reorganized as an engineering school in 1848 by Benjamin Franklin Greene. Here, too, continental textbooks in the French style predominated in engineering training.[18] Of course college training remained exceptional for American engineers until nearly the end of the nineteenth century. But there is at least some evidence that the looser system of apprenticeship and self-study by which most engineers in America were educated was also strongly influenced by French engineering theory. I will cite but one example. Loammi Baldwin, Jr., a Bostonian, was an influential early engineer who regularly took on apprentices. He evidently emphasized French engineering theory, which was well represented in his library. One of his pupils was Charles S. Storrow. After working with Baldwin, Storrow went to France and studied at l'École des Ponts et Chausses. He was one of the engineers who contributed to the great hydraulic and industrial complex at Lowell, and he subsequently created a second complex at Lawrence. He became a founder of a great New England tradition in hydraulic engineering, and he influenced both James B. Francis and Uriah Boyden, the American developers of the French turbine.[19] The influence of French theory on Storrow was evident in a book published in 1835, shortly after his return from France, entitled *A Treatise on Water-Works*.[20] It was made up almost entirely of translations from French engineering writings. Another fruit of the New England school of hydraulic engineering was Joseph Bennett's translation of S. F. D'Aubuisson's *Treatise on Hydraulics* in 1852.[21]

There is no doubt that American engineers of the nineteenth century used and were influenced by the French analytical tradition in engineering. But they attempted to place it within a different philosophical framework, one that was agressively hostile to mathematical theory. This created a tension within American engineering, a tension that helped to shape the American engineering style. On the one hand they wanted to use the fruits of the French and Continental tradition, and in many instances they did so in private. But in public their stance was antitheoretical and empirical.

The tension within American engineering is shown by John C. Trautwine, whose *Civil Engineers Pocket-Book* was enormously influential; it had sold more than 150,000 copies by the early twentieth century. In the preface to this handbook Trautwine denounced all efforts to assimilate engineering within the mathematical tradition. He saw such efforts as "little more than striking instances of how completely the most simple facts may be buried out of sight under heaps of mathematical rubbish." Trautwine, however, needed to use the fruits of the very tradition that he repudiated. He resolved the dilemma, at least to his own satisfaction, by claiming that things such as the laws of mechanics, the principle of virtual velocities, and the like were no more than simple facts of common-sense observation accessible to schoolboys. He pictured two boys playing with a seesaw and deriving the laws of the lever, the concept of moments, and the principle of virtual velocities. He stopped at this point, so the reader was spared the vision of children deriving Galileo's law of fall and Newton's laws of motion. This entire argument is, of course, absurd. The laws of physics from Archimedes to Newton cannot be derived in this manner, much less the principles of engineering. But it allowed Trautwine to adhere to a Baconian, empirical philosophy similar to that expressed by Isherwood, while making free use of whatever results he wanted from mathematical science and engineering.[22]

The attempt to assimilate a Continental tradition of mathematical analysis within an antimathematical tradition led to absurdity and even hypocrisy. But the tension could be creative. It contributed, for example, to an important innovation, the mixed-flow or Francis turbine. Benoît Fourneyron's turbine was eagerly accepted in America, particularly at the great New England hydro-power complexes such as the one at Lowell. But the engineers most involved in the successful assimilation of the Fourneyron turbine, James B. Francis, chief engineer at Lowell, and his collaborator, Uriah A. Boyden, repudiated the theoretical framework within which Fourneyron's work was embedded. Instead, they substituted their own philosophy and, to some extent, their own theory. Their repudiation was, however, complete only in public. In their private correspondence, in private notebooks and the like, it is clear that they used Fourneyron and other European theorists as constant referents for their own approach to the turbine.[23] The result was a sort of creative dissonance. They regarded the turbine from two quite different and conflicting perspectives. The emergence of the Francis design cannot be fully understood without understanding the interaction between these two approaches.

Fourneyron started from a highly abstract conceptualization of an ideal water motor first clearly enunciated by Lazare Carnot.[24] If no internal friction is assumed, Carnot reasoned, complete efficiency would be achieved if the water entered without shock and left the motor with essentially no velocity. In a sense this is a truism: if the water loses none of its energy on entrance or through friction, and if it leaves with no energy, then all the energy must be transformed into work. Fourneyron showed how such an ideal water motor could be approximated in practice. To avoid shock at entrance he invented fixed guides inside the turbine wheel, which directed the water toward the moving wheel at a predetermined angle. Second, by mathematical analysis, he showed how to derive the angle that would allow the water to enter tangent to the surface of the moving blade, thus avoiding shock. Third, he showed how the exit angle that would give minimum velocity could be determined. The mathematical method by which he derived these angles was essentially an extension of classical mechanics to the turbine. But the result was simple formulas for the angles of entrance and exit, and these formulas could be separated from the mathematical theory from which they were derived.[25] (Actually there are two angles to consider for entrance, since with a moving wheel both absolute and relative velocities must be taken into account, but I will ignore this complexity here, though it was essential to Fourneyron's success.)

What is worth noting is that Fourneyron treated the turbine as a black box and concerned himself only with the angles of entrance and exit. This is implicit in the assumption that there are no frictional losses in the turbine. He assumed that the actual path of the water could be ignored, so long as there were no sharp bends or kinks.

Boyden and Francis rejected Fourneyron's theory while using, in private, its practical results. They rejected the assumption of frictionless operation. They developed a less idealized, highly empirical approach by which frictional effects were placed in the forefront. They did not treat the turbine as a black box; on the contrary they traced the absolulte path of the water inch by inch as it passed through the turbine. Their idea was to maintain uniform flow—actually uniformly accelerated flow. As much as possible they kept the absolute flow in a straight line.

Both these approaches are valid and are analogous to methods used today. Treating the turbine as a black box is similar in principle, though not in mathematical detail, to modern control-volume analysis. In essence this involves ignoring all the complex dynamics of what is going on within the control volume, which can be a turbine or other hydraulic machine, and concentrating on the input and output.

Boyden and Francis, with their concern for the particular conditions of flow at each point in the absolute path of the water and their concern for uniform flow were thinking in terms analogous to those of boundary-layer analysis in fluid mechanics. This analogy may appear farfetched; certainly they had no clear conception of laminar flow and boundary layer, much less the mathematical methods needed to deal with them. But while the theory was primitive, the experimental methods were not. Boyden and Francis used methods analogous to those developed earlier by American millwrights. Some of the millwrights within this tradition used glass-walled testing flumes into which they added small pellets in order to observe the detailed flow of the water. The testing flume is, of course, a fundamental experimental tool in fluid mechanics. Indeed, James Leffel, one of the millwrights who perfected the Francis design, used a glass-walled testing flume in an extensive experimental investigation.[26]

Without going into all the details, it is clear that the development of the Francis turbine involved an interaction between these two ways of understanding the turbine. But this fact is by no means evident in Francis's *Lowell Hydraulic Experiments*, in which he published his and Boyden's innovations. Indeed, Francis did not mention Fourneyron by name at all. And amid the many detailed tabulations of data that he presented, Francis left out the three critical angles needed to apply Fourneyron's theory. Francis presented himself and his work as purely empirical, experimental research. Indeed, he denied that the mathematical theory had any real utility. Francis wrote that "the turbine has been an object of deep interest to many learned mathematicians, but up to this time, the results of their investigations . . . have afforded but little aid to the hydraulic engineer."[27] It should be noted that mathematically inclined engineers, such as Fourneyron, were called mathematicians at this time in America.

Francis achieved a stylistic consistency in his book at the expense of the rankest sort of hypocrisy. In his correspondence with Boyden and in his private engineering notebooks Francis made repeated use of precisely the sort of theory that he repudiated publicly. In particular, Francis used a variation on Fourneyron's method of determining the angles of entrance and exit in his private notebooks.[28] He never published these methods, leaving his readers to proceed by cut-and-try or to discover them for themselves in the literature. This is unfortunate, in that Francis made creative additions to Fourneyron's methods that he never published. In particular, his unpublished principle of uniform acceleration of the fluid as it moved through the rotor is fundamental to turbine design.

It is clear that in the case of Francis the reason for suppressing all reference to Fourneyron's theory was stylistic. In other instances economic motives cannot be excluded. But Francis was in a peculiar position; the owners of Lowell were primarily interested in manufacturing textiles. They did not wish to manufacture turbines on a regular basis. They and Francis, as consumers of turbines, hoped that by publishing the results of Francis's experiments they would stimulate others to improve the turbine further. This is, in fact, what did happen. Second, Francis was unusual for that age in his generosity·in acknowledging his debt to others. In the *Lowell Hydraulic Experiments*, for example, Francis makes frequent reference to earlier French experimental research. Indeed, by suppressing all mathematical theory Francis also suppressed some creative achievements of his own. In short, Francis's is a highly unusual case; every motive but the stylistic ones can be excluded.

It is easy to condemn Francis, and indeed, he deserves censure. But we can at least try to understand why he acted as he did. The engineering community in America was in a state of gradual transition. Scientific methods and institutions were only beginning to be absorbed. But these methods and institutions, along with the attendant values associated with the scientific approach, were to some extent alien intrusions. They posed a threat, in the sense that they implied a drastic reorganization of the profession. The American engineering style was a sort of buffer; it admitted scientific methods but insisted that they conform to the existing social structure and values. In America, as in England, engineering was allied with private industry; in France and on the continent engineering had traditionally been the servant of the state. And the institutional structure of governmental engineering led to close alliance between engineering and the scientific community. In both cases there was resistance to change. In France new schools founded to meet the practical needs of industry, beginning with l'École Centrale, tended to resist the practical and utilitarian demands this entailed. They sought to retain the high prestige of the *grandes écoles* and their close alliance with science.

A few French engineers resisted the possibly premature scientification of engineering. Théodore Olivier, one of the faculty of l'École Centrale complained bitterly:

Thus, and I say it with a heavy heart, it was under the baneful influence of Laplace, Poisson, and Cauchy that the École Polytechnique was reorganized in 1816.

These men, who knew no language but algebra [that is, analysis] ... destroyed ... the original organization of studies at the École Polytechnique. Instead of a school of students destined for the public services, they created a school of mathematicians. . . . They suppressed the applied courses. . . . They ... and their disciples have often repeated that the École Polytechnique is destined above all to give the country two or three savants of Analysis each year.[29]

But Olivier was ignored, and the orientation toward mathematical science came to dominate l'École Centrale as well as l'École Polytechnique.

In America the situation was quite different. To American engineers it was the scientific demands of the new engineering that constituted the threat. Even those engineers with college training lacked the mathematical mastery of their French colleagues. More important, attempting to match the Europeans in mathematics would weaken the upward mobility from craftsman to engineer and the close alliance with employers, which followed, in part upon a tradition of on-the-job learning. In short, the French mathematical orientation would be subversive in America. And American engineers found what they thought were good reasons for rejecting it. In response to a young European trained in the mathematical methods of the polytechnic tradition, James B. Francis confessed, "I fear I have a habit of doubting purely theoretical deductions in an unreasonable degree."[30]

Some American scientists attempted to resist the prevailing bias against theory on the part of American engineers. In the case of Francis's own field, water power, a scientific author presented a survey of the mathematical theory of water wheels in a popular encyclopedia in 1852. This author complained that water power was controlled by those who rejected theory: "With few exceptions it has been abandoned to the management of those who recognize in it no principle, and no scope for improvement; and whose practice is not more opposed to improvement than it is empirical and opposite to all true principle."[31] It is in the light of the excessive rejection of theory by American technologists that one should read Joseph Henry's repeated insistence on the priority of theory to practice.[32]

Critics, such as Henry, had a point. What they missed was the fact that a transformation of American engineering would threaten the existing social system of technology. And further, the rejection of theory was exaggerated in both oral and written statements; in actual practice American engineers such as Francis were more in tune with contemporary theory than they were willing to admit in public. But they chose to emphasize a self-image of the engineer as a practical person who met the utilitarian needs of industry by creative design linked to empirical science. To them the polytechnician appeared to be an impractical theorist. In the eyes of American engineers, whether rightly or wrongly, the Continental mathematical tradition represented the wrong way to do engineering. And this wrongness was as much a matter of social structure and values as of method.

1. David Stevenson, *Sketch of the Civil Engineering of North America*, 2d ed. (London: John Weale, 1859).

2. Thomas P. Hughes, "Emerging Themes in the History of Technology," *Technology and Culture* 20 (October 1979): 697–711.

3. Pertti J. Pelto, *The Snowmobile Revolution: Technology and Social Change in the Arctic* (Menlo Park, Calif.: Cummings Publishing Company, 1973).

4. Carroll W. Pursell, Jr., *Early Stationary Steam Engines in America* (Washington, D.C.: Smithsonian Institution Press, 1969); Louis C. Hunter, *Steamboats on the Western Waters* (New York: Harvard University Press, 1949).

5. John H. White, Jr., *American Locomotives: An Engineering History, 1830–1880* (Baltimore: Johns Hopkins University Press, 1968).

6. George S. White, *Memoir of Samuel Slater* (Philadelphia, 1836); see also Gary Kulik, "A Factory System of Wood," in *Material Culture of the Wooden Age*, ed. Brooke Hindle (Tarrytown, N.Y.: Sleepy Hollow Press, 1981), pp. 300–335.

7. John F. Kasson, *Civilizing the Machine* (New York: Penguin Books, 1977); Louis C. Hunter, *A History of Industrial Power in the United States, 1780–1930* (Charlottesville: University Press of Virginia, 1979), vol. 1, *Waterpower in the Century of the Steam Engine*, pp. 204–27, 251–91.

8. Merritt Roe Smith, *Harpers Ferry Arsenal and the New Technology* (Ithaca, N.Y.: Cornell University Press, 1977).

9. Edwin T. Layton, Jr., "The Diffusion of Scientific Management and Mass Production from the U.S. in the Twentieth Century," *Proceedings XIVth International Congress of the History of Science* (Tokyo, 1975), vol. 4, pp. 377–84.

10. Ibid.; see also Patrick Fridenson, "The Coming of the Assembly Line to Europe," in *The Dynamics of Science and Technology*, ed. Wolfgang Krohn, Edwin T. Layton, Jr., and Peter Weingart, Sociology of the Sciences Yearbook, 2 (Dordrecht, Netherlands: D. Reidel, 1978), pp. 159–75.

11. A. L. Kroeber, "Stimulus Diffusion," *American Anthropologist* 43 (January–March 1940): 1.

12. William Sloan, *Benjamin Franklin Isherwood, Naval Engineer* (Annapolis: Naval Institute Press, 1965); Monte A. Calvert, *The Mechanical Engineer in America, 1830–1910* (Baltimore: Johns Hopkins University Press, 1967), pp. 245–61.

13. Benjamin Franklin Isherwood, *Experimental Researches in Steam Engineering*, 2 vols. (Philadelphia: W. Hamilton, 1863), vol. 1, preface and passim.

14. Ibid., pp. xiv–xxv, 139–40.

15. Ralph R. Shaw, "Engineering Books Available in America Prior to 1830," *Bulletin of the New York Public Library*, January 1933, p. 55.

16. Brooke Hindle, "British v. French Influence on Technology in the Early United States," *Actes du XIᵉ Congrès International d'Histoire des Sciences* 6 (Warsaw, 1968): 49–53.

17. George W. Cullum, "Dennis H. Mahan," *Biographical Register of the Officers and Graduates of the U.S. Military Academy at West Point*, 7 vols. (Boston and New York: J. F. Trow, 1891), vol. 1, pp. 319–25; Dennis H. Mahan, *An Elementary Course of Civil Engineering* (New York: Wiley & Putnam, 1837), pp. vii, 44–53.

18. Benjamin Franklin Greene, *The True Idea of a Polytechnic Institute* (Troy, N.Y.: Rensselaer Polytechnic Institute, 1949); "Students Note-Book of Lectures at Rensselaer, 1852–1853," MS volume, Avery Library, Columbia University.

19. "Charles Storer Storrow," *Engineering News* 29 (16 February 1893): 147.

20. Charles S. Storrow, *A Treatise on Water-Works* (Boston: Hilliard, Gray and Company, 1835).

21. J. F. D'Aubuisson de Voisins, *A Treatise on Hydraulics for the Use of Engineers,* trans. Joseph Bennett (Boston: Little, Brown and Company, 1852).

22. John C. Trautwine, *The Civil Engineers Pocket-Book* (Philadelphia: Claxton, Remsen & Haffelfinger, 1872), preface, pp. v–viii.

23. See, for example, Francis to Boyden, 13 March 1851, Papers of Uriah Atherton Boyden, Division of Civil and Mechanical Engineering, National Museum of American History, Smithsonian Institution, Washington, D.C.

24. Charles C. Gillispie, *Lazare Carnot, Savant* (Princeton: Princeton University Press, 1971), p. 60.

25. Benoît Fourneyron, "Mémoire sur l'application eu grande, dans les usines et manufactures, des turbines hydraulique au roues à pallettes courbes de Bélidore," *Bulletin da la Société d'Encouragement pour l'Industrie Nationale* 33 (January 1834): 4–17; (February 1834): 49–61; (March 1834): 85–94; see also Marcel Crozet-Fourneyron, *Invention de la turbine* (Paris: C. Béranger, 1924).

26.. Edwin T. Layton, Jr., "Millwrights and Engineers, Science, Social Roles, and the Evolution of the Turbine in America," in *The Dynamics of Science and Technology,* ed. Krohn, Layton, and Weingart, pp. 68–69, 78–83.

27. James B. Francis, *The Lowell Hydraulic Experiments,* 2d ed. (New York: D. Van Nostrand, 1868), p. 52.

28. See entry "Theoretical Investigations of Turbines," December 22, 1846, "Experiments and Transactions, 1845–1857," pp. 74–82, Francis Engineering Notebooks (MS), Proprietors of Locks and Canals, Lowell, Massachusetts.

29. Quoted in John Hubble Weiss, *The Making of Technological Man* (Cambridge, Mass.: MIT Press, 1982), p. 165.

30. James B. Francis to Luigi D'Auria, 25 February 1878, vol. A-16, Papers of the Proprietors of Locks and Canals, Library of Harvard Business School, Harvard University, Cambridge, Mass.

31. *Appleton's Dictionary of Machines, Mechanics, Engine-Work and Engineering,* 2 vols., s.v. "water-wheels" (New York: Appleton & Company, 1852).

32. Arthur P. Molella and Nathan Reingold, "Theorists and Ingenious Mechanics: Joseph Henry Defines Science," *Science Studies* 3 (1973): 323–51.

International Exchange and National Style: A View of Natural History Museums in the United States, 1850–1900

SALLY GREGORY KOHLSTEDT

When A. B. Meyer, director of the Royal Zoological, Anthropological, and Ethnological Museum of Dresden, concluded after a tour of major museums in the United States in 1900 that "museum affairs in general are on a higher plane [in North America] than in Europe," his somewhat hyperbolic observation reflected a change in ranking among contemporary museums that had occurred during the last third of the nineteenth century.[1] Two aspects of American museums particularly impressed Meyer. He found an emphasis on instruction in public displays nearly everywhere, provided in such a way as to supplement other educational experiences. The administrator, moreover, took detailed notes on technical innovations in museum structures, such as ventilation and fire security. While the comprehensive nature of his survey was unprecedented, Meyer represented a continuing tradition of travel and commentary among museum managers.

Initially American naturalists were the inquirers, introduced to European museums by postgraduate tours and reinforced by intermittent visits and the exchange of specimens and publications. By the end of the century, however, European museum staff joined the stream of travelers who came west across the Atlantic Ocean to visit, among other places, the museums in North America. Obviously directors of late-century institutions in Europe still justifiably prided themselves on the scale and quality of their holdings, but the German visitor's generous estimation of North American museums, published in the Smithsonian Institution's annual report, was a vindication of American innovations. Scholars have yet to investigate, however, the extent to which international exchanges established parallel goals and similar structures and whether familiarity inhibited response to local expectations or led to rivalry. In this essay I shall address such issues in a preliminary way, focusing particularly on the establishment of educational programs in the United States, a characteristic highlighted in Meyer's report.

Museums of natural history were central repositories of information throughout the nineteenth century, as geological explanations were linked to fossil remains as well as to mineralogy, and the theory of evolution was based on evidence documenting the geological distribution of species. Collections that existed initially to support investigation of taxonomical and theoretical research, however, became more and more involved with public audiences and with education, not only in the United States but also in Britain and Europe.[2] Elaborately designed buildings dramatized the importance of the new public museums and galleries, and those well established built additions and added staff in cities eager to demonstrate their cultural acumen. More was happening than expansion, however. The traditional arrangement of specimens and their descriptions were reconsidered in recognition of changing scientific ideas as well as a broader audience. Thus the museum age, from roughly 1860 to 1900, was an expression of reorientation within the circle of practicing naturalists and was simultaneously linked to cultural and educational developments in post–Civil War America.

The practical development of these American museums fell to a group of sometimes voluntary curators and administrators—though gradually more and more of them were paid—who relied heavily on each other for advice. Their contact was necessarily informal because there was no international or even national association of museum staff members. On occasion, museum administrators and curators held sessions on museum-related issues in conjunction with the American Association for the Advancement of Science and, disappointed with the results, helped found the Association of American Naturalists in 1883.[3] The members of the latter had diffuse interests, and only occasionally did they consider museum topics, but it was not until 1906 that staff involved with natural history joined with those in art institutions to found an American Association of Museums.[4] In Britain the Museums Association was organized in 1889 and its early meetings were attended and even intermittently addressed by Americans.[5] These professionalizing efforts took place relatively late, however, so staff involved during the rapid museum development of the last half of the century relied on direct contact with each other. The means they used to exchange ideas and highlight achievements were publications, correspondence, and individual tours of museums.

The rhetoric of nineteenth-century science described an international community of scholars who cooperated to extend the boundaries of knowledge. Museums particularly reflected such reciprocity in the exchange of specimens, publications, and even personnel. Americans subscribed to this notion even as they retained an aspiration to establish parity and perhaps even preeminence.[6] From the founding of Charles Willson Peale's museum in Philadelphia in the late eighteenth century, museums and natural history societies organized specimens and published descriptive literature while unsuccessfully aspiring to the institutional breadth and depth of their European and British correspondents and counterparts. During the last half of the century, however, American museums established their own distinctive institutional identities, forged in response to the opportunities provided by their region and their sponsors. Europeans took increasing note of establishments whose dimensions were ever closer to their own in holdings but whose practices were distinctive. Any comparative work on museum development will need to take into account both the conscious parallels that existed among museums and those characteristics that mark out individual institutions and, it appears, national patterns.

In this essay I shall concentrate primarily on the records maintained by museum curators and directors. They sought self-consciously to achieve the standards set by their European counterparts and at the same time responded to the particular demands and opportunities created by local sponsors, who often insisted on seeing museums as counterparts to libraries, holding a treasure of material and ideas that should be made accessible to a general public. There are, of course, other substantial areas, such as the development of research collections, that could similarly be studied as a product and reflection of international exchange, but they are beyond the scope of this paper.

American Museum Building

British imperialism left its imprint on the collecting patterns of individuals and scientific societies well into the nineteenth century.[7] The early American naturalists who corresponded with Peter Collinson and Sir Hans Sloane, occasionally publishing in the Royal Society *Proceedings*, aspired to little more than recognition for new discoveries and incidental additions to their own cabinets and herbaria.[8] When colonial collectors combined resources, as they did in Charleston during the late eighteenth century, they expressed an eclectic enthusiasm worthy of the young British Museum.[9] In the years surrounding the American Revolution, ambitious young people formed societies and began to acquire books, artifacts, art, and natural history specimens. In time study of the natural environment became the domain of more specialized groups such as the Boston Society of Natural History, the Academy of Natural Sciences of Philadelphia, and the New York Lyceum.[10] Minerals, dried plants, animal skins and skeletons, and Indian objects were donated by members and exchanged among societies for the study of natural history without much attention to the relations among the materials acquired. Local members did indeed emphasize particular subjects for their individual studies, but virtually nothing was refused.

Exchanges and donations were the principal means of accession, since the limited incomes of the societies derived from membership dues and occasional endowments were typically used to rent or to purchase buildings and to publish proceedings and memoirs. Small individual items as well as entire collections were acknowledged in the minute books and correspondence files, and gifts from distinguished honorary and foreign members were especially prized.[11] The deference to foreign notables was appropriate, for there were relatively few publishing naturalists in the country until the 1820s. As cabinets and bookshelves filled, members made plans for larger buildings and public visibility by displaying their collections a few hours each week and sponsoring lecture courses. Optimism outran resources, however, during the 1830s and 1840s, in part because the public found more of interest in lyceum and privately endowed series of public lectures and in the exotic presentations of P. T. Barnum.[12] When public scientific museums were opened nearly a generation later, their directors cautiously borrowed some techniques of promotion and exhibition from the popular entrepreneurs. Like the proprietors of the British museums and amusements discussed in Richard Altick's *Shows of London*, they learned somewhat to their surprise that popular audiences were not as dirty or uncontrolled as expected.[13]

In Europe and Great Britain the principal museums enjoyed government sponsorship by the mid-nineteenth century. Although the French Revolution had opened the museum of the Jardin des Plantes in Paris to a public audience, its curators continued to emphasize the research aspects of their collections and held professorships at the university on government salary.[14] The British Museum in London relied primarily on Parliament for its sustenance and on a self-perpetuating board of trustees for its formal policy. City museums, founded after parliamentary legislation made tax support possible in the 1840s, looked to municipal councils for financial support.[15]

Support of natural history museums in the United States, with the exception of the Smithsonian Institution, whose natural history collections grew under the studied inadvertence of its secretary until the 1870s, was voluntary and private, with occasional grants from state and local authorities.[16] This situation held aspirations in check during the first half century but presented unprecedented opportunities during the last half of the century, when philanthropists identified museums and similar institutions as appropriate beneficiaries. Two characteristics, then, that appear to distinguish the American experience are the lack of sustained support from local and national governments and the absence of a single, central institution in the capital whose practices and policies had a nationwide influence.

Louis Agassiz, the Swiss-born immigrant who arrived in America and took a position at Harvard in 1847, carried with him a vision of museum building that proved irresistible and a background in European museums that took official sponsorship for granted. The enthusiastic naturalist quickly assumed leadership among the Boston and Cambridge intellectuals and persuaded them that a program of research and teaching in the sciences at Harvard would require a

substantial museum collection. Postgraduate work with Georges
Cuvier at the Jardin des Plantes and familiarity with museum holdings throughout Europe because of his travels to study fishes and
fossil remains meant that Agassiz had more than publications in
ichthyology and geology to offer. His views on natural science in
general and on museums in particular were encouraged and promoted by a private clique of friends who called themselves the Laz-
zaroni, by the Cambridge literati in the Saturday Club, and by large
public audiences who attended his lectures in Boston, New Haven,
and elsewhere.[17] The proof of his influence was a $100,000 grant
from the Massachusetts state legislature for his specialized Museum
of Comparative Zoology at Harvard. As a naturalist, Agassiz often
cited his experiences in Germany, France, and Britain to make the
point that those countries which had important museums also led
in scientific inquiry—indeed institutional growth seems to have
been closely linked to scientific aspirations throughout the century.
Agassiz's optimistic prediction that the young United States might
eventually surpass other countries in scientific productivity and his
successful example at the Museum of Comparative Zoology influenced his students significantly. Several of them, including Alpheus
Hyatt, Albert Bickmore, Frederic Ward Putnam, and Edward S.
Morse—and such short-term visitors as Henry A. Ward and George
Brown Goode as well—went on to found or direct museums themselves.[18] Although his museum was intended for advanced research
in comparative zoology, he nonetheless created public displays and
promoted scientific education at all levels.[19]

Perhaps taking a cue from Agassiz, the Boston Society of Natural
History sought public support for its collections, having learned
from the bitter experience of insolvency that their private means
were insufficient to support the facilities and staff required for adequate storage, preservation, and display. They successfully petitioned the state legislature for a grant of land near the proposed public library and institute of technology in the newly filled Back Bay
development.[20] The Boston Museum of Science provided facilities
for research by members while simultaneously opening its doors to
the public. The European and British examples of scientific productivity remained important, and the staff emphasized research priorities. Legislative support, however, albeit intermittent, placed public
access and education in an ever more prominent position.

Large urban museums established after the Civil War were apprised of developments abroad and throughout the United States in a number of ways. The system for international exchange of published proceedings, coordinated by Spencer F. Baird at the Smithsonian Institution, kept mailing costs to Europe at a minimum so that Americans could follow developments abroad as well as at home by reading the annual reports and proceedings of museums and their sponsoring organizations. Simultaneously the effective operation of an American museum supplier in Rochester, New York, who could compete with London and Paris suppliers successfully by eliminating their trans-Atlantic shipping costs, made a broader supply of materials available. Having spent four years visiting and acquiring materials in European museums, Henry A. Ward was well equipped to offer advice on museum structures as well as to mount and sell specimens from his world-recognized Scientific Establishment in Rochester.[21]

Decade by decade civic and scientific leaders built outstanding museums: Boston and New York in the 1860s, Philadelphia in the 1870s, Milwaukee in the 1880s, Chicago and San Francisco (where an earlier building had been destroyed by fire) in the 1890s. During these decades of development and rapid growth the curators turned to each other ever more frequently for additional specimens, for the names of reliable dealers, cabinetmakers, taxidermists, and graphic artists, and for display techniques. Similarities in organization and program among these museums are hardly surprising. Yet there was also an insistent theme of distinctiveness, most evident in the elaborate architectural styles of individual museums but reflected as well in their holdings.

Museum buildings themselves, as centers for public activity, were sufficiently imposing to fit the expectations of businessmen and politicians. Travel and popular literature made educated Americans familiar with the imposing British Museum and with museums on the Continent, with their Roman-arched interiors and Greek-pillared facades. The details might change, particularly on the exterior. The Smithsonian Institution has a look of Norman Gothic and the American Museum in New York City is Romanesque, while the Columbian Exposition Building, which initially housed the Field Museum in Chicago, was a Spanish version of the Ionic.[22] Whatever the style devised for the façade by eclectic Victorian architects, however, museum interiors almost uniformly had high vaulted ceilings, windows placed for design and not lighting, and grand vestibules. As the number of visitors grew and ideas about exhibition were modified to accommodate a more diverse audience, the structures proved to be limited. In an effort to resolve new problems, administrators turned to each other for effective and secure cabinet designs, materials for prominent entrance displays, and measures for the control of visitors.

The British Museum was the single largest influence on American development and was widely commented on by Europeans as well, in part because it went through a major reorganization just at the height of the museum movement.[23] Richard Owen wrote detailed plans for the South Kensington museum, centered around an entry hall, which he thought should have a synoptic exhibit or "museum index," with galleries and wings displaying particular branches of natural history including mineralogy, botany, zoology, and ethnology.[24] Although this plan was never realized, many of its elements found their way into the designs of his successor, William Flower, and those of other museum directors as well. Owen and fellow staff members corresponded with Americans, as well as with numerous curators establishing museums in the far reaches of the British colonial empire.[25] Their holdings were thus unusually comprehensive and rich in materials for international exchange. Cooperation was facilitated by the necessity in smaller museums of concentrating attention on national or regional natural history. While the larger museums might exploit smaller counterparts, competition was only an issue with rare or unique materials. Thus, in anthropology, museum directors such as Frederick Ward Putnam of the Peabody Museum at Harvard would try a variety of strategies to keep North American items from going abroad.[26]

International exchange of specimens and of ideas reflected greater parity among national museums in the late nineteenth century. The Centennial Exposition of 1876 provided a dramatic example of audience enthusiasm. Europeans as well as well-traveled Americans, such as Alpheus Hyatt, were impressed by the Philadelphia displays from throughout the United States and abroad. Hyatt, waxing enthusiastic about his visit, went so far as to claim that the opportunities at the exposition had been "but little short of a year's travel in Europe."[27] A substantial part of these materials became part of the permanent holdings of the Smithsonian Institution after seventy freight-car loads of exposition material helped spur the building of the United States National Museum. The western surveys, the establishment of scientific curriculums in leading public schools and normal schools, and the development of numerous collegiate and public museums even during the depression years of the 1870s contributed to public interest. The new National Museum staff recognized the necessity of highlighting display when it announced that its own operating principle would be to place only objects "capable of teaching and instructing a large proportion of visitors" in public corridors, leaving back rooms for storage and research.[28]

Educational Initiatives

William Barton Rogers, a geologist, an educator, and the author of the proposal for the Boston Museum of Science, had argued that museums deserved public financial support because they, in turn, had a responsibility to "enlighten the public and promote the progress of science."[29] This incentive ran somewhat counter to contemporary efforts to dissociate research and professionalization from popularization in the American Association for the Advancement of Science. As a result, the scientific staff of museums who shared professional aspirations were reluctant to expand public programs, but they were pushed by an alternative set of potential allies. Teachers, school board members, philanthropists, and some naturalists were persuaded that children and adults should learn the rudiments of science. Their motives were mixed. Some were philosophically committed to scientific explanation and method, wanting to teach another generation their assumptions. Others participated in the renewed enthusiasm for nature represented by nature study clubs and popular magazines.[30]

A wide range of individuals and groups expected initially that museums would reflect their interests, and as a result urban public museums enjoyed virtually uncritical support during their formative years.[31] The idea that public museums might offer systematic instruction to a few young scholars studying a specialty and perhaps an even broad range of students was not new, nor did it originate in the United States. The precedent, however, was limited to occasional lectures in museum halls and labeling—with typically limited and technical detail—of specimens.

An argument that caught American attention was presented by Edward Forbes in "The Educational Uses of Museums of Natural History," an essay reprinted, with evident approval, in the *American Journal of Science and Arts* in 1854 and again in the *American Journal of Education* in 1871.[32] Forbes argued that collections were an essential part of effective teaching for advanced students, and his words resounded with the message that Louis Agassiz consistently delivered to school teachers, namely, that they use an "object" method when teaching about natural objects. Forbes's exhortations, however, were somewhat narrowly conceived from his experience with the British geological survey, although in the geological museum Forbes in fact presented innovative displays in lectures that attracted a general audience.

Urban museums began to promote public participation in two distinct but not unrelated ways: first, by ensuring physical and intellectual access to the materials on display, and second, by developing programs for children and adults to supplement these displays. Curators at public museums quickly realized that passive, crowded displays of technically identified fishes, birds, and rocks taught casual visitors almost nothing. A larger, more comprehensive plan was required. In Boston Agassiz's former student formulated a plan to reorganize the interior of the public museum. After Alpheus Hyatt's proposal had been accepted by the society in 1870, he was hired as permanent curator to put it into effect. Using self-instructional labels and exhibit series, he intended to make the collections illustrate "the general laws of nature."[33] Although there is no evidence that the word *evolution* was used in illustrations or labels, Hyatt showed the development of species. Materials were grouped into categories: mineralogy, zoology, botany, paleontology, geography, and anthropology.[34] Collateral material such as maps, graphs, and charts were gradually introduced, and Hyatt willingly exchanged or sold copies of his instructional materials to other museum directors.[35] His goal was to make visitors responsible for learning on their own, after the museum provided basic information. When his aspirations proved too ambitious, he developed handbooks and used volunteers and paid guides to elaborate on displays.[36]

In order to attract people into museums in the first place, promotional activity and some elements of drama were borrowed, cautiously, from popular shows. Albert Bickmore, a former student of Agassiz and onetime collector for P. T. Barnum in Bermuda, combined the traditions of study collections and entrepreneurship. As first director of the American Museum of Natural History, Bickmore was undoubtedly influential in persuading his former employer to lend the museum the skeleton of his famous Jumbo elephant.[37] Visitors to the elephant exhibit increased attendance by several thousand.[38] Other museums turned to supplier Henry A. Ward, purchasing for prominent display a copy of the mastodon, plaster casts of famous fossils, or large mounted elephants like those which have come to characterize the National Museum in Washington and the Field Museum in Chicago.[39] Aesthetically and intellectually, museumgoers appreciated the growing sophistication in the work of taxidermists who created habitat groups and dioramas to show animal groups in typical environment.

Well-designed educational exhibits had broad appeal, and attendance at the museums became larger and more diverse. If earlier the middle and upper classes dominated museum hallways, having paid a fee for the privilege, the working classes began to attend, too, as fees declined and were sometimes dropped altogether. In fact, "free" museum days became the best way for museums to compete with proprietary shows that needed to charge admission. Once museum directors recorded and compared their numbers of visitors, the era of public museums had arrived.

Like schools and libraries, museums were expected to do much more than teach facts: they were to teach values. The audience was intended to discover in a network of cultural institutions "that recreation, entertainment, and educational influence which serves to elevate and ennoble their life and character."[40] As if to make the point, the city of Milwaukee—like the British Museum a century earlier—housed its new museum and library in the same building.[41] In New York, a city with many newcomers to acculturate, the issues of accessibility rose quickly. For several years the museum stood long blocks beyond the end of the subway line—a fact that the press decried and the staff of the museum used to explain low attendance. City officials heeded the complaints and extended public transportation, at the same time suggesting that the weekday and daytime hours kept schoolchildren and members of the working classes from coming, even on free days. Strict sabbatarians on the museum board resisted Sunday hours until 1889, when, under direct pressure from the city council, they agreed to have evening hours two days each week, one on Sunday.[42] When state and city governments assisted museums financially, they influenced policy as well. The Milwaukee Public Museum, whose support from a socialist city government was proportionately larger than that of other large urban museums, was also the first to grant admission without charge throughout the week.[43] Others established free days and discovered, as had the Field Museum, that attendance more than doubled when no payment was required.[44]

Access was important, but neither museum staffs nor many sponsors were satisfied with the response to their displays, even when the displays were supplemented by leaflets or individual guides. Curators, on the negative side, resented the noise, dirt, and distraction of undiscriminating viewers. On the positive side, some staff members enjoyed opportunities for serious teaching. Hyatt at the Boston Museum initiated one of the earliest educational programs, in response to the suggestion of local teachers that the museum offer them instruction in natural history to supplement what they learned in normal school.[45] Saturday classes in geography, botany, and zoology, together with field trips and summer sessions near the ocean, put museum staff in touch with local teachers, and, indirectly, with their students.[46] Educational reformers in Boston helped develop these programs and sustained the demand for new courses. In Philadelphia, where a new museum built to coincide with the Centennial Exposition was sponsored by the Academy of Natural Sciences, the educational programs depended on philanthropists for sustaining support. Within a few years New York, Milwaukee, and Chicago politicians and reformers promoted city and state funding, initially for specified programs.

By the 1880s the educational programs of the Boston and New York museums were so extensive that the trustees designated subcommittees and specific staff to oversee the educational activities, which had expanded far beyond initial expectations.[47] Hyatt continued to manage the museum in Boston as well as direct education, but he had substantial help from school administrator and board member Lucretia Crocker, who mobilized financial support from local women.[48] In New York Albert Bickmore became the head of a newly established Bureau of Instruction.[49] Lectures for techers began in 1881 at the American Museum, and within a few years Bickmore had developed a series of lantern slides and study sets of specimens, which could be lent regularly to teachers for classroom use. By the 1890s his educational program was the largest in the country, with a budget from the New York legislature designated for the cost of programs that served the state normal schools, the training schools in New York City and Brooklyn, the county teachers' institutes, and the common schools of the greater New York City area.[50] It was the educational activity that attracted state support, but the money to pay curators and preparators and to purchase specimens for basic collections still came primarily from private gifts and memberships.

As the programs in Boston and New York became well known, teachers in other cities pressed their local museums to develop similar resources. Almost immediately after the opening of the Field Museum, the president of the Chicago Institute of Teachers and head of the Department of Natural Science at the Chicago Normal School asked that the museum supply information and materials—slides, specimens, and models—that would assist nature study in the classroom and outdoors. The curators at the new museum provided some lectures for teachers and encouraged school classes to visit the museum, but there was no systematic program until a special endowment ensured the availability of instructional staff and materials.[51] The situation in Chicago highlights the tensions felt within large urban museums as their staffs worked to maintain traditional activities of research and display while responding to the demands for public instruction. The situation was quite different at smaller community museums, whose functions were explicitly directed by local teachers and amateurs primarily for educational purposes.[52]

The resistance at the Field and other museums was more than financial. New curators, hired to organize and display large collections, resisted involvement with school children and their teachers. Perhaps Hyatt and Bickmore had felt that their institutions' very survival depended on effective public programs, and certainly many museum adminstrators at the turn of the century shared similar concerns. But other naturalists, trained in specialties and eager to leave some scientific mark, felt that they should not be given instructional responsibility—indeed it was beyond the capacity of some.[53] Thus they might agree to give an occasional lecture, but when "public usefulness" also required the preparation and presentation of natural history objects to elementary school children, some reacted. An assistant in entomology at the American Museum recalled that his supervisor threw a temper tantrum at the suggestion of public lectures, ranting at length about the "injustice heaped upon scientists" in museums.[54] The decision to establish separate offices to coordinate educational projects and to draw lecturers from nearby colleges rather than rely on reluctant curators was perhaps inevitable in such museums.[55]

Thus, despite resistance, the educational programs became the most active and visible part of American museums of natural history by the end of the century. The entire faces of many exhibit areas had been changed by the use of more graphic materials, artistically and scientifically arranged specimens, and readable labels. An individual visitor could learn a lesson from a display series, assisted in some cases by a pamphlet, a volunteer guide, or an introductory lecture. Advertisements brought people into the museum, where exhibits were expected to sustain their interest. By the end of the century, museums used programs as a justification for increasing government support, an important supplement to the less predictable contributions from members and patrons. Distinctions in museum sponsorship between American museums and their European counterparts were becoming less obvious as government funds were appropriated on a regular basis.

Reciprocity and National Identity

Americans had readily acknowledged the educational ideas of Edward Forbes and the structural plans of Robert Owen during the early period of museum development. They had, in fact, turned regularly to European colleagues for exchanges and advice.[56] Within the United States, too, interchange of ideas and materials became common. By the 1880s the urban museums in Boston, New York, and Washington were frequently cited reference points for aspiring Western institutions.

Significant improvements in transportation and communication help explain the growing number of parallels in museum development. Significant, too, were the regular exchange of annual reports and similar publications outlining new programs at museums. Since correspondence was conducted through the office of a paid director rather than by elected officers, visits among museum staff also became more common, extending the long-standing tradition among naturalists of visiting museums to use collections and to investigate their arrangement.

First-hand experience with museums had inspired American visitors from the time of Benjamin Franklin and John Adams, whose early organizational efforts were spurred, in part, by seeing American materials abroad. During the nineteenth century tours of London, Paris, Vienna, Berlin, and St. Petersburg by postgraduate gentlemen inevitably included museums and botanical gardens.[57] For some the transfer of ideas was direct. Henry A. Ward spent six years examining more than eighty museums, and his consequent familiarity with their holdings certified both his stock of museum materials and his credentials as consultant at the Scientific Establishment in Rochester.[58] Alpheus Hyatt went on an extended tour in Europe shortly after the Boston museum opened. His proposal for reorganization relied in part on his assessment of museums there, and his insights contributed to his work during his thirty-year directorship.[59] Albert Bickmore, the promoter behind the founding of the American Museum of Natural History, did not visit foreign museums, but he, too, worked briefly for Agassiz and was incessant in his request for Richard Owen's detailed plans for the natural history collections of the British Museum.[60]

By the time the British Museum (Natural History) opened in South Kensington in the 1870s, it was common practice for American museum staff to examine at first hand the organization and administration of their European counterparts. In 1876 William H. Dall visited Berlin, Stockholm, Copenhagen, London, and other cities with important museums in order to gain information for his colleagues then formulating plans for a new museum building in Washington.[61] Four years later his Smithsonian colleague George Brown Goode used the opportunity provided by his supervision of the American exhibition at the Berlin Fisheries Exposition to investigate European museums as well.[62] Later, as director of the National

Museum, Goode would become the leading spokesman for the theory and practice of museum exposition. Similarly, Otis T. Mason toured collections of ethnography and anthropology in Amsterdam, Copenhagen, Leyden, Leipzig, and Paris in order to determine what arrangement to use on Native American materials. Fascinated by what he learned and by differences he observed, Mason became an outstanding theorist himself and helped shape the debate between those who sought taxonomical arrangement and those who promoted ecological display.[63] Smithsonian curators such as Frederick W. True crossed the Atlantic in order to facilitate exchanges and study administrative practices.[64] Magazines began to reflect this international interest in museum development, and as early as 1877 *Popular Science Monthly* featured Thomas Brewer's detailed review of European museums.[65]

Midwestern and Western museums, established somewhat later, could benefit from the experiences of their predecessors in Europe and the eastern United States. They were able to develop multifaceted institutions within a relatively short time. The Milwaukee Public Museum, which like its eastern predecessors began with the holdings of a local society of naturalists, grew rapidly after its incorporation. A commercial exposition in Milwaukee was a catalyst for individual subscribers and public officials who purchased a large number of specimens from Henry Ward's Scientific Establishment, thus acquiring a core collection prepared for display.[66] The Field Museum in Chicago similarly built on Henry Ward's materials, purchased for about $100,000 by Marshall Field, as well as an anthropological series exhibited at the World's Fair in 1893. For a decade or more Ward also supplied staff members experienced in preparation and display, some of whom advanced from taxidermist to director during the heyday of museum development. Ward lost first William Morton Wheeler, then Carl Akeley, to Milwaukee; twenty years later his son Henry L. Ward became director of that museum. Another Ward alumnus, Frank Baker, became director of the Chicago Academy of Sciences, whose public display was eclipsed when prominent Chicago patrons endowed the larger museum on the other side of town.[67] At Field's Columbian Museum, too, an experienced museum staff created new habitat series, including "deer in four seasons" by Carl Akeley before he moved yet again to the American Museum in New York, while Frederick Ward Putnam added the directorship of the anthropology department to his long list of administrative responsibilities, which included the directorship of the Peabody Museum at Harvard. Such lateral and upward mobility increased the rate of development in new museums, and those inexperienced sought opportunities to visit other museums when they assumed administrative responsibilities. Thus two of the early directors of the Milwaukee Public Museum went on eastern tours preparatory to organizing their museum,[68] and the Field Museum's first director turned down a consultant's offer to go on an information-gathering trip and went instead himself to visit museums in cities on the Atlantic seaboard.[69]

The points of influence shifted somewhat as museums were built farther west, even in Honolulu, Hawaii. In San Francisco a new and larger museum was built by the California Academy of Sciences after the devastating earthquake of 1906, and a former taxidermist from the American Museum in New York was hired to create new displays.[70] When Annie Alexander founded the Museum of Vertebrate Zoology at the University of California at Berkeley, her first director insisted that he must visit other similar museums, at his own expense if necessary, to ensure that the proper principles of cataloguing and organization were established from the beginning.[71] Similarly the naturalists in San Diego and Los Angeles built urban museums in the 1910s, basing aspects of their programs and displays explicitly on successful examples in New York and Chicago. It is therefore not surprising to find that the director of the new Bishop Museum in Honolulu toured mainland museums and attended meetings of the American Association of Museums with similar intentions.[72] The American Association was by 1907 holding annual peripetetic meetings and thus providing regular opportunities for members to learn of other museums at first hand. The association was formed too late to assist in the initial development of all but a few museums, but it did help standardize technical aspects of presentation in the twentieth century.

It was not only in the American West, however, where museum staff observed and emulated the principal urban museums from New York to Chicago and Boston to Washington. The growing European awareness of developments across the Atlantic was marked in 1884 by Karl A. Zittle's article in *Science,* "Museums of Natural History in the United States," in which the Munich professor of paleontology suggested that European museums might well imitate the display techniques and educational programs he had observed.[73] Attendance of Americans such as George Brown Goode, who gave a talk entitled "Principles of Museum Administration" at the Newcastle meeting of the British Museums Association in 1895, certainly familiarized the English and some Europeans with activities on the other side of the Atlantic. A few years later H. Lebrun of the Royale d'Histoire Naturelle de Belgique, contemplating the writing of a book on American museums, planned to focus on the methods used for teaching natural history, which he found to be the most distinctive characteristic of American museums.[74] Travelers on general tours had long commented on museums with less practical concerns in mind, such as the unaffiliated naturalist Alfred Russell Wallace, who was particularly taken as well with the educational facilities.[75] Thus A. B. Meyer was basically repeating and elaborating themes in his extensive overview in 1903.[76] His assessment of museums in the Eastern United States suggested it was time for stocktaking, in a mood that characterized the *fin de siècle*. The great age of museum building was nearly over, and most institutions had taken the basic shape they would sustain into the twentieth century.[77]

The tendency to measure achievement by an international set of standards, made possible by the mutual familiarity of museum administrators, was widely in evidence by the end of the century. Liberal exchange of material and ideas had characterized their development. The result, however, was not a dull conformity. Museums, in their efforts to attract support and provide the progressive education sought by patrons, typically highlighted local specimens and phenomena. If the entry halls of many of the larger museums boasted mounted elephants from Africa or India, the habitat groups in the gallery were likely to display local birds and geological formations that would be discussed in relation to phenomena represented in the region of the museum. International standards, then, were more concerned with methodology and programs than with content. The characteristics of American museums that seemed significant to Meyer and that caught international attention were such well-executed educational displays and activities. In comparative context American museums had achieved the parity self-consciously sought at mid century and, moreover, a distinctive identity.

1. A. B. Meyer, "Studies of the Museums and Kindred Institutions of New York City, Albany, Buffalo, and Chicago, with Notes on Some European Institutions," *Report of the National Museum . . . for 1903* (Washington, D.C., 1904), appendix 2, p. 323. Meyer admitted that he offered his review in a spirit of appreciation and gratitude, which makes the glowing comparison somewhat suspect. Nonetheless, his emphasis on education and technical features, particularly lighting, ventilation, fire prevention, and security, demonstrates what he found most useful and perhaps distinctive about the American institutions.

2. Alma S. Wittlin, *Museums: In Search of a Usable Future* (Cambridge: MIT Press, 1970), p. 119. Wittlin argues that natural history museums initially grew more rapidly in the United States than did art museums because of the interest in natural resources. It could also be argued that the natural specimens were more readily available and affordable than fine art. The outdated but still standard survey of American museums is Lawrence Vail Coleman, *The Museum in America* (Washington, D.C.: American Association of Museums, 1939). For the pattern of colonial development see the essay of Susan Sheets-Pyenson, and her background notes on bibliography, in this volume.

3. Alpheus Hyatt to Spencer F. Baird, 2 January 1884, RU 189, Smithsonian Institution Archives (hereafter SIA), Washington, D.C. Originally the Society of Naturalists of the Eastern United States, the name had been changed by 1885 to the more geographically inclusive American Society of Naturalists. The title of their *Records* varied with name changes. For a different perspective on the purposes of the society, see Hamilton Cravens, *The Triumph of Evolution: American Scientists and the Heredity-Environment Controversy* (Philadelphia: University of Pennsylvania Press, 1978), pp. 28–29.

4. American Association of Museums, *Proceedings* 1 (1908): 1–3.

5. Museums Association, *Proceedings* 1 (1890): passim.

6. Discussion of the tension between nationalism and internationalism is to be found in Bruce Sinclair, "Americans Abroad: Science and Cultural Nationalism in the Early Nineteenth Century," in *The Sciences in American Context: New Perspectives*, ed. Nathan Reingold (Washington, D.C.: Smithsonian Institution Press, 1979), pp. 35–54.

7. An interpretive overview of the politics of botanical exchange is to be found in Lucile Brockway, *Science and Colonial Expansion: The Rule of the British Royal Botanic Gardens* (New York: Academic Press, 1979).

8. The most detailed study of British-American exchange during the colonial period is Raymond P. Stearns, *Science in the British Colonies of America* (Urbana: University of Illinois Press, 1970).

9. Caroline M. Borowsky, "The Charleston Museum" *Museum News* 41 (1963): 11–21.

10. There are no modern book-length studies of these organizations. The best guide to the primary and older secondary accounts remains Max Meisel's, *Bibliography of American Natural History*, 3 vols., (New York: Premier Publishing Company, 1924). Also see examples on Albany (by James Hobbins), on Cincinnati (by Henry Shapiro), and on Philadelphia (by Patsy Gerstner) in *The Pursuit of Science in the Early American Republic: American Scientific and Learned Societies from Colonial Times to the Civil War*, ed. Alexandra Oleson and Sanborn C. Brown (Baltimore: Johns Hopkins University Press, 1976).

11. This remained true at major museums throughout the nineteenth century, as reflected in the files of museums from Boston to San Francisco. The minute books

and correspondence files are, for most institutions, extant, although rarely organized archivally, with the notable exception of the Academy of Natural Sciences of Philadelphia, whose early records are also available on microfilm. The Smithsonian Institution has now acquired and organized many of the records of the National Museum in its archives. The American Museum of Natural History's records are held by the librarian and available with the permission of the Board of Trustees. The Field Museum records are housed with the registrar, together with a brief list, prepared by this author, of files held elsewhere in the museum. The Boston Society of Natural History's papers are in the Boston Museum of Science library. Most of the files of the California Academy of Sciences were destroyed by the San Francisco earthquake and fire of 1906, but those records saved by Alice Eastwood and others are now in the library of the academy. The librarians of the San Diego Museum of Natural History and the Los Angeles Museum also have the early records of their institutions.

12. Neil Harris, *Humbug: The Art of P. T. Barnum* (Boston: Little, Brown and Company, 1973).

13. Richard Altick, *The Shows of London* (Cambridge: Harvard University Press, Belknap Press, 1978).

14. Camille Limoges, "The Development of the Museum d'Histoire Naturelle of Paris, c.1800–1914," in *The Organization of Science and Technology in France, 1808–1914*, ed. Robert Fox and George Weisz (Cambridge: Cambridge University Press, 1980), pp. 211–40.

15. The centenary of the opening of the British Museum (Natural History) in South Kensington produced a spate of literature, including A. E. Gunther, *A Century of Zoology at the British Museum through the Lives of Two Keepers* [J. E. Gray and Albert Gunther] (London: British Museum, 1975), and William T. Stearn, *The Natural History Museum at South Kensington: A History of the British Museum (Natural History) 1753–1980* (London: Heinemann, 1981). On the network of museums see Henry Miers, *A Report on the Public Museums of the British Isles* (Edinburgh: Constable, 1928), pp. 9–10.

16. For one view of the original director of the Smithsonian see Wilcomb E. Washburn, "Joseph Henry's Conception of the Purpose of the Smithsonian Institution," in *A Cabinet of Curiosities: Five Episodes in the Evolution of American Museums*, ed. Whitfield J. Bell, Jr., and others (Charlottesville: University Press of Virginia, 1967), pp. 106–66.

17. The best analytical biography is Edward Lurie, *Louis Agassiz: A Life in Science* (Chicago: University of Chicago Press, 1960).

18. In a number of essays by Ralph W. Dexter are documented the subsequent careers of Agassiz's students; see, for example, "Frederick Ward Putnam and the Development of Museums of Natural History and Anthropology in the United States," *Curator* 9 (1966): 151–55.

19. For Agassiz on public education see Joan Burstyn, "Early Women in Education: The Role of the Anderson School of Natural History," *Journal of Education* 159 (1977): 50–64; see also Mary Pickard Winsor, "Agassiz and the Species Question," *Studies in the History of Biology* 3 (1979): 89–117.

20. Sally Gregory Kohlstedt, "From Learned Society to Public Museum: The Boston Society of Natural History," in *The Organization of Knowledge in Modern America, 1860–1920*, ed. Alexandra Oleson and John Voss (Baltimore: Johns Hopkins University Press, 1979), pp. 186–406.

21. The idea of the public museum projected by Etienne Boullée in 1783 was a "secular pantheon," as well as an expression of the Age of Enlightenment, according to Ludwig Glaeser, *Architecture of Museums* (New York: Museum of Modern Art,

1969), unpaged museum catalogue. See also pictures of museums at the turn of the century in A. B. Meyer's "Studies of the Museums" and Alma Wittlen's *Museums,* pp. 140–43.

22. For a detailed account of the physical evolution of a research institution and museum see Kenneth Hafertepe, *America's Castle: The Evolution of the Smithsonian Building and Its Institution, 1840–1878* (Washington, D.C.: Smithsonian Institution Press, 1984).

23. See Adrian Dollfus, *Le Museum de Londres* (Paris, 1889), and A. Yaschenko, *The British Museum and Other Zoological Institutions of the West of Europe* (St. Petersburg, 1894).

24. See references in note 15; see also J. Mordaunt Crook, *The British Museum* (London: Oxford University Press, 1972), and Albert Gunther, *The History of the Collections Contained in the Natural History Departments of the British Museum* (London: Longmans & Co., 1912).

25. Some idea of the nature of the correspondence can be found in the bound volume of A. E. Gunther's "American Correspondents, 1851–1913" in the Natural History Library, British Museum, London.

26. Putnam warned Ernest Theodore Hamy, for example, that he was likely to buy fakes if he negotiated with dealers rather than with museum personnel; see letter of 8 October 1883, Hamy MSS, Jardin des Plantes, Paris. The theme of European exploitation persisted, and in 1909 the Los Angeles City Council voted $500 to excavate the paleontological remains at Rancho LaBrea, stating a worry that the numerous foreign scientific bodies that had applied for permission to excavate this wonderful deposit would carry away "fossils which should remain in a home museum"; see Gretchen Sibley, "Seventieth Anniversary of Academy Excavations at Rancho LaBrea," Southern California Academy of Sciences *Bulletin* 78 (1979): 151–62.

27. Hyatt to Audella Hyatt, 14 June 1876, Hyatt MSS, Syracuse University, Syracuse, New York (hereafter SU).

28. United States National Museum, *Proceedings . . . 1881* (Washington, D.C.: 1882), appendix 16. For a discussion of historical materials on the period see Sally Gregory Kohlstedt, "Institutional History," in *Historical Writings on American Science,* ed. Sally Gregory Kohlstedt and Margaret Rossiter, *Osiris,* 2d ser. 1 (1985): 17–36.

29. Rogers to Bickmore, 20 December 1864, Bickmore MSS, American Museum of Natural History Archives, New York City (hereafter AMNH).

30. See, for example, *Popular Science Monthly* 8 (December 1875): 240–43; 10 (February 1878): 475–76; and 17 (September 1880): 698–99. See also Matthew D. Whalen and Mary F. Tobin, "Periodicals and the Popularization of Science in America, 1860–1910," *Journal of American Culture* 3 (1980): 195–202.

31. Discussion of urban cultural movements, including museums, can be found in Helen Lefkowitz Horowitz, *Culture and the City: Cultural Philanthropy in Chicago from the 1880s to 1917* (Lexington: University Press of Kentucky, 1976); Douglas Sloan, "Science in New York City, 1867–1907," *Isis* 71 (1980): 35–76; and Joseph Ewan, "San Francisco as a Mecca for Nineteenth Century Naturalists," in *A Century of Progress in the Natural Sciences, 1853–1953* (San Francisco: California Academy of Sciences, 1955).

32. *American Journal of Science and Arts,* 2d ser. 18 (November 1854): 340–52; *American Journal of Education,* January 1871: 117–30. Forbes, apparently for financial reasons, held a chair of botany at Kings College, London, while also serving as curator of the geological museum.

33. Records of the society, including a published copy of Hyatt's "Proposed Plan of

Organization," (see p. 2 for quotation) are maintained in the Library of the Boston Museum of Science. See also "Alpheus Hyatt" in *Biographical Dictionary of American Scientists, the Seventeenth through the Nineteenth Centuries*, ed. Clark A. Elliott (Westport, Conn.: Greenwood Press, 1978), p. 134.

34. Hyatt to William H. Dall, 4 March 1870 and 13 January 1871, Dall MSS, RU 7073, SIA.

35. Requests even came from abroad. See Han Reusch, director of the Geological Survey of Norway, to Hyatt, 1 November 1890, Hyatt MSS, SU.

36. See Kohlstedt, "From Learned Society to Public Museum."

37. Albert S. Bickmore, "An Autobiography with a Historical Sketch of the Founding and Early Development of the American Museum of Natural History," typescript, AMNH. An undocumented account by *New Yorker* columnist Geoffrey Hillman, *Bankers, Bones and Beetles: The First Century of the American Museum of Natural History* (Garden City, N.Y.: Natural History Press, 1969), is lively and in general accurate. A survey particularly attuned to political influences is in John M. Kennedy, "Philanthropy and Science in New York City: The American Museum of Natural History, 1868–1968" (Ph.D. dissertation, Yale University, 1968).

38. Morris K. Jesup to P. T. Barnum, 20 April 1889, Barnum MSS, AMNH.

39. Sally Gregory Kohlstedt, "Henry A. Ward: The Merchant Naturalist and American Museum Development," *Journal of the Society for the Bibliography of Natural History* 9 (April 1980): 647–61.

40. Undated petition, Laws and Legislation folder in vault, AMNH.

41. Walter B. Hendrickson, "The Forerunners of the Milwaukee Public Museum," *Lore* 22 (Summer 1972): 89–103.

42. L. P. Gratacap, "History of the American Museum of Natural History," typescript, AMNH. A similar debate was waged in London during the 1880s; see "Newspaper Clippings, 1879–1902," in the British Museum (Natural History) Library, London.

43. The free admission policy was built into the state law (330A, Chapter 329, approved 31 March 1882) establishing the museum. The formal records, including the minutes of the Board of Trustees, are at the museum, while some uncatalogued correspondence, primarily for the period after 1904, are housed in the Milwaukee Public Library.

44. Annual Report 1895–96, p. 111.

45. Sally Gregory Kohlstedt, "In from the Periphery: American Women in Science, 1830–1880," *Signs* 4 (Autumn 1978): 93–94; for a comprehensive analysis see Margaret Rossiter, *Women Scientists in America: Struggles and Strategies to 1940* (Baltimore: Johns Hopkins University Press, 1982).

46. Pamphlets and other literature related to the "Teachers' School of Science" are held in the library of the Boston Museum of Science.

47. Gratacap, "History of the AMNH," especially chap. 4.

48. Kohlstedt, "From Learned Society to Public Museum," passim.

49. Gratacap, "History of AMNH," chap. 4.

50. "Laws and Legislation affecting the Museum," in vault, AMNH.

51. Wilbur Jackman, a pioneer in the nature study movement, in a series of letters in 1896, especially that of 12 November 1896, discussed the museum's program to provide duplicate specimens and slides to local teachers. John H. Tear, 25 August 1895, offered to help sponsor weekend lectures by curators for school teachers. Letters to F. J. V. Skiff, Director, Field Museum of Natural History (FMNH). "Fifty Years of Progress," *Field Museum News* 14 (September–October 1943): 22–25.

52. Sally Gregory Kohlstedt, "Collections, Cabinets and Summer Camp: Natural History in the Public Life of Nineteenth Century Worcester," paper presented at West Virginia University, January 1984.

53. See, for example, C. F. Millspaugh, 16 January 1897, and O. C. Farrington, 16 January 1897, to F. J. V. Skiff, FMNH.

54. Raymond Ditmars, *The Making of a Scientist* (New York: Macmillan, 1937), p. 21. The antagonism between the educational and research departments of the museum is discussed in Clark Wissler, "Survey of the American Museums of Natural History, Made at the Request of the Management Board in 1942–1943," AMNH.

55. Curators also demanded responsibility for their own departments. Charles Cory wrote to Edward Ayer, 27 June 1894, that he would allow "no unscientific man to dictate the management or arrangement of the collection." FMNH.

56. Attention to European practices continued throughout the period, in an ever more analytical, even critical, tone. See O. C. Farrington, "Notes on European Museums," *American Naturalist* 33 (October 1899): 763–81; and Edmund Otis Hovey, "Notes on Some European Museums," *American Naturalist* 32 (September 1898): 697–715.

57. Jeffries Wyman to Albert Bickmore, 22 December 1864, Bickmore MSS, AMNH.

58. Ward's extensive correspondence, which includes a series of letters from his European contacts, is in the Rush Rhees Library, University of Rochester, Rochester, New York.

59. Hyatt's letters to his wife, Audella, detail his tour; they are in the Hyatt MSS, SU.

60. See letters from Owen, 14 February 1870, and from G. R. Gray, 10 August 1871, in Bickmore Letterbook, AMNH. Owen had directed the Museum of the College of Surgeons before his appointment in 1856 as superintendent of collections at the British Museum; see *Dictionary of Scientific Biography* 10 (1970–80): 260–63.

61. Dall to Spencer F. Baird at the Smithsonian Institution in 1876, describing Dall's experiences in Berlin, Stockholm, Copenhagen, London, and other cities, RU 7002, SIA.

62. G. Carroll Lindsay, "George Brown Goode," in *Keepers of the Past*, ed. Clifford L. Lord (Chapel Hill: University of North Carolina Press, 1965), p. 133.

63. Mason, "The Educational Aspect of the United States National Museum," in *Studies in Historical and Political Science*, Eighth Series, 10, Supplement (1890), p. 508. A typescript of letters from Mason to George Brown Goode and to Mason's family, 17 July to 7 October 1889, are in RU 7086, SIA. For a general overview, see Curtis Hinsley, *Savages and Scientists: The Smithsonian Institution and the Development of American Anthropology, 1846–1910* (Washington, D.C.: Smithsonian Institution Press, 1981).

64. A notice of True's tour was reported in *Science* 3 (25 January 1884): 11.

65. "A Run through the Museums of Europe," *Popular Science Monthly* 2 (1871): 472–81.

66. Nancy Oestreich Lurie, *A Special Style: The Milwaukee Public Museum, 1882–1982* (Milwaukee: The Museum, 1983).

67. Walter B. Hendrickson and William J. Beecher, "In the Service of Science: The History of the Chicago Academy of Sciences," *Bulletin of the Chicago Academy of Sciences* 2 (1972): 211–68.

68. Carl Doerflinger, first curator of the Milwaukee Public Museum, turned to the U.S. National Museum for assistance and also toured Eastern museums; see his let-

ters to Spencer F. Baird, 1 February, 27 March, and 18 April 1883, RU 189, SIA. Later, Henry Nehrling produced a typescript, "Report on Eastern Museum," in which he found the American Museum most worthy of emulation; copy (1892) in the library of the Milwaukee Public Museum.

69. Charles Walcott to F. J. V. Skiff, 5 February 1894, FMNH.

70. Theodore Hittell's typescript "History of the California Academy of Sciences" traces the history to the earthquake of 1906; copy in Academy Library, San Francisco. The subsequent history must be followed in the reports produced annually for the Board of Trustees.

71. Joseph Grinnell to Annie Montague Alexander, 6 December 1970, Alexander MSS, Bancroft Library, Berkeley, California.

72. Some of the early efforts of the San Diego museum are traced in the letters of Charles R. Orcutt, horticulturist and founder of *West American Scientist*, in the library of the San Diego Natural History Museum and the Huntington Library, San Marino, California. On Hawaii, see *A Museum to Instruct and Delight: William T. Brigham and the Founding of the Berenice Pauahi Bishop Museum* (Honolulu: Bishop Museum Press, 1980). Eventually this interchange reached Australia as well, and directors included North America on their international tours. See Sally Gregory Kohlstedt, "Australian Museums of Natural History: Public Priorities and Scientific Incentives in the Nineteenth Century," *Historical Records of Australian Science* 5 (1983): 17.

73. "Museums of Natural History in the United States, *Science* 3 (1884): 190–96; see also "Karl Alfred von Zittle," *Dictionary of Scientific Biography* 14: 626–27.

74. Lebrun to F. J. V. Skiff, 14 March 1906, Registrar's Office, FMNH.

75. Alfred Russell Wallace, *My Life: A Record of Events and Opinions*, 2 vols. (New York: Dodd, Mead, 1905); chaps. 30–32 detail his tour of the United States and Canada in 1886 and 1887.

76. Meyer's theme echos in subsequent reports. E. E. Lowe's report of 1928, sponsored by the Carnegie United Kingdom Trustees stressed the size of the buildings, the attractive, "even spectacular," exhibits, and the extensive work with school children in "A Report on American Museum Work," to be found as an appendix to Henry Miers, *A Report on the Public Museums of the British Isles* (Edinburgh: Constable, 1928). Later Lawrence Vail Coleman's classic survey, *The Museum in America*, called educational programs "America's contribution to museum practice," p. 341.

77. When Frederick A. Lucas of the American Museum toured the British Museum, he was bemused by the fact that "the aesthetic side of exhibits, including the question of proper color backgrounds, labels, and cases" were being considered in the same ways they were in the United States. Lucas to Henry L. Ward, 21 September 1912, records of the Milwaukee Public Museum in the Milwaukee Public Library.

The Maturation of American Medical Science

RONALD L. NUMBERS

JOHN HARLEY WARNER

"What does the world yet owe to American physicians or surgeons?" an essayist for the *Edinburgh Review* asked contemptuously in 1820.[1] Offended Americans, unable to claim any medical heroes of their own, responded defensively by arguing that they excelled in the practice of medicine. Nevertheless, they resented their obviously dependent status in the medical sciences and yearned for the respect that would come with scientific achievement. It is time, proclaimed one patriotic surgeon in 1856, "to declare ourselves free and independent of our transatlantic brethren, as we did eighty years ago declare ourselves free and independent of the British crown."[2] In an effort to spur the medical community into action, the *Philadelphia Journal of the Medical and Physical Sciences* for several years displayed the insulting Edinburgh query prominently on its title page.

The American struggle for medical independence lasted for nearly a century after the 1820 incident. As Table 1 indicates in a crude, quantitative way, the United States continued to lag behind Europe in contributing to the medical sciences until late in the nineteenth century. The Americans overtook the English in the 1880s, however, the French in the 1890s, and the Germans in the 1910s. By 1920 they led the world in medical research.[3]

It is important in tracing the maturation of American medical science to distinguish between the history of the so-called basic medical sciences, such as anatomy, physiology, biochemistry, pathology,

Table 1.   The Number of Discoveries in the Medical Sciences,
by Country, 1820–1919

Decade	United States	England	France	Germany
1820–29	1	12	26	12
1830–39	4	20	18	25
1840–49	6	14	13	28
1850–59	7	12	11	32
1860–69	5	5	10	33
1870–79	5	7	7	37
1880–89	18	12	19	74
1890–99	26	13	18	44
1900–09	28	18	13	61
1910–19	40	13	8	20

Source: Joseph Ben-David, "Scientific Productivity and Academic Organization in Nineteenth Century Medicine," American Sociological Review 25 (1960): 830.

and pharmacology, and the development of the clinical sciences, particularly therapeutics. The former reached maturity during the years between 1890 and 1920, when the United States created a self-sustaining institutional base for medical research. Medical therapeutics followed a much different course, achieving maturity during the last third of the nineteenth century, when American clinicians abandoned their insistence on a distinctively American practice in favor of therapies based on the principle of medical universalism. It is also important to bear in mind that even during the period when American physicians failed to keep pace with their European colleagues in using hospitals and laboratories for medical research, it was not uncommon for them to engage in armchair theorizing about the functions of the human body or investigate the relation between climate and disease empirically. Regarded retrospectively, such activities may have contributed little to the advancement of medical science, but they were scientific nonetheless.

The Basic Sciences

Nineteenth-century American physicians tended to attribute their meager scientific output to the relative immaturity of their country. "In the great family of enlightened nations, we are the last born," explained one doctor. "In our youth we must be sustained."[4] The United States may indeed have been a youth among nations, but, as Richard H. Shryock long ago pointed out, it lacked neither the population nor the wealth to support scientific research. By 1860 the United States claimed eight cities with populations in excess of 150,000; its per capita income exceeded that of any European country; its industry led the world in mechanization.[5] Thus whatever the reasons for the country's failure to contribute to medical science, they did not stem from either poverty or the absence of an urban culture.

Some American physicians, embarrassed by being "the mere recipients" of European knowledge, blamed the scientific inactivity of their colleagues on the availability of foreign literature and on feelings of national inferiority. Harvard's Oliver Wendell Holmes, for example, discerned a "fatal influence" to the growth of indigenous science emanating from the indolence created by the "fairest fruits of British genius and research [being] shaken into the lap of the American student."[6] The much-discussed American custom of pirating foreign works and selling them well below the cost of the original editions, which continued until the United States recognized international copyrights late in the century, encouraged this parasitical tendency.[7]

American reliance on foreign works reflected what some contemporaries diagnosed as "a morbid feeling of inferiority to our transatlantic brethren," a paralyzing fear that the humble efforts of Americans would elicit nothing but scorn from the scientific capitals of Europe.[8] Professional leaders repeatedly chastized American physicians for being slaves to foreign authority and urged them "to interrogate nature and experience more, and European opinions less," but little progress was made as a result.[9]

Although a proclivity for borrowing and a sense of intellectual inferiority may have contributed indirectly to America's poor record in the medical sciences, a far more basic cause was the commercial system of medical education that prevailed in the United States. Before the nineteenth century medical schools had traditionally stressed the dissemination rather than the production of scientific knowledge; they had frequently provided institutional homes for medical scientists, but had tended to leave organized research to individual initiative or to scientific academies. During the first half of the nineteenth century this arrangement changed as educational reformers turned European medical schools, particularly German schools, into patrons of laboratory-based medical research. At the University of Berlin, for example, state-paid professors were expected by 1810 to conduct research as well as teach.[10] In America most medical schools, even those nominally affiliated with a college or university, remained proprietary institutions, run for prestige and profit by ill-equipped local practitioners, many of whom could not have qualified for matriculation as medical students in Europe. Unlike most European governments, which regulated and supported medical education, the state legislatures in America granted charters virtually upon request—an estimated 457 by 1910—and allowed schools to set their own standards.[11] Such legislative liberality may have provided the expanding country with an ample supply of medical practitioners, but it did little to promote medical science.

American medical schools derived their income almost solely from the students' fees, which the professors divided among themselves. This scheme virtually guaranteed mediocrity, since high standards would inevitably have reduced the number of fee-paying customers. Because medical schools generally required less of their matriculants than liberal arts colleges, they often enrolled "*the leavings of all the other professions*" (italics in original). Most medical students never attended college, and some barely knew how to read and write. The college boys who did go into medicine, complained one educator, were often those "who, from various causes— ill-health, poor scholarship, bad conduct and general discouragement—fall by the wayside and after one or two years of study, leave college without a diploma."[12] In view of such conditions, it is not surprising that contemporary critics frequently identified inadequate preliminary education as the highest barrier to the cultivation of medical science in America. "Our physicians and other professional men have genius enough," observed a Boston medical journal in 1833; "their defect is in mental discipline, which was not acquired during their preparatory studies in such a degree as to make the daily acquisition of knowledge, and the habitual exercise of the mental powers, become a primary object of pursuit, and a principal source of their highest enjoyment."[13]

Students not only entered medical school ill prepared for a scientific career, they left in the same condition. In contrast to the leading European schools, which at midcentury required attendance for four years and devoted from thirty-seven weeks, as at Edinburgh and Paris, to forty-one weeks a year, as at Berlin and Pavia, to lectures, the medical school of the University of Pennsylvania, one of America's best, required only twenty-five weeks a year for two years, and most American schools offered annual terms of only sixteen weeks. To make matters worse, American medical students until the last quarter of the century customarily repeated the same courses during their second year that they had taken during their first.[14] In eight to twelve months of formal training they were expected to learn anatomy, physiology, chemistry, and medical botany. Instruction in the basic sciences, except for anatomy, consisted of didactic lectures. As the historian John B. Blake pointed out, "until late in the century, anatomy was traditionally the only laboratory course in medical school. It was, however, generally taught simply for its practical value, chiefly for surgery, and, unlike the other medical sciences, gross human anatomy had very limited potential for stimulating original research."[15] American medical students may have picked up the vocabulary of science, but unlike German students, for example, they had little opportunity to learn its methodology.

Although the quality of American medical schools and their graduates varied greatly, by and large they produced craftsmen, not scholars. "In Europe an educated physican is presumed to be an accomplished *belles lettres* and professional scholar," noted the American Medical Association Committee on Medical Education in 1863. "In this country . . . a doctor has no special prominence, and, because a graduate, is not therefore regarded as educated or learned."[16] (It should be noted, however, that both American physicians and foreign visitors occasionally observed that individual physicians enjoyed higher social standing in the United States than in Europe.[17]) Given their cultural environment and training, American physicians understandably valued practice above science, wealth over scholarly reputation. William Beaumont, the frontier physician-physiologist, observed these traits during a visit to New York in 1833. "The professional gentlemen of this City have quite too much personal, political and commercial business on hand to permit them to turn their attention to animal and physiological chemistry, whose

high honours and rewards to them are to be the results," he wrote to
a friend. "Their curiosity once gratified, they are silent and aloof
from the subject."[18]

The American obsession with practice and "getting ahead" de-
terred even scientifically inclined physicians from engaging in research. "You will lose a patient for every experiment you make in
the laboratory," one medical professor warned a student contemplating a scientific career.[19] This attitude helps to explain why the
nearly 700 American physicians who studied in Paris between 1820
and 1861 failed to establish a research tradition in America. An eminent American physician, upon hearing that his Paris-trained son
wished to devote several years to clinical research before entering
practice, explained to his son's French mentor why he could not
approve of such plans. "In this country," he wrote, "his course would
have been so singular, as in a measure to separate him from other
men. We are a business doing people. We are new. We have, as it
were, but just landed on these uncultivated shores; there is a vast
deal to be done; and he who will not be doing, must be set down as
a drone."[20]

American independence in the basic medical sciences did not
come until American medical schools freed themselves from dependence on student fees and acquired endowments sufficiently large
to allow them to raise standards for admission and provide professors with the time and facilities to undertake scientific work. Since
almost no medical professorships generated sufficient income from
fees to provide a decent living, American professors customarily
supported themselves by practicing medicine. In fact, observed one
young physician, "a professorship in a medical college is generally
sought as an advertisement in acquiring practice, rather than as an
opportunity for study and investigation."[21] In 1878 the president of
the American Medical Association contrasted conditions in the
United States, where "the names of those who have made undeni-
able and valuable additions to the common stock" of medical
knowledge could be counted on the fingers of two hands, with those
in Europe, where scientists were supported by the hundreds. Americans, he said, must recognize

*that pure science, while it is a mine of wealth to the state, cannot
remunerate the investigator; that it cannot live upon itself; that*

those who consecrate themselves to the pursuit of it must isolate themselves from the money-getting world around them; must be relieved from all care and anxiety as to their daily bread; and must be supplied with every necessary appliance while with concentrated thought and patient toil they seek to penetrate as it were with a diamond drill the flinty barriers which separate the known from the unknown. This is particularly true of those engaged in biological research.[22]

Experience in Europe demonstrated that medical schools could provide a home for science, but as long as American institutions remained primarily business enterprises, they stood little chance of attracting the necessary governmental and philanthropic assistance. The "peculiar commercial organization of medical colleges," explained John D. Rockefeller's chief philanthropic advisor, accounted for the reluctance of the wealthy to support medicine "while other departments of science, astronomy, chemistry, physics, etc., had been endowed very generously."[23] It also helped to explain the preference of American millionaires for theology over medicine. In 1890 American seminaries claimed 171 endowed chairs, whereas medical schools had only 5, and none of the latter was adequate to pay even one professor's salary. The combined capital funds of all medical schools amounted to less than a quarter of a million dollars, approximately one forty-eighth of what theological schools, with half the number of students, possessed. This disparity, grumbled one jealous physician, existed despite the fact that Edward Jenner's discovery of smallpox vaccination "saves the community more dollars in one year than all the endowments of all the theological schools in all time."[24]

The absence of salaries and laboratories that endowments could have provided influenced not only individual careers but the general pattern of activity in the biomedical sciences. Americans, noted one physician, displayed "a bias toward systematizing and utilizing the already existing knowledge, rather than the exploration of yet un-travelled routes of investigation."[25] Those with scholarly inclinations often channeled their energies into writing financially remunerative textbooks and reference works rather than conducting basic research. John Call Dalton, for example, who in the late 1840s studied in Paris with the French physiologist Claude Bernard, achieved his greatest fame as a teacher and author of texts, not as a researcher. An admiring colleague commented on this unfortunate outcome:

This eminent physiologist is by mental constitution evidently qualified to hold the position in America which in Paris is occupied by M. Bernard. He should be exploring the dark and untravelled regions of physiology instead of leading undergraduates along its beaten track; his pen should be occupied in tracing new provinces of thought added by his genius to the ever-spreading map of discovered biological science, instead of writing text-books for students. . . . His own proclivities would lead him to produce original monographs, circumstances confine him to the systematic routine of writing a college text-book.[26]

Practicing physicians—and virtually all biomedical scientists in America until the last quarter of the century did practice medicine—found little time for systematic scientific investigation. As one medical journal pointed out, "A man, fatigued with the details of practice, and whose time is never at his own disposal, can rarely do more than keep himself acquainted with the existing condition of medical science."[27] The experience of S. Weir Mitchell, a Philadelphian who studied the physiological effects of snake venom, illustrates the difficulties facing those who combined research and practice. "It was my habit," he wrote, "to get through work at three or four o'clock; to leave my servant at home with orders to come for me if I was wanted, and then to remain in the laboratory all the evening, sometimes up to one in the morning, a slight meal being brought me from a neighboring inn."[28] As more than one investigator discovered, such self-financed research could also be expensive. William Beaumont, whose experiments on digestion in the 1820s and 1830s won international acclaim, calculated his out-of-pocket expenses at more than $3,000, and he suffered less than most because his position as a salaried surgeon in the United States Army provided a steady income and considerable free time.[29]

When the first Americans began returning from German laboratories in the 1870s, they, too, experienced difficulty in finding full-time employment as scientists. One of the earliest returnees, Henry Pickering Bowditch, who in 1871 established the first laboratory for experimental physiology in the country, was able to devote full time to research and teaching only because family money supplemented his Harvard salary.[30] T. Mitchell Prudden and William H. Welch, America's pioneer pathologists, were not so fortunate. Upon returning to the United States in 1878 both reluctantly practiced medicine for a period before finding institutional homes where they could continue their research. In a letter to his sister Welch described the frustration he experienced in trying to launch his career as a professional scientist:

I sometimes feel rather blue when I look ahead and see that I am not going to be able to realize my aspirations in life. . . . I am not going to have any opportunity for carrying out as I would like the studies and investigations for which I have a taste. There is no opportunity in this country, and it seems improbable that there ever will be.

I was often asked in Germany how it is that no scientific work in medicine is done in this country, how it is that many good men who do well in Germany and show evident talent there are never heard of and never do any good work when they come back here. The answer is that there is no opportunity for, no appreciation of, no demand for that kind of work here. In Germany on the other hand every encouragement is held out to young men with taste for science.

All these evils, he continued, derived from the fact that "the condition of medical education here is simply horrible."[31]

But even as Welch penned these words, the reformation of American medical education was beginning. Although the proliferation of substandard schools continued unabated, the best institutions were grading and lengthening their curricula to three years, requiring evidence of preliminary education, and, led by the Harvard Medical College, abandoning proprietary status to become integral parts of universities.[32] The dramatic growth of laboratory-based medical science in the latter half of the century encouraged such reforms, as did the German training of approximately 15,000 Americans between 1870 and 1914. Although only a minority of the total specialized in the basic sciences, men such as Bowditch, Prudden, and Welch succeeded in transplanting the research laboratory to American soil.[33] When it became apparent that students' fees alone could not support such expensive facilities, medical schools began trading proprietary autonomy for the financial security of university connections.

A further prod to educational reform came from the state legislatures, each of which passed some kind of medical licensing act between the mid 1870s and 1900. The state licensing boards influenced medical education in two ways. First, most of them required a candidate to hold a diploma from a reputable medical school—that is, one requiring evidence of preliminary education and, in some instances, offering a three-year course of study, a six-month term, and clinical and laboratory instruction. This forced any school hoping to compete for students to upgrade its curriculum, at least superficially. Second, many states, especially during the late 1880s and 1890s, revised their statutes to require all candidates, even those holding medical degrees, to pass examinations. Although some of the weaker schools quickly learned how to coach students to pass these tests, graduates of strong institutions had much better chances of passing. Medical commercialism, observed Abraham Flexner, thus "ceased to pay."[34]

No single event contributed more to the reformation of American medical education than the opening in 1893 of the Johns Hopkins School of Medicine under the leadership of Welch. At a time when, according to Welch, no American medical school required a preliminary education equal to "that necessary for entrance into the freshman class of a respectable college," the Hopkins faculty, at the insistence of its patron, demanded a bachelor's degree.[35] Modestly following the Hopkins example, more than twenty schools had by 1910 raised their entrance requirements to two years of college.[36] As Robert E. Kohler pointed out, this reform, more than any other, "stretched the financial resources of the proprietary school beyond the breaking point. . . . Higher entrance requirements disrupted the established market relation with high schools, diminished the pool of qualified applicants, and resulted in a drastic plunge in enrollment. Medical schools could not survive on fees."[37]

Blessed with a large endowment, Johns Hopkins became the first real center for medical science in the country. In addition to creating chairs in anatomy, physiology, pathology, and pharmacology, it provided their occupants—recruited nationally—with well-equipped laboratories and salaries sufficient to free them from the burdens of practice. Before long Hopkins students were spreading across the land, similarly transforming other medical schools. "It is no exaggeration to say that the few teachers who manned these [Hopkins] departments . . . revolutionized within a single decade the status of anatomy, physiology and pathology in America," reported a national body in 1915.[38] Welch, who in 1878 despaired of ever finding employment as a pathologist, was able less than a quarter of a century later to write:

Today, pathology is everywhere recognized as a subject of fundamental importance in medical education, and is represented in our best medical schools by a full professorship. At least a dozen good pathological laboratories, equipped not only for teaching, but also for research, have been founded; many of our best hospitals have established clinical and pathological laboratories; fellowships and assistantships afford opportunity for the thorough training and advancement of those who wish to follow pathology as their career . . . and as a result of all these activities the contributions to pathology from our American laboratories take rank with those from the best European ones.[39]

By the turn of the century the medical schools at Harvard, Pennsylvania, Chicago, and Michigan had joined Hopkins as important medical research centers, but the country still lacked an institution comparable to the Koch Institute in Berlin or the Pasteur Institute in Paris. In 1901, however, the United States Congress provided funds for a national hygienic laboratory to investigate infectious and contagious diseases, and, more important, John D. Rockefeller, the oil magnate, donated the first of millions of dollars to create an institute that would become "the crown of medical research in this country."[40] The Rockefeller Institute for Medical Research not only freed its staff from practicing medicine but from teaching as well, allowing them to devote their entire lives to medical science. This environment brought the United States its first Nobel Prize in medicine—awarded to the French-born Alexis Carrel in 1912—and helped to reverse the flow of medical science and scientists from east to west. Its success soon inspired the creation of other American institutes for medical research and provided a model for the Kaiser Wilhelm Gesellschaft, which opened in Berlin in 1911.[41] By 1920 snide Europeans no longer asked what the world owed to American physicians and surgeons.

The Clinical Sciences

In clinical medicine, which involved diagnosing and treating diseases, the process of maturation did not always parallel the transition from colonial dependence to independence that characterized the development of the basic sciences. Although Americans admired the superior clinical facilities of Europe, they commonly believed that singular circumstances in the United States demanded uniquely American responses to disease and made European knowledge suspect and in certain respects irrelevant. Thus, in such areas as medical therapeutics American physicians never established a traditional colonial relationship, and they achieved maturity not by declaring independence, but by abdicating it.

The development of diagnostic tools and knowledge about morbid natural history flourished in the great hospitals of nineteenth-century Europe. Particularly in Paris, easy access to large numbers of diseased bodies made possible systematic clinical observation of disease processes, pathoanatomical correlation of these clinical patterns at autopsy, and statistical portraits of diseases based upon such studies. From the early 1820s, hundreds of American physicians were drawn to study in Paris both by its clinical facilities and by an environment conducive to medical research. "Merely to have breathed a concentrated scientific atmosphere like that of Paris," one physician studying in the French hospitals wrote home to Boston, "must have an effect on any one who has lived where stupidity is tolerated, where mediocrity is applauded, and where excellence is defied."[42]

Americans, however, lacked comparable institutions until late in the century, and physicians who studied in France found little opportunity at home to apply what they had learned abroad. William Wood Gerhard, who in the 1830s successfully employed Parisian methods to distinguish between typhoid and typhus fever, lamented the conditions at his "own little hospital" in Philadelphia. "I regret much the slender materials I possess and the difficulties wh[ich] seem inseparable from observation in this country," he wrote. Despite his inferior facilities for research, he optimistically expected that American physicians would place greater faith in his modest statistical studies on *American* patients than in conclusions drawn

from manyfold more Parisians. Many of his countrymen nevertheless remained skeptical of his conclusions, which contradicted European opinion, until Sir William Jenner confirmed them in the
1840s.[43] American contributions to differential diagnosis and the
natural history of diseases increased as the century progressed and
large hospitals became common in American cities.[44]

It was in surgery that American clinical medicine attained its
highest level in the eyes of both Americans and Europeans during
the nineteenth century. Attributing their surgical skills to native
mechanical ingenuity and frontier resourcefulness, Americans celebrated such pioneering work as Ephraim McDowell's 1809 operation in Kentucky for an ovarian cyst and J. Marion Sims's operation
for vesicovaginal fistula, which he perfected in operations on slave
women while practicing in a small Georgia town during the 1840s.
The first successful application of ether anesthesia for surgery, at the
Massachusetts General Hospital in Boston in 1846, greatly inflated
the American medical ego and convinced Americans that they no
longer needed to apologize for their medical backwardness. American medical men "may not dive so deeply into abstract sciences, or
linger there so long as in the old and somewhat *senile* establishments of Europe," declared one American surgeon, "but . . . as skillful operators, and practical men, they are the equals to any in the
world, and second to none whatsoever."[45] Although a few exceptional achievements like the application of ether may not have war-
ranted such pride, surgery—and the similarly mechanical field of
dentistry—did represent the most accomplished branches of clinical medicine in nineteenth-century America. The mechanical nature of dental and surgical therapeutics made them largely immune
to arguments of American particularism, which were prominent in
discussions of medical therapeutics.

The beliefs that therapeutic knowledge gained from experience
with European patients and diseases might not be suitable for American practice and that therapeutic principles, unlike the tenets of
the basic sciences, might not apply to all environments had deep
roots in American medical thought. Although based in part on cultural nationalism, this conviction derived chiefly from the pivotal
importance assigned by American physicians to specificity in treatment of patients with different backgrounds and in different set-

tings. Prevailing therapeutic theory stressed the necessity of tailoring therapy to the patient's age, sex, ethnicity, and habits, as well as to climate, topography, and population density.[46]

This commitment to specificity suggested, for example, that the therapeutic needs of the immigrant poor differed from those of native-born patients and, consequently, that information gained by observing pauperized Irish immigrants in a large urban hospital might be deceptive as a guide to the treatment of middle-class private patients or even the hospitalized native-born. Thus physicians at the Commercial Hospital of Cincinnati in the mid-nineteenth century prescribed rest for many patients "just from Ireland via New Orleans" while treating many of the other inmates with full depletive regimes of bleeding and purging.[47] A clinical lecturer at the Massachusetts General Hospital identified a phthisical woman to his students as "one of the cases of broken down health so often met with among her class";[48] another physician at that hospital, reflecting on the consequences of this sort of class-specific constitution, noted that although copious bleeding and purging were appropriate for hale constitutions, hospital practice provided few opportunities to employ these remedies.[49] The notion was widespread that although active depletion might not be tolerated by degenerate urban dwellers who lived sedentary lives in vitiated surroundings, the robust farmer required a forcefully depletive therapeutic strategy.[50]

The stress on specificity fostered the notion that, because different regions of the country required distinctive therapeutic practices, physicians should be educated where they intended to practice. The peculiar features of the South, for example—its characteristic diseases, large population of Negroes, and warm climate—all argued for southern students studying medicine at southern medical schools. "Anatomy, Physiology, General Therapeutics, and Chemistry, may be studied to perfection in the Capitols [sic] of Europe and the United States," explained one proponent of this view; but, he warned, before practicing in the South such students would either have to unlearn the practical precepts their northern teachers had taught them or fail miserably in their efforts to heal the sick. In his opinion, an ill southerner would be

better in the hands of some Planter or overseer who had long resided in this region, and who was perfectly familiar with the dis-

ease, than he would be in the hands of the ablest Physician of London or Paris, who had never practiced beyond their precincts, and who would be guided in his treatment solely by the general principles of Medicine.[51]

Just as northern therapeutic practices could be inappropriate for southerners, so, too, might European practices be invalid or even dangerous for Americans. Charles Caldwell, a medical professor in Louisville, Kentucky, returned from a European tour in the early 1840s to warn his classes that "the climate of London and Paris were entirely different from our own; the diet and habits of the people altogether different; and that these with other circumstances so modified the constitutions of the people and the character of the diseases, as to make the latter totally different from the diseases of this country." He expressed the common American suspicion of therapeutic knowledge generated in European clinics:

The Hospitals of those great cities were very extensive and filled with persons laboring under great varities of diseases; but they were from the very dregs of society [,] a class whose constitutions have been depraved by intemperance and want, and modified by vice, habit and climate until they possess no analogy in constitution or disease to any class in our own country. From this class or this kind of cases is the student of medicine to derive his knowledge and experience in visiting the Hospitals of London & Paris.

Caldwell concluded that European "constitutions and diseases are so modified and so totally different from those in our country, the knowledge of Pathology and Therapeutics to be gained by visiting these hospitals can be of but little advantage to the practice of Medicine in the United States."[52]

Physicians generally believed, moreover, that climate modified the influence of remedial agents on the body just as it influenced disease actions. One physician who held this view cited as his evidence "the different aspects of hyosciamus in England and Italy; of nitrite of silver in Naples and England; of the eau medicinale in Russia and France; [and] the vastly different effects of mercury in different climates."[53] If American practitioners remained "satisfied with the imbecility of European practice," they would, according to the estimate of one Boston physician, "undoubtedly lose a third or half of our patients."[54]

Americans criticized not only specific European therapies but also more general therapeutic philosophies. The therapeutic skepticism characteristic of the Paris clinical school, which argued for discarding any treatments whose clinical value had not been established by empirical observation, found an even more extreme expression in Vienna during the 1840s as therapeutic nihilism—that is, the complete rejection of therapeutic intervention in certain cases.[55] American physicians, who regarded active intervention as a crucial element in professional identity and legitimacy, found such inactive postures impractical and perhaps immoral. "The temporalizing course pursued by the French renders their therapeutics often inefficient," argued a Cincinnati practitioner, referring to the French inclination to leave the patient's cure to the healing power of nature.

In anatomy, physiology, and pathology, they stand unrivaled; but beyond this they seem scarcely to look. Having made a diagnosis, *the next most important matter is to prove its correctness; and as this can only be verified in the* dead body, *more enthusiasm is manifested in a post mortem examination than in the administration of medicine to cure disease.* The triumph of these physicians is in the dead-room.[56]

As this quotation suggests, adherence to therapeutic localism did not imply a belief in the relativity of all medical knowledge. Among the medical sciences, therapeutics was largely exceptional. Although medical therapeutic knowledge did not function equally in all environments, the tenets of such basic sciences as chemistry and anatomy were universally applicable. American physicians also admitted the possibility that European therapeutics might have some applicability in the United States, but they insisted that each therapy be validated independently for the American market. Most believed, however, that therapeutic knowledge grounded upon American experience held far more promise than knowledge of foreign provenance.

In proclaiming their distrust of European clinical knowledge, American physicians assumed the burden of investigating American diseases and cures. A Kentucky student emphasized in his medical thesis the broad gap that existed between the clinical principles set forth by European medical writers and the requirements of American circumstances. Writing in the 1830s, he suggested that this

imposes on us the greater necessity of observing for ourselves, and of culling, from among the useless rubbish of their productions, something worthy of an extensive and enlightened nation. Thus, although it has been vauntingly asked by one of their writers, "what does medicine owe to America," her sons may yet explore her wilds, and collect the materials, to rear upon the ruins of Eastern speculation, an edifice both complete and durable.[57]

The American environment provided both the opportunity and the responsibility for the reconstruction of medicine to meet American needs.

The program for medical research implied by this emphasis on the American environment was clearly localistic, drawing from a region knowledge to be applied within that region. Among the most active areas of research in early- and mid-nineteenth-century American medicine was a species of natural historical investigation that linked together meteorological, epidemiological, and therapeutic observations. Perhaps the most original medical theses were of this genre. While most theses were merely derivative exercises, many a student elected to write an original essay based on his own investigations of the topography, climate, diseases, and therapeutic practices of his home county.[58] Daniel Drake's massive treatise *On the Principal Diseases of the Interior Valley of North America* (1850–54) was, in many respects, only a singularly ambitious expression of the same endeavor.[59] Studies of prevailing diseases, weather conditions, and appropriate treatments also thrived in the discussions of local medical societies, whose meetings were otherwise thin in scientific content.[60] Although American physicians did not excel in those branches of medical science that held universal interest or application, they did actively conduct research in a sort of environmentally oriented clinical natural history that was of considerable local import.

American allegiance to therapeutic localism and knowledge gained from direct clinical observation could be seriously challenged only during the last third of the nineteenth century, when a new therapeutic epistemology took its grounding in experimental laboratory science. Growing interest in this way of generating medical knowledge was both reflected in and fostered by the return from Germany of American physicians eager to exploit laboratory science as a means of transforming medical practice and elevating the status of the profession. Central to their program of reform was the idea that the laboratory was a legitimate arbiter of therapeutic knowledge.

From the early 1870s a number of prominent American physicians began arguing forcefully that the laboratory should join the bedside as an appropriate locus for the generation and validation of therapeutic knowledge. The ensuing clash between the advocates of the laboratory and the defenders of empirical clinical observation did not pit science against art, but entailed two largely incommensurable conceptions of the proper boundaries of therapeutic epistemology. During the mid 1870s, for example, Alfred Stillé, a Philadelphia practitioner committed to clinical observation and environmental specificity, argued that the intrusion of laboratory science into the realm of therapeutics was presumptuous and destructive:

The domain of therapeutics is, at the present day, continually trespassed upon by pathology, physiology, and chemistry. Not content with their legitimate province of revealing the changes produced by disease and by medicinal substances in the organism, they presume to dictate what remedies shall be applied, and in what doses and combinations. Their theories are brilliant, attractive and specious. . . . When submitted to the touchstone of experience, they prove to be only counterfeits. They will neither secure the safety of the patient nor afford satisfaction to the physician.[61]

Roberts Bartholow, an American enthusiast for the therapeutic promise of laboratory physiology, denounced Stillé's views to the members of a Baltimore medical society as "reactionary." "Modern physiology," he asserted

has rendered experimental therapeutics possible, and has opened an almost boundless field which is being diligently cultivated. . . . It is obvious that no science of therapeutics can be created out of empirical facts. We are not now in a condition to reject all the contributions to therapeutics made by the empirical method, but a thorough examination of them must be undertaken by the help of the physiological method.[62]

Bartholow and like-minded physicians rejected clinical observation as the principal way of gaining therapeutic knowledge—and as insufficient for the creation of a science of therapeutics—but they did not rule out the clinic as a source of therapeutic progress; rather, they advocated a new role for clinical observation in the testing of laboratory-generated therapeutic principles and practices.

The ascendance of this new view fundamentally altered the relation between American and European therapeutics. Experimental science investigated disease processes and the practices that altered them. A tacit assumption that animated the rising vogue of vivisectional research during this period was that some fundamental tenets of physiological and therapeutic knowledge could be transferred profitably from the lower animals to man; medical scrutiny was focused on physiological processes, and it was to a certain extent irrelevant whether these processes took place in a laboratory animal or an Irish immigrant. In this context, the heretofore crucial differences between northerner and southerner, immigrant and native, and American and European grew small indeed. The new experimental science, gradually taken up during the next few decades, prescribed in principle standardized treatments for diseased bodies, and considerations based on national variations—other than incidental ones—became stigmata of inferior medical practice.

Recognition of the therapeutic relevance of knowledge generated in the laboratory—abstracted from both the patient and the patient's environment—meant that therapeutic knowledge could be transferred freely between Europe and America. Thus, at the same time that American physicians acquired their own institutions for clinical research, they also freed themselves from their commitment to a distinctive "American" practice. Maturity in this context implied international reciprocity grounded upon allegiance to medical universalism, not national independence.

Numbers is primarily responsible for the section on the basic medical sciences, Warner the section on the clinical sciences. The former wishes to thank Mark Shale for his research assistance. The latter gratefully acknowledges the support of part of the research for this paper by NSF Grant SES-8107609.

1. [Sydney Smith], Review of *Statistical Annals of the United States of America*, by Adam Seybert, *Edinburgh Review* 33 (1820): 79.

2. S. D. Gross, "Report on the Causes Which Impede the Progress of American Medical Literature," American Medical Association *Transactions* 9 (1856): 348. On American excellence in medical practice, see, for example, the prospectus, *Philadelphia Journal of the Medical and Physical Sciences* 1 (1820): ix.

3. By the late 1870s Americans were publishing more articles on surgery, obstetrics and gynecology, and diseases of the nervous system than persons of any other nationality; see Mary E. Corning and Martin M. Cummings, "Biomedical Communications," in *Advances in American Medicine: Essays at the Bicentennial*, ed. John Z. Bowers and Elizabeth F. Purcell, 2 vols. (New York: Josiah Macy, Jr., Foundation, 1976), vol. 2, pp. 731–33.

4. A. B. Palmer, "Report of the Committee on Medical Literature," AMA *Transactions* 11 (1858): 231.

5. Richard H. Shryock, *American Medical Research: Past and Present* (New York: Commonwealth Fund, 1947), p. 28; see also Robert William Fogel and Stanley L. Engerman, *Time on the Cross: The Economics of American Negro Slavery*, 2 vols. (Boston: Little, Brown and Company, 1974), vol. 1, pp. 248–50; Thomas C. Cochran, *Frontiers of Change: Early Industrialism in America* (New York: Oxford University Press, 1981), p. 114.

6. Oliver Wendell Holmes and others, "Report on the Committee on Literature," AMA *Transactions* 1 (1848): 286–87. The phrase about being "mere recipients" appears in Samuel Jackson and others, "Report of the Special Committee Appointed to Prepare 'A Statement of the Facts and Arguments Which May Be Adduced in Favour of the Prolongation of the Course of Medical Lectures to Six Months,'" ibid. 2 (1849): 365.

7. See, for example, Gross, "Report," pp. 344–46; and Alfred Stillé and others, "Report of the Committee on Medical Literature," AMA *Transactions* 3 (1850): 181. For British reaction to this practice, see "Report on the Progress of Midwifery and the Diseases of Women and Children," *Half-Yearly Abstract of the Medical Sciences*, no. 22 (July–December 1855): 208.

8. Thomas Reyburn, "Report of the Standing Committee on Medical Literature," AMA *Transactions* 4 (1851): 493.

9. Usher Parsons, "Address," ibid. 7 (1854): 48–49. For references to American physicians as "slaves" and "toadies," see S. D. Gross, "On the Results of Surgical Operations in Malignant Diseases," ibid. 6 (1853): 157; and Gross, letter to the editor, *Medical Record* 4 (1868): 191. For a counteropinion, see *American Journal of the Medical Sciences*, n.s. 33 (1857): 389–90.

10. Hans H. Simmer, "Principles and Problems of Medical Undergraduate Education in Germany during the Nineteenth and Early Twentieth Centuries," in *The History of Medical Education*, ed. C. D. O'Malley (Berkeley and Los Angeles: University of California Press, 1970), p. 189; Theodor Billroth, *The Medical Sciences in the German Universities: A Study in the History of Civilization* (New York: Macmillan Company, 1924), p. 27. French medical schools only belatedly supported laboratory-based science; see Erwin H. Ackerknecht, *Medicine at the Paris Hospital, 1794–1848* (Baltimore: Johns Hopkins University Press, 1967), pp. 123–26.

11. Abraham Flexner, *Medical Education in the United States and Canada* (New York: Carnegie Foundation, 1910), p. 6; Alfred Stillé, "Address," AMA *Transactions*

22 (1871): 83; William O. Baldwin, "Address," ibid. 20 (1869): 75. On the teaching of the various medical sciences, see Ronald L. Numbers, ed., *The Education of American Physicians: Historical Essays* (Berkeley and Los Angeles: University of California Press, 1980).

12. "American vs. European Medical Science Again," *Medical Record* 4 (1868): 182–83.

13. "Medical Improvement—No. 1," *Boston Medical and Surgical Journal* 9 (1833): 92. On inadequate preliminary education, see, for example, Stillé and others, "Report," p. 173; and N. S. Davis, "Report of the Committee on Medical Literature," AMA *Transactions* 6 (1853): 125.

14. F. Campbell Stewart and others, "Report of the Committee on Medical Education," AMA *Transactions* 2 (1849): 280; see also E. Giddings, "Report of the Committee on Medical Education," ibid. 22 (1871): 137.

15. John B. Blake, "Anatomy," in *Education of American Physicians*, ed. Numbers, pp. 39–40.

16. Charles Alfred Lee, "Report of the Committee on Medical Education," AMA *Transactions* 14 (1863): 84.

17. See, for example, Stewart and others, "Report," p. 344; "American Surgery," *Boston Medical and Surgical Journal* 92 (1875): 21.

18. Ronald L. Numbers and William J. Orr, Jr., "William Beaumont's Reception at Home and Abroad," *Isis* 72 (1981): 598.

19. S. Weir Mitchell, "Memoir of John Call Dalton, 1825–1889," National Academy of Sciences, *Biographical Memoirs* 3 (1895): 181. See also "The Scarcity of Working Medical Men in America," *Medical Record* 2 (1867): 277; and John S. Billings, "Literature and Institutions," in *A Century of American Medicine, 1776–1876* (Philadelphia: H. C. Lea, 1876), pp. 363–64. Allegiance to medical practice helped to kill the short-lived Philadelphia Biological Society; see Bonnie Ellen Blustein, "The Philadelphia Biological Society, 1857–61: A Failed Experiment?" *Journal of the History of Medicine and Allied Sciences* 35 (1980): 188–202.

20. James Jackson, *A Memoir of James Jackson, Jr., M.D., with Extracts from His Letters to His Father, and Medical Cases, Collected by Him* (Boston: I. R. Butts, 1835), p. 55. On Americans in Paris, see Russell M. Jones, "American Doctors and the Parisian Medical World, 1830–1840," *Bulletin of the History of Medicine* 47 (1973): 40–65, 177–204.

21. Simon Flexner and James Thomas Flexner, *William Henry Welch and the Heroic Age of American Medicine* (New York: Viking Press, 1941), p. 85.

22. T. G. Richardson, "Address," AMA *Transactions* 29 (1878): 96–97.

23. George W. Corner, *A History of the Rockefeller Institute, 1901–1953: Origins and Growth* (New York: Rockefeller Institute Press, 1964), p. 579.

24. Shryock, *American Medical Research*, p. 49. On support for theological and medical schools, see Flexner and Flexner, *William Henry Welch*, p. 237.

25. Henry F. Campbell, "Report of the Committee on Medical Literature," AMA *Transactions* 13 (1860): 773.

26. Ibid., pp. 774–75.

27. "American Medicine," *Philadelphia Journal of the Medical and Physical Sciences* 9 (1824): 405.

28. Edward C. Atwater, "'Squeezing Mother Nature': Experimental Physiology in the United States before 1870," *Bulletin of the History of Medicine* 52 (1978): 330. Atwater emphasizes the importance of financial support for the progress of physiology.

29. Numbers and Orr, "William Beaumont's Reception," p. 596. In this instance, Beaumont generously padded his expense account. On self-supporting science, see

also the *Autobiography of Samuel D. Gross, M.D.*, 2 vols. (Philadelphia: George Barrie, 1887), vol. 1, pp. 96–97.

30. W. Bruce Fye, "Henry Pickering Bowditch: A Case Study of the Harvard Physiologist and His Impact on the Professionalization of Physiology in America" (M.A. thesis, Johns Hopkins University, 1978), p.78.

31. Flexner and Flexner, *William Henry Welch*, pp. 112–13; *Biographical Sketches and Letters of T. Mitchell Prudden, M.D.* (New Haven: Yale University Press, 1927), p. 32.

32. See Martin Kaufman, *American Medical Education: The Formative Years, 1765–1910* (Westport, Conn.: Greenwood Press, 1976); and Robert P. Hudson, "Abraham Flexner in Perspective: American Medical Education, 1865–1910," *Bulletin of the History of Medicine* 56 (1972): 545–61.

33. Thomas Neville Bonner states that "German study, especially in the basic sciences, was probably the most important factor in explaining the remarkable progress in medical studies in this country after 1870"; see his *American Doctors and German Universities: A Chapter in International Intellectual Relations, 1870–1914* (Lincoln: University of Nebraska Press, 1963), p. 137. Robert G. Frank, Jr., Louise H. Marshall, and H. W. Magoun, in "The Neurosciences," in *Advances in American Medicine*, ed. Bowers and Purcell, p. 557, identify study in Germany as the "essential ingredient" in the maturation of the neurosciences.

34. Flexner, *Medical Education*, p. 11; Martin Kaufman, "American Medical Education," in *Education of American Physicians*, p. 19.

35. Flexner and Flexner, *William Henry Welch*, pp. 219, 222–23.

36. Flexner, *Medical Education*, p. 28.

37. Robert E. Kohler, "Medical Reform and Biomedical Science: Biochemistry—a Case Study," in *The Therapeutic Revolution: Essays in the Social History of American Medicine*, ed. Morris J. Vogel and Charles Rosenberg (Philadelphia: University of Pennsylvania Press, 1979), p. 32.

38. Richard H. Shryock, *The Unique Influence of the Johns Hopkins University on American Medicine* (Copenhagen: Ejnar Munksgaard, 1953), p. 22; see also Edward C. Atwater, "A Modest but Good Institution . . . and Besides, There Is Mr. Eastman," in *To Each His Farthest Star: University of Rochester Medical Center, 1925–1975* (Rochester: University of Rochester Medical Center, 1975), p. 6.

39. Flexner and Flexner, *William Henry Welch*, pp. 266–67.

40. Kohler, "Medical Reform," p. 53; Corner, *Rockefeller Institute*, p. 149; A. Hunter Dupree, *Science in the Federal Government: A History of Policies and Activities to 1940* (Cambridge: Harvard University Press, 1957), pp. 267–68.

41. Corner, *Rockefeller Institute*, pp. 76, 150–51; Shryock, *American Medical Research*, p. 93.

42. Oliver Wendell Holmes to his parents, Paris, 13 August 1833, reprinted in John T. Morse, Jr., *Life and Letters of Oliver Wendell Holmes*, 2 vols. (Cambridge, Mass.: Riverside Press, 1896), vol. 1, pp. 108–9.

43. William Wood Gerhard to James Jackson, 1 January 1835, James Jackson Papers, Francis A. Countway Library of Medicine, Boston; Dale C. Smith, "Gerhard's Distinction between Typhoid and Typhus and Its Reception in America, 1833–1860," *Bulletin of the History of Medicine* 54 (1980): 368–85; see also Ackerknecht, *Medicine at the Paris Hospital*, and Jones, "American Doctors and the Parisian Medical World."

44. Phyllis Allen Richmond, "The Nineteenth-Century American Physician as a Research Scientist," in *History of American Medicine: A Symposium*, ed. Felix Marti-Ibañez (New York: MD Publications, 1959), pp. 142–55. On American hospi-

tals, see Morris J. Vogel, *The Invention of the Modern Hospital: Boston, 1870–1930* (Chicago: University of Chicago Press, 1980).

45. Valentine Mott, quoted in Courtney R. Hall, "The Rise of Professional Surgery in the United States, 1800–1865," *Bulletin of the History of Medicine* 26 (1952): 234. American surgical excellence is discussed in "American vs. European Medical Science," *Medical Record* 4 (1869): 133–34; S. D. Gross, "American vs. European Medical Science," ibid., pp. 189–91; and John Eric Erichsen, "Impressions of American Surgery," *Lancet*, 21 November 1874, pp. 717–20.

46. For a particularly useful analysis of nineteenth-century American medical therapeutics, see Charles E. Rosenberg, "The Therapeutic Revolution: Medicine, Meaning, and Social Change in Nineteenth-Century America," in *The Therapeutic Revolution*, pp. 3–25.

47. Casebooks for Medical Ward Female, 30 May 1848–7 March 1850, Cincinnati General Hospital Archives, History of the Health Sciences Library and Museum, University of Cincinnati Medical Center, Cincinnati, Ohio.

48. John Ware, Clinical Lectures, 1830, John Ware Papers, Francis A. Countway Library of Medicine, Boston.

49. George Cheyne Shattuck, Diary Notes on Patients, vol. 2, entry for 12 December 1832, Francis A. Countway Library of Medicine, Boston.

50. "Effects of Breathing Impure Air," *Boston Medical and Surgical Journal* 6 (1832): 14; Northern Medical Association of Philadelphia, "Discussion on Bloodletting," *Medical and Surgical Reporter*, n.s. 3 (1859): 271–74, 495–500, 515–21; 4 (1860): 34–39, 486–97, 517, 518.

51. "Introductory Address," *New Orleans Medical Journal* 1 (1844): ii–iii; see also Jas. C. Billingslea, "An Appeal on Behalf of Southern Medical Colleges and Southern Medical Literature," *Southern Medical and Surgical Journal* s.2, 12 (1856): 398–402; and John Duffy, "A Note on Ante-Bellum Southern Nationalism and Medical Practice," *Journal of Southern History* 34 (1968): 266–76.

52. Courtney J. Clark, Notes on the Medical Lectures of Charles Caldwell, Medical Institute of Louisville, Kentucky, 1841–1842, Courtney J. Clark Papers, Manuscripts Department, Duke University Library, Durham, North Carolina.

53. Edward H. Barton, *Introductory Lecture on the Climate and Salubrity of New-Orleans and Its Suitability for a Medical School* (New Orleans: E. Johns and Company, 1835), p. 17.

54. Celsus, "Treatment Demanded by Malignant Diseases," *Boston Medical and Surgical Journal* 6 (1832): 141; "Public Medical Information," ibid., p. 336.

55. Ackerknecht, *Medicine at the Paris Hospital*, pp. 129–38; Erna Lesky, *The Vienna Medical School in the Nineteenth Century*, trans. L. Williams and I. S. Levij (Baltimore: Johns Hopkins University Press, 1976). I do not suggest by this that Viennese nihilism was fully derived from Parisian skepticism; see Erna Lesky, "Von den Ursprüngen des therapeutischen Nihilismus," *Sudhoffs Archiv für Geschichte der Medizin und der Naturwissenschaft* 44 (1960): 1–20.

56. Review of "Lectures on the Theory and Practice of Physic.—by William Stokes . . . and John Bell . . ." *Western Lancet* 1 (1842–43): 354–57. On American attitudes toward the healing power of nature and its associations with therapeutic skepticism, see John Harley Warner, "'The Nature-Trusting Heresy': American Physicians and the Concept of the Healing Power of Nature in the 1850's and 1860's," *Perspectives in American History* 11 (1977–78): 291–324. See also "Andral's Medical Clinic," *Western Lancet* 2 (1843–44): 148; John P. Harrison, "On the Certainty and Uncertainty of Medicine," ibid. 3 (1844–45): 118; and "Modern Practice of Medicine," *Boston Medical and Surgical Journal* 12 (1835): 351–52.

57. William Wood, "An Inaugural Dissertation on the Causes of Epidemics" (M.D. thesis, Transylvania University, 1834), Special Collections and Archives, Transylvania University, Lexington, Kentucky.

58. Typical of such theses are Robert H. Hanna, "An Inaugural Dissertation on the Medical Topography and Epidemic Diseases of Wilson County Kentucky" (M.D. thesis, Transylvania University, 1835); and Thomas Hunter, "A Dissertation on the Topography of South Alabama and the Diseases Incident to Its Climate" (M.D. thesis, Medical College of the State of South Carolina, 1843), Waring Historical Library, Medical University of South Carolina, Charleston, South Carolina.

59. Daniel Drake, A Systematic Treatise, Historical, Etiological, and Practical, on the Principal Diseases of the Interior Valley of North America, 2 vols., vol. 1 (Cincinnati: Winthrop B. Smith and Company, 1850); vol. 2, ed. S. Hanbury Smith and Francis B. Smith (Philadelphia: Lippincott Crombe and Company, 1854).

60. The proceedings of one such local society are recorded in the Minutes of the Union District Medical Association, vol. 1, 1867–1880, Walter Havighurst Special Collections Library, Miami University, Oxford, Ohio.

61. Alfred Stillé, Therapeutics and Materia Medica, 2 vols. (Philadelphia: Henry C. Lea, 1874), vol. 1, p. 31. Stillé discusses the influence of such factors as climate, season, and occupation on the actions of medicines on pp. 33, 90–94. The pairing of this and the following quotation is suggested by Alex Berman, "The Impact of the Nineteenth Century Botanico-Medical Movement on American Pharmacy and Medicine" (Ph.D. dissertation, University of Wisconsin, 1954), pp. 36–37.

62. Roberts Bartholow, Annual Oration on the Degree of Certainty in Therapeutics (Baltimore, 1876), pp. 12–14. For an assessment of the relation between experimental physiology and therapeutics in the mid-nineteenth century, see John Harley Warner, "Physiological Theory and Therapeutic Explanation in the 1860s: The British Debate on the Medical Use of Alcohol," Bulletin of the History of Medicine 54 (1980): 235–57; Warner, "Therapeutic Explanation and the Edinburgh Bloodletting Controversy: Two Perspectives on the Medical Meaning of Science in the Mid-Nineteenth Century," Medical History 24 (1980): 241–58.

Other Perspectives

On Visiting the "Moving Metropolis": Reflections on the Architecture of Imperial Science

ROY MACLEOD

The last decade has seen a growing interest in the role of science in the history of imperial expansion. In England this has mirrored a growing interest in the "imperial relations" of British science. In this country, as earlier in the United States and Canada, historians have begun to describe the ways in which national scientific traditions developed from imperial precedents, often modifying imperial structures in the process. The pioneering work of Ann Moyal and Michael Hoare has opened the field by mapping Australasian activity from the voyages of Banks and Cook to the emergence of an independent Australian scientific identity. Abroad, scholars in Canada and India have explored different models of colonial scientific development that appear within the experience of transplanted Britons. This work has revealed enormous gaps in our knowledge, particularly of those general features of innovation and consolidation that characterized the "heroic age" of "imperial science," whether in India, Africa, Asia, or the settler colonies. At this stage one is prompted to reflect on the nature of "metropolitan" and "peripheral" relations and on the wider architectural principles that gave British science, in its imperial setting, its special and enduring quality. In the experience of Swift's Lemuel Gulliver, prevailing views of court and metropolis assumed an intriguing perspective when seen from a distant point on the periphery; and from Gulliver's island, where that famous castaway found his bearings, familiar notions of the history of science and statecraft invite careful revision.[1]

I have borrowed my title from Professor W. K. Hancock, to whom I owe much.[2] With him, let me immediately deflect criticism for any misuse of the terms *empire* or *imperial, colonial* or *colonialism, capital* or *capitalism.* As Hancock reminds us, *imperialism* is no word for scholars; neither is the word *model,* and both rest especially uncomfortably upon historians of science.[3] This is because Western science has traditionally been regarded as benevolent, apolitical, and value neutral; its extension, a value-free aid to material progress and civilization. Western science, since the seventeenth century, has had, it is argued, little more than a contingent relationship to conquest. Trade follows the flag, and science may improve the prospects of trade, but this imposes no responsibility upon science. The civilizing, improving advantages of new knowledge, in moral and material progress, surely cannot be questioned. If the imperial idea is accepted, if the complex association of commercial, humanitarian, and ideological motives underlying empire is understood, then science has only an incidental function in its articulation.

Closer inspection, however, reveals certain flaws in this reasoning. The creation of a free market based on economic hegemony, the control of the seas, the provision of communication and the protection of transport, and the glorification of progress as a civilizing ideal, all raise questions in which new knowledge has quite specific application. The control of that knowledge became critical. The way that knowledge was controlled, the "metropolitan" forces to which it refers, may have moved and changed, but the bonds forged through science are indissolubly linked to political development. Through science comes a language—conveniently the language of the mother country; through this language, neatly conveying the instrumental rationality of Western knowledge, comes control—in the imperial context, control often without accountability to the people who are governed, and knowledge "marginalized"—directed to the limited purposes of government, in such a way that the great majority of people remain far from enjoying the "relief of man's estate." This condition of life, familiar to science educators and development economists, reveals the contradiction, familiar to all historians of empire, that improved means do not necessarily imply improved ends.

This fact raises a further question: How did the pursuit of natural knowledge become a part of statecraft, both directly as part of policy and indirectly as part of government-aided economic development? To this may be added the question, How did arguments alleged to have been derived from scientific views of the world and its purposiveness come to inform naturalistic models of economic and political development, and arguments for the introduction of techniques that have proved at best economically palliative, but far more often socially and economically divisive? Finally, what happens when imperial relations become so intertwined that the metropolis *depends* upon the periphery for both economic and intellectual resources. Can the Empire strike back?

Once these questions are asked, the central issue becomes no longer science *in* imperial history, but science *as* imperial history. "Imperial science" thus becomes an expression of a will and a purpose, a mission, a vocation, often inarticulate, but enormously powerful. At one point in *Britain's Imperial Century*, Ronald Hyam refers to the conventional wisdom that "England's writ ran no further than the range of her ship's guns." [4] In fact, her writ ran where English institutions survived, and of English imperial institutions, among the most durable and enduring were those of science. I have undertaken to understand this better, as part of what Laurens van der Post has called the "honest reappraisal of the meaning of the vanished British Empire . . . one of the most urgent historical tasks of our time."

In what follows, I want first to consider selected models of imperialism and science, in what they tell us both about the history of science and the outlook of historians, to illustrate some of the difficulties inherent in this form of analysis, to offer a slightly different framework within which old questions may acquire different emphasis, and, finally, to discuss some analytical implications and limitations of this and almost any other general framework that seeks to embrace the range and variety of experience now available to historians of colonial and postcolonial development.

Let me make one further apology for offering frameworks in preference to facts, systems instead of surgical analysis. Actually, there is little alternative. If the history of the British Empire has passed into an Alexandrian age, where criticism has overpowered creation, this has not yet applied to the history of the sciences and their role in the building and binding of empire. This is partly because there is so little agreement about the boundaries of the subject, or about the definition of its terms. Common words commonly acquire new layers of meaning, and the resultant definitions, such as they are, are porous in the extreme. For example, it is important not to confuse "colonial science," with its sense of structures, institutions, precepts and boundaries, with "scientific colonialism," a term that implies a process, even a deliberate policy, with objectives and means to achieve those objectives. To confuse is to conflate an interactive ideology with a particular historical form of its experience. Indeed, scientific—versus unscientific—colonialism may simply mean a version of colonial policy. Similarly, I take "imperial science" to be a set of structures, staff, and legal and corporate institutions, serving many areas of policy and embracing quite different philosophical and political principles at different times, while "scientific imperialism" describes a version of imperialism that flourished in the late nineteenth century and that in large measure reflected the "scientization" of social theory. Finally, there is the "imperialism of Western scientific ideas," a phrase that implies the incarnation of the Baconian program, a proposal to subject everything to the acceptance of scientific and technological development.

In practice imperial science and scientific imperialism are two sides of the same coin, but failure to keep them analytically distinct confuses historians of all persuasions. In the imperfect glossary that I shall use, *metropolitan science* means not just the science of Edinburgh or London, or Paris or Berlin, but a way of *doing* science, based on learned societies, small groups of cultivators, certain conventions of discourse, and certain theoretical priorities set in eighteenth-century Western Europe. The Transit of Venus expeditions were thus "metropolitan," not just because they were launched from London or Paris, but because they implied a set of intellectual structures and questions common to the metropolis.[5] Metropolitan science *was* science.

Colonial science, on the other hand, was, by definition, done at a remove from Europe. It was, to paraphrase Thornton, "imperial science seen from below."[6] It meant different things: to those at home, recognizing their dominion over palm and pine, it meant derivative science, done by lesser minds working on problems set by savants in Europe. It was, looked at from the metropolis, "low science," identified with fact gathering. The work of theoretical synthesis would take place elsewhere. This division of labor fitted epistemologically well with the requirements of natural history and astronomy and conceptually well with a philosophical climate predicated upon inductive discovery.

In the colonies, "colonial science" could mean something else— merely science as practiced in the colonies, for example. This could be intellectually derivative. It could also be "metropolitan," in Hancock's sense, with values confirming the rule of "Britannia in another world," looking to local or international mercantile interests in minerals and wheat, concerned as much with the generation of capital as with colonization. From this usage would grow the concept of imperial science, embodying intellectual and institutional rivalries that reflected political divisions.

Imperial science again took on a different coloration, when regarded from England, or from the periphery. From London, it embodied a "spirit of power and service" in science, expressed organically as a nervous system whose functional cerebrum reposed in Burlington House, the Athenaeum, or eventually in the Oxbridge-London triangle. From the perimeter, however, "imperial science" implied a cooperative spirit, in which the appropriate analogy was not that of a single organism, moving to instruction, but a family, coexisting in a common linguistic ecology, whose existence was necessary to maintain a balance of nature and culture.

In this language, our definitions suffer from historiographical confusion and neglect. The phrase "imperial science," like "imperialism" itself, gained currency in the 1870s and the 1880s as an expression of the wish to strengthen the links between Great Britain and the British Empire, through what Lord Rosebery later called a larger patriotism. Merely an implicit part of the expansion of Europe to Sir John Seeley and Goldwin Smith, science and technology were given a more definite place in imperialism by V. I. Lenin and J. A. Hobson, in describing the successive stages of capitalist development.[7] But when imperialism came to be understood principally as a shorthand for the development overseas of finance or monopoly capital, with the associated expansion and redivision of world markets and the growth of state intervention, militarism, and colonial annexation, the contribution of scientific activity was relegated to the status of a necessary but unproblematic given. Only with later elaborations of Marxist theory—often focused on colonial annexations and the search for markets, on stimuli to industrialization, and on the relations between industrial countries and the third world—has the importance of scientific activity, as instrument and ideology, become prominent again.[8] In this context, the history of science in European imperialism has acquired fresh significance.[9]

To illustrate the variety and assumptions of standard approaches to science and imperialism, let me consider four related models. The first, which I shall for convenience call instrumentalist, is embodied in Carlo Cipolla's *European Culture and Overseas Expansion*. Cipolla's method, in *Guns and Sails*, is to focus on *technique*, to explore the reasons Europeans gained better technology and why these achievements enabled Europe to make conquests overseas. Multiple social, economic, and technical factors are reflected in different periods: an early Age of Innovation is followed in the sixteenth century by an Age of Reconnaissance, when increased use of the compass and decreasing supplies of galley slaves prompted sailing and shipbuilding, which, with commercial impetus provided by new trade routes to the East, provided Europeans—especially the Portuguese and Spanish—with a competitive advantage. Reconnaissance was followed by religion, religion by riches. When Vasco da Gama dropped anchor at Calicut and was asked what the Portuguese were looking for in Asia, he replied: "Christians and Spices." Or, as Cipolla puts it, "Religion supplied the pretext and gold the motive. The technological progress accomplished by Atlantic Europe during the fourteenth and fifteenth centuries provided the means."[10]

The consequences, in Cipolla's terms, are later seen in a phase of maritime expansion overseas, and in the eighteenth century in the support of industrial revolution at home.[11] Cipolla, however, confines himself to technique; this is all important, he says, "while philosophy and social and human relations are degraded to the role of means."[12] Phases of imperial development are determined by the technologies needed to bring them about; the history of empire is also the history of technology.[13]

This instrumental approach has an interesting parallel with the cultural independence model fashionable among historians of

American science during the late 1950s and early 1960s. I. Bernard Cohen of Harvard referred in 1959 to the prevailing attitude of inferiority to European science until the Second World War, in a paper revealingly entitled "The New World as a Source of Science for Europe."[14] His model was bimodal: a long period of colonial science, extending "far into the mid-nineteenth century," slowly giving way, following the Civil War, to a view of university and industrial science, which, drawing on European, especially German, experience, eventually produced traditions identifiably American. Cohen was followed, in 1962, by Donald Fleming of Harvard, who drew Australia and Canada into an extended comparison, and subsequently added a third phase—the Flight of the Muses from Hitler during the 1930s.

Both Cohen and Fleming were concerned with two particular problems: the accuracy or inaccuracy of Tocqueville's prophecy that egalitarian countries could never—or could hardly ever—reproduce the "aristocratic," "high science" of metropolitan Europe, and the failure of the American public to patronize cultural activity for its own sake, and the special priority given "practical" science in America and the factors that impelled creative effort toward the fostering of an independent science-based technological tradition, as part of America's urge to national self-assertion. Reflecting their interest in American history, both Fleming and Cohen saw the principal intellectual dilemma of the colonies as one of *enforced* provincialism. This provincialism was highlighted by the emphasis, in the colonies, of the natural history tradition. This tradition was important to Europe for commercial and philosophical reasons, but was adopted willingly by colonists. First, because it served "a fundamental part of the quest for a national identity in societies where the cultural differentiation from Britain was insecure, and the sense of the land correspondingly important for self-awareness." Second, because it "coincided with the primary national purpose of mastering [the] environment and canvassing its economic potentialities."[15]

To both Cohen and Fleming, however, the natural history tradition was a two-edged sword; winning mastery over a new continent, but for whom? In fact, the process proved also to be an instrument of control. Colonial scientists in America were relegated to the status of mere collectors. European savants received their tribute from the ends of the earth and honored by eponymy their colonial servants. This model of "absentee leadership" placed colonists at a disadvantage; the "gatekeepers" in Europe determined the prizes, and the result was "intellectual colonialism," a "psychology of abdication," making over to Europeans "the highest responsibilities in science"—a view that ostensibly overcast American science for a hundred and fifty years.[16] Fleming and Cohen were both concerned with the alleged inferiority of American science and the connection between political and cultural independence. The accuracy of their interpretation, in American terms, has prompted much constructive criticism. Inadvertently, perhaps, Fleming and Cohen exposed a paradox. As Europeans moved abroad, as the Empire grew, cultural dependence was an unavoidable, even necessary consequence. One could maintain a place in metropolitan circles by accepting the role allotted by those circles. Colonists differed in the extent to which they found this role chafing; in America, "classification without recognition" too closely resembled "taxation without representation." Eventually the tables were turned, and colonial science would hold the metropolis in its debt. But within the surviving Empire, both black and white, any metropolitan recognition that this debt existed was long avoided, for reasons that I shall presently explore.

In 1967, these issues of dependency and dominion were subsumed into a diffusionist perspective, when George Basalla turned to the spread of Western science throughout the world. In an influential article in *Science*, Basalla proposed a three-phase model,[17] applying an evolutionary pattern of cultural progression to the diffusion of science. During phase one, what he calls the nonscientific society, be it China, India, or pre-Columbian America, provides, borrowing Cohen's phrase, "the source for European science"—nonscientific, in his terms, meaning non-Western. In this phase of exploration by travelers or diplomats, flora and fauna are described and classified. Observers, all from Europe, settle and report their findings. In this model, the duration of phase one varies from region to region: consider the Spanish and the Portuguese in the Americas in the sixteenth century, the British in America in the seventeenth century, the French and British in India and the Pacific in the eighteenth century; both in Africa in the nineteenth century, and the Germans in the Pacific in the twentieth century. In phase one the scientific interests are those of the mother country, invariably a maritime power. As Thomas Sprat wrote in his *History of the Royal Society,* it was the maritime nations that were "most properly seated, to bring home matter for new sciences, and to make the same proportion of Discoveries . . . in the Intellectual Globe, as they have done in the Material."[18]

In his new philosophy, with commercial interests closely at issue, this matter of science was the stuff of which more than dreams were made. Eventually, this phase, argued Basalla, gave rise to a second phase, which he calls colonial, or dependent, science—not a pejorative phrase, he adds, nor one implying a relation of servility. Colonial science is essentially the same, he continues, whether it be within the Spanish sphere of influence or the British. It is inferior in a technical sense; it is subcritical in size; it lacks unity; colonial scientists cannot share in the "invisible college" of Europe or of twentieth-century America. Its strength lies in producing facts and field-workers. Its speed of development can be accelerated by favorable circumstances, as in America, Canada, and Australia, or retarded by an unfavorable environment, as in Africa and Latin America. Ultimately, says Basalla, when certain conditions are reached, a sense of nationalism pervades the colonial community; national institutions are created, and absolute reliance upon the external scientific culture is ended. This leads to phase three, of "independent science," in which a country's scientists are trained at home, rewarded at home, for work done at home and overseas, and published in national and international journals.

Basalla goes on to describe a set of barriers that must be overcome before phase two can give way to phase three: (1) resistances to science, especially religious or philosophical, must be overcome (China being a notable case); (2) the social status of scientists must be clarified and raised—that is, science must be seen as socially prestigious; (3) state aid to science must be forthcoming; (4) science education must be advanced; (5) scientific organizations must be founded and journals established; and (6) a "proper technological base" should be made available, with adequate scientific instrumentation as vectors of technical skill.

This model presents a program for scientific development familiar to students of bilateral and international aid policies in the 1970s. Basalla is careful to say it is not "necessarily causal," but the program has strong necessitarian overtones. In fact, let us examine its difficulties. Perhaps the greatest of these are:

1. It generalizes all societies, regardless of cultural context, into a single scheme.

2. The scheme is linear and homogeneous, assuming that there is a single Western scientific ideology which is disseminated uniformly. It does not take into account "south-south," or intercolonial movement, or movement between the colonies of one country and other European countries.

3. It aludes to, but does not explain, the political and economic dynamics within a "colony" that make for change and that occupy a shaded area between phases one and two or between two and three.

4. It fails to account for the relations among technological, social, and economic developments and the part played by science in bestowing legitimacy on political forces that may arrest development in other spheres.

5. It does not take into account the cultural dependence that lingers long after formal colonial political ties have been thinned or cut.

6. It does not take into account the wider economic interdependencies that have, since empire, contributed to the plight of the third world, for which science alone scarcely offers consolation.

Indeed, Basalla's model conveys a sense of economic unreality. Regarding with impatience this liberal, ostensibly value-free picture, the Argentine economist Francisco Sagasti rejected its implicit belief in an "invisible hand" guiding the responsible to independence. He sees in cultural imperialism instead a series of phases leading to economic control by Europe, the United States, and the multinationals and producing a marginalization of research effort and the erosion of democratic development.[19] In his "countermodel," there was a colonial, or preindustrial, stage of metropolitan science from the eighteenth century to the early nineteenth century, followed by a stage of Western science from the 1850s to the 1930s, during which dependent economies were manipulated to increase colonial integration through world markets and in which the extension of knowledge, through education, remains dependent on the metropolis. This gave way to a slow progress of independent industrialization, after World War II, by means of the substitution of imports—a process which, since the 1960s, has been overtaken by the concerted economic strategies of Western governments and multinationals, rendering dependent economies vulnerable to wider patterns of international trade. From Sagasti's perspective, one form of dependency, based upon the import of manufactured goods, has merely been replaced by another, more powerful, dependency, based upon the import of capital, machinery, and technical knowledge. Either system denies the illusion of independence. The image of the West diffusing science for the greater glory of the human spirit diminishes, even disappears, when thus closely examined.[20]

Finally, the Basalla model—almost inescapably, given its context—fails to account for the *political* character of science within this process of "intellectual colonization." Even if consideration is limited to the British Empire, at least four questions immediately arise:

1. Why is there no *analysis* of the proposition that knowledge is controlled or manipulated by the mother country? If there is an independence movement, when and how are alternative institutions established, and what extrascientific factors bring them about? What are the agents of change between phases? Is there an indissoluble affinity between political and cultural dependence, or independence, which it may be in the interest of both parties to preserve? In India, science was arguably an instrument of control, to which Indians were denied access.[21] What were the implications of this policy for India in the postcolonial world?

2. What use is made of colonial science, precisely, by the mother country? Possessions, notably India and Ireland, were consciously used as "social laboratories" for metropolitan policies.[22] New information, drawn from the colonies, was used to support European theoretical positions, whether concerning race, the structure of the earth, or the movement of the heavens. Colonial science, forcing open held positions to the incontestable evidence of the senses, may paradoxically hold a vital, not a subordinate position in institutions at home.

3. What features characterized the "scientific colonists"? Basalla's model is too broad to articulate with any precision the particular influences at work upon individuals within *different* settings. To explain imperialism in science as an irradiation of barbarism by civilization can be to offer an explanation so general as to be distorting. At the same time, it can neglect the traditions of different *corps d'élite*—"generalists" and geologists, astronomers and engineers—of which the stuff of history is made. Just as history based exclu-

sively on "explorers" neglects the wider social and political implications of discovery, so history without individuals neglects the importance of anomaly, and makes men mere mannequins in the
fashions of political theory.

4. Finally, what connection, if any, can be found between patterns
of economic and political evolution and patterns of intellectual development in the various sciences? While it would be too sweeping
to suggest that the conceptual content of specific fields was *determined* by the fact of imperial expansion, it would not be unreasonable to explore the dynamic interaction occurring between the two.

So far, this discussion may have provoked the recollection of many
long and contentious debates about the utility of historical models—debates that no historian since Marx or Toynbee has been able
to ignore or resolve. There are many models of imperial development: *sociological models,* dwelling on the relationships of culture
and distinctions between center and periphery; *economic* models,
dealing in labor supply and capital investment; and *diplomatic-
administrative* models, based on international relations, and the
"right to rule." Each poses difficulties which any description of imperial science, or of imperialism in scientific ideas and institutions,
tends to highlight.

Among social historians, the use of models is often confusing and
imprecise. We commonly confuse explanatory models, for example,
with predictive or heuristic models. Historical blades seem at times
too dull for philosophical chopping blocks. Since Collingwood's corrective, and the arid debates of the 1930s on the nature of historical
causation, we have often delegated "cause" to the philosophers, and
have got on with the job. But at the same time, we may have lost
sight of the damage this delegation can do to our own analytical
purposes. In particular, we commonly let our language confuse the
ordinary act of imposing structure upon chronology or events for the
purpose of explanation and interpretation with the more unusual
act of identifying "natural processes" abstracted from particular
contexts or periods. The latter tendency inclines us toward a kind
of naturalistic fallacy, in which social systems are interpreted as reflecting regularities observable in the natural world. There follows
from this a tendency to assume that what is highly valued is natural,
rational, and therefore unproblematic. This is the difficulty with
Basalla's model. In fact, this kind of model seems to acquire a som-

nambulist historicism, which not only can do inadvertent injustice to the density of contextual variation, but also can lend itself to misleading uses.

In this case, and for the purpose of my argument, it is necessary to recast Basalla's pioneering effort, perhaps considering in a different light some of the important evidence he assembled. Of course, creating new categories is no virtue in itself. But any improved model should reflect, rather than remove, the importance of controversy and conflict and should allow for the possibility that stages of underdevelopment are not necessarily transitional or intermediate, awaiting some eventual, revolutionary call to independence.

It is obviously impossible to reduce easily into a simple scheme the diversity of the experience that reflected different versions of imperial development in different parts of different European empires. It is all the more difficult to superimpose upon the history of empire this history of the natural sciences and of technologies that served different functions at different times and that underwent quite distinct patterns of development. Nonetheless, some attempt at a wider synthesis may be helpful, if only to focus criticism and to encourage the cultivation of counterexamples.

Accordingly, I would like to offer an impressionistic taxonomy to describe some characteristics of the main phases of British imperial science between c.1780 and 1939 and to outline in passing some of the structural relations which, whether necessary or contingent, appear to have strengthened the connection between political, economic, and technical developments. Examples will be selected principally from the experience of Australia, although comparisons with other parts of the British Empire may well be relevant and worth pursuing. In this speculative blueprint there are five phases or "passages" (see table 1).

Phase One: Metropolitan Science

The first period, extending from 1780 until the mid 1820s, might be described as the Banksian era or, just as well, the Laplacian or Cuvierian era. In this period the selection of problems, the ordering of nature, was dictated from the Royal Society—or the Académie des Sciences.[23] This policy continued the European tradition of discov-

Table 1. *Passages in British Imperial Science*

Aspect of scientific practice	Metropolitan	Colonial	Federative	Efficient imperial	Empire or commonwealth
Institutional ethos	Explorative; Banksian; internationalist; systematist; centrist; individualist	Envelopment by metropolis; degagement by periphery; autochthonous societies; individual research; scientific services	Cooperative, imperial research; university development and professional legitimation; extended scientific services	Expert, official, public; codified science; specialization of disciplines; expansion of tertiary education;	Metropolitan trusteeship; coordinated fundamental research; delegated research responsibility from London
Social and political characteristics	Monarchist; centripedal	Colonial ascendancy; intercolonial rivalry; responsible government	Intercolonial and interimperial association	Optimistic liberalism; defensive imperialism; application of scientific methods to government	Imperial unity; self-determination
Economic and technological functions	Expansion of maritime trade; discovery of raw materials and new markets	Primary products; pioneer, adaptive technology; local markets	Improved technology (communications, agriculture); extended participation in world markets	Management of land, resources, industry; early government regulation of mode of production	State encouragement of applied science; early growth of secondary industry

ery and explanation, in part with the hope of completing a world view determined by Europeans: geographically, to settle for all time questions concerning the shape and texture of the earth; botanically and zoologically, to confirm systematic views of the continuity, linearity, and continuous gradation of species; astronomically, to complete the Newtonian world picture—to "fix the frame of the world." Thus, the Transit of Venus expedition of 1769 was to determine the "astronomical unit," the "celestial meter stick." During this period there was a clearly internationalist ethos in science, given force by instructions from the admiralty against interfering. So, too, were there indignant protests against the confinement of Matthew Flinders in Mauritius by the French.[24]

With the spirit of common endeavor came a vigorous sense of cultural competition. Even amid the stresses of war, France and England had enjoyed an almost uninterrupted flow of scientific and philosophical influences. After Waterloo military rivalries were replaced by commercial and scientific competition. If fears of French invasion had died, the specter of French domination in cultural and material progress remained. Anxiety and caution, suffused with national pride, resounded through the pages of the "heavy quarterlies" and informed the reform movements in English science and education. This invocation of domestic motives in imperial maneuvers became a standard practice, neglected by historians of science today at their peril. To an earlier French historian of empire Britain was explicable only as a country in which "l'esprit commercial s'allie curieusement à l'esprit de découverte . . . et les appétits du sport."[25]

After 1780, the end of Britain's first empire and the consolidation of her second did not interrupt this process of cultural expansion. Economically, the metropolis at first exported little and invested little. But by the late eighteenth century Britain was seeking new sources of raw materials and the development of new markets. Politically and commercially, activity was directed from London through chartered companies in Hudson's Bay, India, and the East Indies and through plans to explore the commercial utility of newly discovered products. In all this activity strategic considerations were never far distant. As Geoffrey Blainey points out, the Royal Navy surveyed the possibility of cultivating flax for sails and pines for masts on Norfolk Island well before permanent settlements were envisaged. No astute observer could ignore the importance of exploration to the prestige of learned societies of London and Edinburgh. New reports of discoveries created new intellectual capital and strengthened the sterling of English science.

Phase Two: Colonial Science

After the loss of the American colonies, the greatest imperial investments, in science as in capital, were in India. But, around the 1830s, metropolitan science, consolidating its position in Calcutta and Bombay, began to move beyond India. During this period, and for about thirty-five years, is seen the first permanent extension of the metropolis, at once enlarging and ensuring the domain of what became known as colonial science. Cosmopolitanism among metropolitan peers was an indulgence to be enjoyed from a position of strength. Hence the pressure to occupy greater intellectual space, from Herschel's explicit efforts to annex the Southern Skies from his observatory at the Cape of Good Hope to Sir Thomas Brisbane's zeal for astronomy in New South Wales; so, too, Sir Edward Sabine's proposed ring of magnetic stations from Bangalore to Baffin Bay.[26]

So, also, the importance of Kew—botanical specimens, cinchona, rubber—all sought for wider imperial purposes of political and economic expansion.[27] So, finally, the need for fresh voyages of discovery. The search for the Northwest passage, long a fascination of maritime nations, became by the 1830s important to the security of Britain's trade routes.[28] Hydrographic surveys and the voyages of HMS *Beagle* and HMS *Rattlesnake* drew their impulse from this simple fact of geopolitics. The navy's surveyors and surgeons, and often its regular officers as well, were, as before, the handmaidens of science.

Where there was trade, there was the navy, and where the navy sailed, or the army rested, the natural sciences benefited.[29] But so did the newer, more specialist and metropolitan scientific societies, which, in Banks's memorable phrase, threatened to dismantle the Royal Society, not leaving "the old lady a rag to cover her."[30] Against this fragmenting pressure of new knowledge, the imperial epicenters of metropolitan science fought to buttress themselves. In 1833, Vernon Harcourt used the same imperialist metaphor at the British Association, when he warned of the dangers of specialization, as colony after colony separated itself from the declining Empire of the Royal Society.[31] The politics of science were embedded in the politics of the wider Empire. If, by degrees, the commonwealth of science was under threat in the colonies there was also danger of imperial fragmentation, with local interests becoming paramount. Many of these interests were vital to Britain—from the minerals and forests of Canada to India and Ceylon and their markets for textiles and teas, from the plantation products of the West Indies to the agriculture and livestock of Australia and New Zealand.

In the event, traditional British institutions, pressed at home by economic recession and demands for constitutional reform, adapted to these portending breezes of change by assuming a wider vision of empire and creating a new conceptual framework for colonial science. To London, this represented a convenient "political envelopment"; in the colonies, among transplanted Britons, the same impulse produced a mythology of devolved responsibility, even a division of labor, which seemed equitable enough.

In this conceptual structure there were two principal, and sometimes contradictory, tenets, both reflected in Fleming's account of colonial American science. First, there was an emphasis on practical versus theoretical knowledge. Thus, Kathleen Fitzpatrick comments on Sir John Franklin's innovations in Tasmania:

Bushlife demands the virtues of action—initiative, hardihood, quickness in decision and improvisation. The most blessed word in the colonial vocabulary is 'practical' as the need for the virtues appropriate to pioneering conditions passes away, the tradition is tenacious. A contemptuous tolerance is the best that the scholar, the artist and the pure research scientist can . . . hope for.[32]

This emphasis was accompanied by a spirit of deference to the metropolis, a spirit that reinforced the colonial mentality decried by imperial historians today. This fact-gathering, derivative mentality was not, one would think, a source of pride. But, pace Fleming, it glorified the character of colonial science. From the Asiatic Society of Bengal to the Philosophical Society of Australasia, colonial scientists saw their task as one of ascertaining "the natural state, capabilities, productions and resources of [theirs] and the adjacent regions . . . for the purpose of publishing from time to time such information as may be likely to benefit the world at large."[33] Thus, information arising from P. D. King's maritime geography was processed by John Grey at the British Museum.[34] In the colonies, the fact of intellectual dependence underscored the practical realities of frontier life. As the *Sydney Morning Herald* editorialized in 1830, "Zoology, Mineralogy, Astronomy and Botany are all very good things, but we have no great opinion on an infantile people being taxed to support them. An infant colony cannot afford to become scientific for the benefit of mankind."[35]

Unavoidably, these two colonial characteristics—the practical and the deferential—obscured deeper currents. In fact, in North America, India, and Australasia, colonial science could be highly theoretical and highly dispositive, especially in the Humboldtian sciences of natural history, geology, astronomy, and meteorology. Exploration in Africa and the tropics revealed features that bore directly on central debates and important reputations in geology and zoology. Michael Adanson, visiting Senegal during the 1750s, recalled that "botany seems to change face entirely as soon as one leaves our temperate countries."[36] Australia, especially, caught the metropolis off guard. Sir James Smith agreed that "When a botanist first enters . . . so remote a country as New Holland, he finds himself . . . in a new world. He can scarcely meet with any fixed points from whence to draw his analogies."[37] As John Oxley wrote in 1821, referring to continental theorists, "Nature has led us through a mazy dance of intellectual speculation, only to laugh at us . . . on this fifth continent."[38] More than providing mere exceptions to European systematics, Australia was the "land of contrarieties," a strange place, wrote Barron Field, "where the laws of Nature seem reversed: her zoology can only be studied and unravelled on the spot, and that too only by a profound philosopher."[39] In botany, Robert Lawrence, collecting for W. J. Hooker, found evidence for external proofs of the new "natural system" of Jussieu, which was ultimately to question the system of Linnaeus. In astronomy, as Sir Thomas Brisbane recognized, the skies were no limit. "Avec un ciel vierge," he wrote the Colonial Office, "what may not be achieved."[40]

It was perhaps inevitable that, as the metropolis moved, the same controversies would march in step. The same beliefs in the wisdom of God read through the Book of Nature; the same debates between idealists and associationists, and the same zeal to find evidence confirming, rather than disputing, the received view of creation.[41]

That all these would migrate to the colonies, transmitting a "carrier conservatism" in the knapsacks of the fossicking frontiersman, is not surprising. But to dismiss colonial scientists as blinkered conservatives misses a wider reality. In this connection, three separate questions arise: whether the Empire was important to metropolitan science; whether colonial scientists appreciated their importance; and if so, whether they did anything about it. How did the center actually deal with the new insights available only from the periphery?

By the 1820s, there was already wide recognition of the central role that colonial science would play in metropolitan debates. Robert Knox, the Edinburgh anatomist, admitted to Sir Thomas Brisbane that Australian specimens revealed "the most wonderful deviations from the usual types of forms which nature employed in the formulation of the animal kingdom."[42] Certainly, in the tradition of Cuvierian comparative morphology, these "deviations" from a "European standard" were of first importance. In the organization of English science, they bore directly upon the prestige of that handful of men whose fortunes rested upon the tenets of catastrophism, Neptunism, and *Naturphilosophie*.

But the threat was swiftly contained. This rush to maintain metropolitan primacy is reflected spectacularly in the expedition of Matthew Friend, who arrived in Sydney in April 1830 bearing a combined commission to find correspondents for the Royal Society, the Zoological Society, the Geological Society, the Medico-Botanical Society, and the British Museum (Natural History). As he rather patronizingly told John Henderson's remarkable Scientific Society in Van Diemen's land:

Your country is still a land of mystery, supposed to abound with anomalies, which, if verified and ably described, would tend much to illustrate many of the most abstruse and important questions in the history of organic life. The transition forms—the animals intermediate betwixt different orders where the diagnostic marks are mixed with each other, are of the utmost consequence in physiology.[43]

The colonists, he advised, had a duty to collect placental organs and crania and ship them home, where there were "greater experience [and] more numerous and better fitted appliances." The mother

country and they, he added, would "divide the honours" between
them.

The political implications of this were lost neither on the colonists—though their protests were muted—nor on London, where
the power brokers of metropolitan science rushed to assimilate the
new revelations from Australian discoveries, just as the Royal Society had done for the American colonies and India a century before.
These were very numerous in the middle decades of the nineteenth
century; indeed, they accounted for about two thirds of the articles
in the *Annual Magazine of Natural History* in the 1850s and 1860s.
This process of manipulation, and its implications, are easy to trace.
As William Hooker was to botany and George Airy to astronomy, so
Richard Owen (at the Royal College of Surgeons, 1826, and the British Museum [Natural History], 1856–84) was to zoology and morphology. Owen had succeeded Banks as the Czar of English natural
history in the colonies. During the 1840s, as we know from Ann
Moyal's pioneering work, he moved particularly quickly, in Banksian manner, to seize the results of new discoveries, particularly associated with marsupials, recognized as the key exception to most
governing principles in the "great chain of being."[44]

Owen, following the idealist tradition in Cuvier and Oken, had
fashioned a theory of a unitary archetype, to which all animal creation approximated and from which all animal creation may have
drawn, in the mind of the Creator, its governing plan. The existence
of this archetype was supported by interpretations of analogy of
function and homology of structure that connected all sections of
the animal kingdom with each other, but which required continuous and separate creations. The archetype required consistent interpretations, which required a confident grasp of periods and processes
of creation and an ability to overlook, or dismiss as uncertain, dis-
cordant evidence. But fossil deviances, and the analysis of organ and
reproductive structure among living animals of Australia, threatened this position. In 1827, Peter Cunningham returned from two
years in New South Wales to announce that "The dissimilarity of
the animal and vegetable diluvial remains [in Australia] to what we
see in a state of living existence, proves that all the products of the
earth were quite different to what they are now."[45] And when, in
1829, the Wellington Cave Fossils were taken as evidence of Hutto-
nian principles, soon to be elevated to a broader position of unifor-
mitarianism by Lyell's *Geology* in 1830–33, it was clear that new

and unbeatable forces were massing against him. Normally, new ideas do not gain ground by the logic of their advocates but by the death of their opponents. In this instance it was different. From the 1830s Owen was forced into a slow, bitter retreat before the advancing forces of scientific naturalism. His defeat, anticipated in 1884 by William Caldwell's discovery of the oviparous nature of the platypus, was a crisis in a drama that touched every corner of Britain's cultural empire.

From the 1840s to the 1870s, this spirit had been anticipated by a steady flowering of colonial scientific enterprise, celebrated in the expansion of learned societies and museums and in scientific surveys from India to British North America.[46] The colonial scientific movement was not without its martyrs. Thus, Edmund Kennedy, the young surveyor of the York Penninsula, fell to Aboriginal arrows in 1846, "a sacrifice," as recorded in St. James's, Sydney, "to the cause of science, the advancement of the colony and the interests of humanity." John Gilbert, the ornithologist, and a casualty of the Leichhardt expedition, was similarly commemorated in marble above Sydney's leading congregation, with the revealing Horatian paraphrase: *Dulce et decorum est pro scientia mori.* Coinciding with anti-imperialist sentiment in England and with the policies associated with Aberdeen, Russell, and Derby and interrupted only by Palmerston, this "colonial ascendancy" in science was not resisted by London through the 1860s, especially since the administrative importance of colonial science for imperial rule could not be denied. As Sir William Denison told Admiral Beaufort in 1849, an astronomer in a colony was a guarantee that a certain amount of science was there at the disposal of the government.[47]

Eventually, this spirit suffused the new universities at Sydney, Melbourne, and Cape Town and kept alive, under difficult circumstances, learned societies from New South Wales, Victoria, and New Zealand to the Cape Colony, Ontario, and India. Still, however, the twin principles of intellectual deference and practical service operated to sustain dependence upon the metropolis.

Phase Three: Federation

In 1875, in order to study the tropical diseases, Patrick Manson had, typically, to return to England to "drink at the fountain of science."[48] But these artesian sources were not to be confined to England indefinitely. By the early 1880s, accompanying the arrival of new political factors in the government of metropolitan science, a "federative" language begins to replace the language of colonialism in science. In 1884, Seeley's *Expansion of England* led to a belief in the merits of encouraging a "greater Britain" in cultural and political affairs. In part, this was accompanied by the victory of T. H. Huxley and his allies, many of them reflecting sympathies and experience in the colonies.[49] In London, new men were succeeding to the czardom of English science. In natural history, the "Winter Palace" moved its court from Kew, the Linnaean Society, and the Natural History Museum to Cambridge, Rothampstead, and the Imperial Institute; in astronomy, Greenwich shared influence with the Solar Physics Observatory at South Kensington. Facing increasing competition from Germany and the United States, arguments for containing "by federation" the skills of Britons overseas needed little justification. Thus, the British Association, the "Parliament of Science," met in Montreal in 1884, in the first steps toward what would become a virtual beating of the bounds of Empire, reminding the colonies of their status in respect to the Mother of Parliaments.[50] In 1885, Huxley, as president of the Royal Society, urged all colonial learned societies to "associate" with each other, through the Royal. This federative strategy was not immediately successful, but it established a clear precedent. It is with the late 1880s, moreover, that imperial science also became an explicit political program, accompanying Disraeli's vision of empire and the opportunism of Chamberlain's Unionists.[51] As always, there was Ireland, and Home Rule contributed to both the division of the Liberal Party and the political separation of scientific friends along imperial lines.[52] In many ways, just beginning to be explored, federation arose as a solution to such fissiparous tendencies that seemed both neutral and attainable. All shades of political interest could be united, with economic science and public sentiment sustained by "scientific" social Darwinism and given force by voluntary interimperial cooperation.[53]

Regarded from scientific London, federation was a policy of promoting Britain as primus inter pares. Within the Royal Society, as in government, a species of institutional condescension ran in counterpoint to expressed desires for cooperation. From India to Canada, the passing of "colonial science" saw the reinforcement of "imperial science," in the reassertion of imperial interests in geology, botany, meteorology, astronomy, agriculture, and forestry, all now staffed by a new army of "scientific soldiers."[54] As the tribulations of Manson and Ross in India revealed, "constructive imperialism" was a policy of containment. The font of honors remained securely unchallenged in Burlington House and in Whitehall.

But the prospect of federation was seen differently at the periphery. It was local pride that prompted the government astronomer, H. C. Russell, F.R.S., to tell the Royal Society of New South Wales in 1888 that "there were many objects for investigation which men coming from the civilised world took the honour and credit of studying what might otherwise belong to the colony."[55] To support that contention, the same year saw the creation of the Australasian Association of the Advancement of Science. Michael Hoare argues that the establishment of this daughter association (ANZAAS, of today) was the first concrete step toward political federation. Indeed, Sir James Hector told the association as much in 1891.[56] Certainly, the movement to federate local societies on a colonial level—and colonial societies on an imperial level—provided a tantalizing precedent. In fact, by federation the colonies strengthened their intellectual position and their loyalty to British science in the same stroke. The *Sydney Morning Herald* in August 1888 welcomed the first meeting of the Australasian Association for the Advancement of Science, in Huxleyan metaphor, as presaging a new imperial scientific army marching together under the Southern Cross. But make no mistake, England had accommodated, assimilated, and kept control. The Southern Cross was still quartered by the Union Jack.

Phase Four: "Efficient" Imperialism

By the turn of the century, the institutions of imperial science, in common with most British institutions, were badly shaken by the Boer War. In the aftermath, the directorates of Kew, South Kensington, Bloomsbury, and Burlington House found allies in Liberal Imperialism and Conservative Unionism. The Empire was an important trust and resource. Its efficient administration was central to the idea of national efficiency. Under Churchill at the Colonial Office, that quarter of the globe covered red would, as W. S. Blunt put it, be kept "in part by concession, in part by force, and in part by the constant intervention of new scientific forces to deal with the growing difficulties of imperial rule."[57]

If imperial unity was the desired end, scientific unity was the one universally acceptable means. In Britain, the equinoctial symbol of efficiency was found in the British Science Guild (BSG), founded in 1905 by Sir Norman Lockyer, editor of *Nature*. The BSG welded the ambitions of science and the purposes of politics, through the rational use of scientific method.[58] The universal applicability of scientific method to domestic politics, reasoned both the Fabians and the BSG, had equally universalist application to the Empire. Men of science agreed, as William Ramsay put it, "The best way of fitting your men for the manifold requirements of Empire is to give them the power of advancing knowledge."[59] The exaltation of scientific method served to bring fresh attention to the prevention of epidemic disease, whether plague in Sydney or cholera in India. As Chamberlain said at a luncheon to raise funds for a proposed school for tropical hygiene in 1898,

The man who shall successfully grapple with this foe of humanity and find the cure for malaria, for the fevers desolating our colonies and dependencies in many tropical countries, and shall make the tropics liveable for white men . . . will do more for the world, more for the British Empire, than the man who adds a new province to the wide Dominions of the Queen.[60]

From this to the spirit that sanctioned the Imperial Universities Conference in 1903 was a short step. The encouragement of the universities would in practice buttress empire. Indeed, if Lord Rosebery had had his way, the University of London would have become the center of that empire. At the same time, the enshrinement of science in the universities of the Empire would make the world safe for English liberal values. Science would also, it was alleged, in time improve the management of agriculture in India, Africa, and the West Indies and the success of manufacturing industry in India and the Crown Colonies. Indeed, scientific method would, it was argued, unite empire in unity of truth, of tradition, and of leadership, from Curzon's India and Lorne's Canada to Smuts's South Africa. In its identification of the rule of law with scientific method, this policy was pleasingly nonpartisan. Science was the balm of Gilead; so the British Association preached when it visited Pretoria and Johannesburg in 1905 to heal the wounds of war. Imperial science was the route to a wider patriotism that transcended party, national, and even imperial politics.

This implicit political meaning of imperial science was voiced in 1907, when Alfred Deakin, then Prime Minister of Australia, told the BSG and the British Empire League that the most urgent task of empire was "the scientific conquest of its physical, and shall we not be bold to say, ultimately its political problems."[61] Deakin, in 1910, welcomed the prospect of a British Association meeting in Australia, which came to pass in 1914, as securing the country in the "brotherhood of nations . . . visibly and before the eyes of the world [bringing] us into notice as a portion of Europe . . . united by intellectual ties as well as those of patriotism and blood . . . an added step towards Imperial unity . . . and one likely to be of great value to the Commonwealth."[62]

Until the First World War, imperial bonds between Britain, India, and the white settler colonies stressed this image of imperial unity. For Lord Milner and Lord Amery, imperial communication and research were the twin keys to imperial development. In this development, Britain was intended to retain leadership. As Tom Mboya remarked, "Efficiency is the last refuge of the imperialist."[63] Wider social programs involving holistic, preventive models, or the improvement of living standards, as proposed by Ronald Ross for West Africa, were not easily accepted, or even understood, by the Colonial Office. Throughout the Empire, at least until 1914, the application of technical knowledge was principally limited to the purposes of government. The direction of science for metropolitan self-interest continued well into the new century. The idea of a neat division of labor—cultivating mines and forests in the colonies and theoretical physics in the metropolis—was threatened from the day Ernest Rutherford won a scholarship to Cambridge. But it sustained a spirit of cooperation in the market of scientific ideas within the Empire, transacted by the scientific "city," with the colonies providing more and more of the merchant capital. As we know, this proved to be of decisive benefit to Britain in 1914 and beyond.

Phase Five: Empire and Commonwealth

Efficiency might be considered ironical, given the waste of the Great War. In the event, the fellow suffering of the war saw the mechanistic language of efficiency give way before the organic language of coordination. A new phase emerged in imperial science, one in which London was regarded as the "honorary secretary" of a voluntary association, the "home" of the imperial "family." *Deference* was, if anything, now defined as *loyalty*. The Empire paid embassy to Britain in science and confirmed, after all was said, the idea of imperial unity that Banks had assumed as natural and inevitable a century earlier. As A. G. Butler, writing of the Australian wartime medical effort, put it, the organization of science witnessed the same "uncertainties, compromise, cooperation without compulsion, the union that is organic rather than formal, which characterises the relation between the various parts of the Empire."[64] In fact, imperial cooperation in science provided an exemplary model for allied and imperial cooperation in other directions. In the process, debts were imposed on all sides. Arguably, British recognition of Indian independence began not with 1947, but with the first meeting in Calcutta of the Indian Science Congress in 1914.

In this coordinative phase, continuing from the First World War until well after the Second, intellectual leadership in British science was increasingly shared with the dominions. Indeed, in many fields—such as agriculture, entomology, and nutrition—the leadership of British science was in fact no longer in Britain. In the postwar decades, models of the British Department of Scientific and Industrial Research (DSIR), transmitted throughout the Empire, contained the residual seeds of centralism.[65] But the architecture of "imperial science" altered to incorporate a timely sense of cooperation, especially in economic policy. As *Nature* put it, in 1924, at the time of the Empire Exhibition, there was "a by no means imaginary connection between the spirit of science and the political ideals of Empire."[66]

During the interwar years, the spirit of science was redefined to suit these political ideals. In the circumstances of the 1920s, that meant cooperation, not metropolitanism. This was exemplified particularly well in the new pattern of scientific cooperation between Australia, New Zealand, and Canada and was preached widely throughout East Asia and South Africa. Imperial self-sufficiency through science was the goal of which the Imperial Bureaux, the Empire Marketing Board, the Colonial Research Committee, and the Imperial Economic Conferences were the collective symbols.[67]

As a political strategy, it had difficulties. But the scientific model survived.[68] By 1930–31, following the Ottawa conference, coordination gave way to a species of delegation in which metropolitan scientific research was conceived as an aid to local economic development and self-determination, particularly in Africa, Malaya, and the West Indies. One consequence was the slow but steady encouragement of "indigenous" scientific activity, often, as it transpired, as a prelude to political independence. That dependence was, of course, to prove illusory, as the old Empire began, with the aging metropolis, to suffer the new imperialism of the superpowers, the international agencies, and the multinational corporations. The passages of imperial science reached an end, perhaps sometime in the 1950s, and a beginning, as a new set of dependencies replaced the old, with Britain no longer at the center. In a psychological sense, "dependency" would remain a characteristic of science as practiced in the Empire for many years. But in many ways, by then the spirit of innovation, long resident at the metropolis, had moved to the periphery, and the fixed certainties of power and competition were replaced by fresh tests of partnership and commonwealth.

Conclusion

In this essay I have not proposed to examine the complex effects of empire on the development of individual disciplines; this is too large a universe for a single exploration. I have suggested, however, that certain mentalities do operate in the relations of science and empire, which may affect the conduct of those disciplines, if at one remove. Indeed, if it is possible at all to distill a generalized conception of science and empire, it is certain that this must be as much concerned with political as with technical issues. I use the word *political* in two senses. First, the changing connotations of what I have described as metropolitan science, colonial science, and imperial science reflect and mediate the changing perceptions of vested interests, both in Britain and in the colonies. By observation, it is clear that imperial science, looked at from the center, was an integral part of the changing policies of colonial development, that problems chosen as important were determined by the interests of the imperial power, and that the practice of science in the Empire was itself influenced by changing ideologies of empire. Changing patterns of trade, incentives to development, all had direct effects on scientific activity, and that activity was more and more to shape the direction of policy well outside the laboratory and the learned society. At the periphery, the particular relations distinguishable between science and politics in any country—contrast the differences, for example, between the elite, mock-metropolitan institutions of Canada, torn between America and Britain with the more egalitarian, isolated scientific societies of Australasia—do vary, as did the corresponding histories of scientific development and political leadership. There are also obvious differences in the requirements of official science in various places, and thus in the overall complexion of imperial science in the political configuration of the settler colonies, India, the "occupied" empire of Africa and the West Indies, and the informal empire of Latin America. Of course, the *precise* political dimension of science in any particular context remains to be revealed by national comparisons. But it is clear that in its most general political usage, science became a convenient metaphor for empire itself—or, more exactly, for what the Empire might become. The ethos, methods, and organizational strategies of science were all used by imperial spokesmen in the discussion of federation, of coordination, and of cooperation. The British Association, the universities, and the machinery of economic policy gradually replaced the metropolitan learned societies in fostering the sense of an imperial mission, but the universal appeal of this mission was never a phantom. We have only to recall the eagerness of colonial governments and societies to copy and preserve metropolitan models and the testimony of those hosts of imperial and colonial botanists, surveyors, astronomers, zoologists, and geologists who, in serving science and empire, helped expose the problems and opportunities that political independence would bring.

"Science and empire" is also a political expression in cultural terms. Unquestionably, there were vital issues of cultural and economic domination involved in the pursuit of natural knowledge. What *is* striking is that so few colonials were aware of the influence they enjoyed over the center. That famous telegram sent in 1884 by William Caldwell from the Burnet River to Archibald Liversidge, relayed to the British Association in Montreal: "Monotremes oviparous, ovum meroblastic," signaled the beginning of a revolution against the czardom of Owen. It was to the scientific world as a signal from the battleship *Aurora*, giving fresh support to the anti-Owenite sentiment gathering in Cambridge and South Kensington. It remains to us to discover more instances in which the institutions and leadership of Britain were *dependent* upon colonial discovery and enterprise. Ironically, ex-colonials—American, Australian, Canadian, and perhaps South African—may have been looking at the metropolis upside down and cosseting unnecessarily defensive and deferential attitudes toward the diffusion of metropolitan science.[69]

Finally, I have suggested that a dynamic conception of imperial science gives a fresh outlook to the study of imperial history. There is no static or linear extrapolation of ideas; there are multiple autochthonous developments that have reverberating effects. I have suggested, following Hancock, that the idea of a fixed metropolis, radiating light from a single point source, is inadequate. There is instead a moving metropolis—a *function* of empire, selecting, cultivating intellectual and economic frontiers. In retrospect, it was the peculiar genius of the British Empire to assimilate ideas from the periphery, to stimulate loyalty within the imperial community without sacrificing either its leadership or its following.

This flexible formulation of imperial science may also afford a new perspective on the study of underdevelopment in the postcolonial world. V. G. Kiernan once remarked upon the tendency today to ascribe to British rule a far less forceful effect, for good or ill, than used to be supposed by friends and foes alike. When the history of imperial science is written, it may well demonstrate how pervasive, yet how unobtrusive, that influence could be.

This essay, previously published in *Historical Records of Australian Science* 5, no. 3 (1982), was given originally as a lecture, delivered, in different versions, at the Australian National University in August 1980 and at the University of Melbourne in May 1981. The author is grateful to the Research School of Social Sciences of the Australian National University, Canberra, for generous support of the work from which it is derived. For helpful comments he is particularly grateful to Oliver MacDonagh, Rod Home, Lloyd Evans, Pat Moran, and Ann Moyal.

Since this essay was published, a number of studies have appeared which bear on the subject of "imperial science," particularly within Australian and Canadian contexts. These are to be found in the *Historical Records of Australian Science* and in the proceedings of the Kingston (Ontario) Conferences. Qualifications and applications of the interpretation advanced in this essay form the basis of two major studies currently underway in Sydney.

1. According to Swift, Gulliver was driven to shore at 32° 2′S, northwest of Van Diemen's Land—a position that might have placed him on the Sydney side of the Murray River.

2. Keith Hancock, "The Moving Metropolis," in *The New World Looks at History*, ed. A. R. Lewis and T. F. McGann (Austin: University of Texas Press, 1963), pp. 135–41.

3. For a discussion of historical models in imperial history see J. M. S. Careless, "Frontiers, Metropolitanism, and Canadian History," *Canadian Historical Review* 35 (1954): 1–21.

4. Ronald Hyam, *Britain's Imperial Century, 1815–1914: A Study of Empire and Expansion* (London: Batsford, 1976), p. 23.

5. Harry Woolf, *The Transits of Venus: A Study of Eighteenth Century Science* (Princeton: Princeton University Press, 1959).

6. A. P. Thornton, *The Imperial Idea and Its Enemies: A Study of British Power* (London: Macmillan, 1959).

7. See J. A. Hobson, *Imperialism: A Study* (London: Nisbet, 1902); and V. I. Lenin, *Imperialism, the Highest Stage of Capitalism* (Moscow: Foreign Language Publishing House, 1974).

8. Standard accounts, which vary widely in their treatment of science and technology, include M. Barratt-Brown, *After Imperialism* (London: Heinemann, 1963); A. P. Thornton, *Doctrines of Imperialism* (London: John Wiley, 1965); R. Owen and B. Sutcliffe, eds., *Studies in the Theory of Imperialism* (London: Longman, 1972); P. D. Curtin, *Imperialism* (London: Macmillan, 1972); see also Peter Worsley, *The Third World* (London: Weidenfeld & Nicholson, 1968).

9. Cf. Anis Alam, "Imperialism and Science," *Race and Class* 19, no. 3 (1978): 1–13; Deepak Kumar, "Patterns of Colonial Science in India," *Indian Journal of History of Science* 15 (May 1980), pp. 104–13; and "Racial Discrimination and Science in Nineteenth-Century India," *Indian Economic and Social History Review* 19 (1982): 63–82.

10. Carlo Cipolla, *European Culture and Overseas Expansion* (Harmondsworth: Penguin Books, 1970), part 1, "Guns and Sails," p. 100.

11. See Eric Hobsbawm, *Industry and Empire* (London: Weidenfeld & Nicholson, 1968).

12. Cipolla, *European Culture*, p. 108.

13. J. Allen, "The Technology of Colonial Expansion," *Industrial Archaeology* 4 (1967): 111–37; François Crouzet, "Trade and Empire: The British Experience from the Establishment of Free Trade until the First World War," in *Great Britain and Her World, 1750–1914: Essays in Honour of W. O. Henderson*, ed. Barrie M. Ratcliffe (Manchester: Manchester University Press, 1975).

14. I. B. Cohen, "The New World as a Source of Science for Europe," *Actes du IXᵉ Congrès International d'Histoire des Sciences*, Madrid, 1–7 September 1959, pp. 96–130; cf. R. P. Stearns, *Science in the British Colonies of America* (Urbana: University of Illinois Press, 1970).

15. Donald Fleming, "Science in Australia, Canada, and the United States: Some Comparative Remarks," *Proceedings of the 10th International Congress of the History of Science* 18 (1962): 180–96; and "Emigré Physicists and the Biological Revolution," *Perspectives in American History* 2 (1968): 152–89.

16. Cohen, "New World as a Source"; Fleming, "Science in Australia," p. 181.

17. George Basalla, "The Spread of Western Science," *Science* 156 (5 May 1967): 611–22.

18. Thomas Sprat, *History of the Royal Society of London* (London, 1667), p. 86.

19. F. Sagasti and M. Guerno, *El desarrollo científico y technológico en America Latina* (Buenos Aires: Instituto para la Integración de America Latina, 1974).

20. Cf. T. W. Keeble, *Commercial Relations between British Overseas Territories and South America, 1806–1914* (London: Athlone, 1970).

21. R. MacLeod and R. Dionne, "Science and Policy in British India, 1858–1914: Perspectives on a Persisting Belief," *Proceedings of the Sixth European Conference on Modern South Asian Studies*, Colloques Internationaux du CNRS, Asie du Sud: Traditions et Changements (Paris: CNRS, 1979).

22. W. L. Burn, "Free Trade in Land: An Aspect of the Irish Question;" *Transactions of the Royal Historical Society*, 4th ser., 31 (1949): 61–74.

23. The most recent life of Banks is by Charles Lyte, *Sir Joseph Banks: Eighteenth Century Explorer, Botanist, and Entrepreneur* (London: David & Charles, 1981).

24. Cf. Hunter Dupree, "Nationalism and Science: Sir Joseph Banks and the Wars with France," in *A Festschrift for Frederick B. Artz*, ed. D. H. Pinckney and Theodore Ropp, (Durham, N.C.: Duke University Press, 1964).

25. F. de Langle, *La Tragique Expedition de la Pérouse et de Langle* (Paris: Hachette, 1954), p. 16; cf. C. W. MacFarlane and L. A. Triebel, eds., *French Explorers in Tasmania and the Southern Seas* (Sydney: Australasian Publishing Company, 1937).

26. David S. Evans and others, *Herschel at the Cape: Diaries and Correspondence of Sir John Herschel, 1834–1838* (Austin: University of Texas Press, 1969). European astronomy had, of course, a well-established political pedigree. Matthew Turner, radical lecturer in physics, compared the discovery of Uranus by Herschel's father in 1781 to an intellectual conquest. "It is true we had lost the *terra firma* of the thirteen colonies in America, but we ought to be satisfied with having gained in return by the generalship of Dr. Herschel a *terra incognita* of much greater extent *in nubibus*"; J. T. Rutt, ed., *The Theological and Miscellaneous Works of Joseph Priestley* (London, 1817), pp. i, 76.

27. Lucile Brockway, *Science and Colonial Expansion: The Role of the Royal Botanic Gardens* (New York and London: Academic Press, 1980).

28. Louis Becke and Walter Jeffrey, *The Naval Pioneers of Australia* (London: John Murray, 1899).

29. See G. F. Lamb, *Franklin: Happy Voyager* (London: Ernest Benn, 1956), chaps. 3 and 4.

30. Quoted in Hector Charles Cameron, *Sir Joseph Banks* (London: Batchworth Press, 1952).

31. William Vernon Harcourt, Objects and Plans of the Association, *Report of the First and Second Meetings of the British Association for the Advancement of Science, held at Cambridge in 1831, and at Oxford in 1832* (London: John Murray 1833), p. 28.

32. Kathleen Fitzpatrick, *Sir John Franklin in Tasmania, 1837–1843* (Melbourne: Melbourne University Press, 1849), p. 51.

33. *Minutes of the Philosophical Society of Australasia*, 4 July 1821.

34. On P. D. King, see Michael Hoare, "Science and Scientific Associations in Eastern Australia, 1820–1890" (Ph.D. dissertation, Australian National University, 1974), p. 27 and passim. I am indebted to Hoare's work for several Australian examples used in this essay.

35. Quoted in Ann Mozley Moyal, "Sir Richard Owen and His Influence in Australian Zoological and Palaeontological Science," *Australian Academy of Science* 3 (November 1975): 41–56.

36. F. A. Stafleu, in *Adanson: The Bicentennial of Michel Adanson's "Fawiller des Plantes,"* ed. George H. M. Lawrence (Pittsburgh: Hunt Botanical Library, 1963), vol. I, p. 179, quoted in Basalla, "Spread of Western Science," p. 613.

37. B. Smith, *European Vision and the South Pacific, 1768–1850* (London: Oxford University Press, 1960), p. 6.

38. J. Oxley, quoted in Hoare, "Science and Scientific Associations," p. 8.

39. B. Field, *Geographical Memoirs* (London, 1825), p. viii.

40. Brisbane to Bruce, 28 March 1822, *Brisbane Papers,* National Library of Australia; see also Sir Thomas Brisbane, *Reminiscences of General Sir Thomas Mac-Dougall Brisbane* (Edinburgh: Constable, 1860).

41. Cf. Moyal, "Sir Richard Owen."

42. Quoted in Hoare, "Science and Scientific Associations," p. 37.

43. Matthew Friend, quoted ibid., pp. 72–73.

44. Moyal, "Sir Richard Owen," pp. 45f.

45. Peter Cunningham, *Two Years in New South Wales,* 4 vols (London, 1827), quoted in Hoare, "Science and Scientific Associations," p. 50.

46. Cf. Ronald Strahan, *Rare and Curious Species: An Illustrated History of the Australian Museum, 1827–1929* (Sydney: The Australian Museum, 1979); S. C. Ghosh, "The Utilitarianism of Dalhousie and the Material Improvement of India," *Modern Asian Studies* 12 (1978): 97–110.

47. Sir W. T. Denison, *Varieties of Vice-Regal Life,* 2 vols. (London, 1870), p. 107; see also Stephen G. Foster, *Colonial Improver: Edward Deas Thomson, 1800–1879* (Melbourne: Melbourne University Press, 1978); and Frederick M. Johnston, *Knights and Theodolites: A Saga of Surveyors* (Sydney; Edwards and Shaw, 1962).

48. Quoted in M. Worboys, "The Emergence of Tropical Medicine," in *Perspectives on the Emergence of Scientific Disciplines,* ed. G. Lemaine and others (The Hague: Mouton, 1976), p. 81.

49. In this context, the Royal Gardens at Kew, the India Office and the Colonial Office deserve reappraisal; see Brockway, *Science and Colonial Expansion.* For the intellectual circle of Huxley and Booker, see R. MacLeod, "The X-Club; A Scientific Network in Late-Victorian England," *Notes and Records of the Royal Society* 24 (1970); 305–22.

50. Cf. R. MacLeod, "On the Advancement of Science," in *The Parliament of Science: Essays in Honour of the British Association for the Advancement of Science,* ed. R. MacLeod and P. Collins (London: Science Reviews, 1981).

51. Cf. M. Worboys, "Science and Colonial Imperialism in the Development of the Colonial Empire, 1895–1940" (D. Phil. dissertation, University of Sussex, 1979); V. de Vecchi, "Science and Government in Nineteenth Century Canada" (Ph.D. dissertation, University of Toronto, 1978).

52. On the conflict between John Tyndall, Huxley, and others in the X-Club, see A. S. Eve and C. H. Creasey, *The Life and Work of John Tyndall* (London: Macmillan, 1945).

53. B. Semmel, *Imperialism and Social Reform* (London: George Allen & Unwin, 1960).

54. Cf. R. MacLeod, "Scientific Advice for British India: Imperial Perceptions and Administrative Goals, 1898–1923," *Modern Asian Studies* 9 (1974); 343–84.

55. H. C. Russell, F.R.S., Government Astronomer, *Minutes of the Australasian*

Association for the Advancement of Science, 12 April 1888, MSS 988/1, Mitchell Library, Sydney.

56. Sir James Hector, F.R.S., Presidential Address, Australasian Association for the Advancement of Science, Christchurch, 1891.

57. Winfred Scawen Blunt, *My Diaries* (1909), vol. 2, pp. 287–95, quoted in Ronald Hyem, *Elgin and Churchill at the Colonial Office, 1905–1908: The Watershed of the Empire-Commonwealth* (London: Macmillan, 1968), p. 506. During the 1890s, departments of botany and agriculture were created in the West Indies and in the British possessions of Africa; for the development of the colonial scientific service, see G. B. Masefield, *A History of the Colonial Agricultural Service* (Oxford: Oxford University Press, 1972), and C. J. Jeffries, *The Colonial Empire and Its Civil Service* (Cambridge: Cambridge University Press, 1938).

58. For the British Science Guild, see W. H. G. Armytage, *Sir Richard Gregory: His Life and Work* (London: Macmillan, 1957), chaps. 6–11.

59. Sir William Ramsay, *Essays, Biographical and Chemical* (London: Constable, 1908), p. 239.

60. Quoted in R. V. Kubicek, *The Administration of Imperialism* (Durham, N.C.: Duke University Press, 1972), p. 143.

61. Report of speech by Alfred Deakin, "Science and Empire," *Nature* 76 (9 May 1907): 37.

62. Federal Parliament, *Hansard* 55 (1910): 1266–67.

63. Quoted in Robin Winks, "On Decolonization and Informal Empire," *American Historical Review* 81 (1976): 540; cf. G. Jones, *The Role of Science and Technology in Developing Countries* (Oxford: Oxford University Press, 1971).

64. See A. G. Butler, *The Australian Army Medical Service in the War of 1914–18* (Canberra: Australian War Memorial, 1943) vol. 1, p. vii.

65. R. MacLeod and K. Andrews, "The Origins of the DSIR: Reflections on Ideas and Men," *Public Administration* 48 (1970): 23–48.

66. C. W. Hume, "The British Empire Exhibition," *Nature* 113 (1924): 863.

67. G. Currie and J. Graham, "Growth of Scientific Research in Australia: The Council for Scientific and Industrial Research and the Empire Marketing Board," *Records of the Australian Academy of Science* 1 (November 1968): 25–35.

68. R. MacLeod and K. Andrews, "The Committee of Civil Research: Scientific Advice for Economic Development, 1925–1930," *Minerva* 7 (1969): 680–705; A. G. Church, *East Africa: A New Dominion* (London: Wetherby, 1927).

69. For alternative models, see Careless, "Frontiers, Metropolitanism, and Canadian History."

The Limits of Scientific Condominium: Geophysics in Western Samoa, 1914–1940

LEWIS PYENSON

Throughout the twentieth century, pure sciences have figured in the foreign policies of imperialist powers. Research installations assume special importance as metropolitan councilors seek to legitimate their control over distant land that can be defended only with great difficulty. Prior discovery of new domains notwithstanding, the presence of scientist emissaries has come to reinforce territorial claims. To function as political legitimizers, of course, men of learning must appear to be engaged in abstract and disinterested research. Lending support to economic exploiters—for example, by combatting tropical disease or surveying a railway—can compromise a scientist who claims to be acting in the interests of all countries. Although not all students of tropical medicine and not all railway engineers in settings beyond the North Atlantic world see themselves as agents of imperialism, it can hardly be denied that much of their activity is related to making peripheral territory secure for metropolitan investment.

Among the pure sciences, geophysics has been especially well suited for supplying researchers to staff outposts of empire. Because the discipline of geophysics covered all terrestrial phenomena, a prima facie case existed for geophysicists to want to erect observatories in distant lands. Furthermore, unlike their geologist colleagues, geophysicists have not until recently sought to discover mineral wealth; unlike agronomers, they have not tried to tamper with local forms of life or introduce new crops. At least before about 1940, in branches of geophysics such as seismology and terrestrial magnetism data were collected to advance global theories. Geophysicists directed their attention toward observing, instead of transforming, the environment.[1]

The German geophysical observatory at Apia, Western Samoa, was one of the premier institutions enlisted in the service of cultural imperialism. By 1914 it had achieved distinction as the finest research center of its kind in the Southern Hemisphere and one of the most advanced beyond the North Atlantic world. As interesting as the observatory's rise to prominence is its survival, under various forms, after the fall of Western Samoa to a New Zealand expeditionary force. With the conquest of Apia, the government at Wellington found itself in possession of a scientific institute far in advance of anything it had previously known. During the war and into the 1920s, New Zealand and its allies financed and defended German-directed geophysics in Samoa. After the departure of the German staff in 1921, New Zealand and the rest of the English-speaking world tried, and failed, to maintain the standard set by German scientists. By 1940, the former geophysical institute had for all practical purposes ceased attending to original research.[2]

In the following pages, I shall trace the course of events surrounding the Apia observatory under New Zealand stewardship. I shall consider the tacit, wartime condominium between Allied administrators and the German director, the friendly postwar relations among New Zealand, American, and German researchers, and the long-term financial commitments made by American foundations and the British admiralty. The evolution of geophysics in Samoa, I shall contend, forces a reevaluation of the usual interpretation concerning a cold war between scientists situated in countries on either side of the European trenches.

The Extramural Professor Arrives

Late in July 1914, Gustav Angenheister arrived in Apia, Western Samoa, with his bride, the daughter of a distinguished professor of chemistry at Göttingen. The thirty-six-year-old physicist had begun a nine-year contract with the Göttingen Scientific Society as director of the Samoan geophysical observatory. According to the terms of the contract, he would spend eighteen months alternately at Apia and at Göttingen. Three years earlier he had been made an honorary professor at the University of Göttingen, and now his salary—commensurate with that of the German director of the naval observatory at Tsingtau on the north China coast—surpassed remuneration accorded many *ordentlichen*, or regular, full professors in Prussia. Angenheister was, in a real sense, the first extramural professor in the history of his university.[3]

Angenheister was coming to the Pacific island chain for the third time. He had first directed the observatory from 1907 to 1909, only several years after completing his doctorate at Berlin and then becoming postdoctoral assistant to Emil Wiechert at Göttingen. In 1907, the observatory was still only a collection of temporary huts overseen by a single, junior scientist. Angenheister supervised a significant expansion of both real estate and research programs. After returning to Göttingen and participating in a geophysical expedition to Iceland, he took up a second tour of duty at Apia in 1911. By then he commanded a postdoctoral assistant and a trained machinist, and his observatory had become an important research center. Angenheister's special interest lay in terrestrial magnetism, but he carefully maintained the scientific work of greatest value, seismology. Data on Pacific earthquakes taken and reduced at Apia proved essential to Wiechert's seismology group at Göttingen as it pioneered a new conception of the interior of the earth.[4] Angenheister succeeded in keeping the seismographs and chronometers in working order and in supervising vast networks for collecting data on tides and meteorological phenomena. Angenheister was indeed an old Samoa hand.

On the long trip to Samoa, one that generally involved passing through Australia even if use was made of the recently completed Panama Canal, Angenheister would no doubt have talked with his wife about their destination. He would have described a European outpost at the back of a natural harbor, a settlement not much different from the Apia detailed by Somerset Maugham in his story "The Pool." The main street ran parallel to the beach for a mile and a half. There was a new dock to facilitate landing and embarkation. Frame structures at the center of town included a court house, a Roman Catholic cathedral, government offices, a public school, a general store, a printer, commercial offices, and a few hotels. A kilometer or so away was the hospital, a number of white buildings valued at well over £50,000 (M 1 million) and endowed by a German colonist. Just outside Apia was Papaseea, the lovely waterfall and pool described by Maugham. Some kilometers inland, on a hill over-looking Apia, lay Robert Louis Stevenson's grave, located near his estate, Vailima, which had become the residence of Governor Erich Schultz-Ewerth. Nearby, too, were a number of missionary outposts and schools, the largest of which was run by the London Missionary Society. As this amalgam of cultures suggests, Western Samoa was unique in modern times in having a bilingual English-German government; the local newspaper used both languages.[5]

The origins for these circumstances are found in European economic and political designs on the Samoan archipelago. Volcanic in origin, the principal islands of Savaii, Upolu, and Tutuila are graced with unusually rich soil, plentiful rainfall, and serviceable harbors. By the middle of the nineteenth century, Samoa came to be seen by the imperialist powers as a desirable acquisition, for great profits were to be won by creating plantations for cocoa, coffee, and copra in the hospitable and accessible terrain. Only the last item—dried coconut meat—was native to Samoa, and around 1900 Europeans and North Americans had developed an enormous craving for it. Copra made fortunes for the exporting companies. The farmers who saw to its cultivation dreamed, too, of retiring in splendor. By the 1880s a handful of settlers, principally German and British, had set down roots, and others followed with the usual secondary institutions providing civilized amenities and retailing civilized religion.[6]

Despite many generations of Western involvement, the islands served no European masters until around 1900. Foremost in fore-stalling imperialist domination were the native Samoans themselves, a proud people who spoke the most ancient of the Polynesian tongues. During the nineteenth century, the Samoan islands were ruled by a monarch and his council. No different from their policies in Africa, India, or China, Western countries sought to persuade the Samoan king to sell the freedom and patrimony of his people. When the shrewd monarch resisted, his rivals were solicited. Into the Samoan "entanglement" rushed the foreign offices of Germany, Great Britain, and the United States, the three powers whose nationals had by 1880 substantial financial stakes in the place. Each eager to steal a march on its rivals, the imperialists tried to strengthen their positions by supporting various factions during a civil war. Toward the principal Samoan harbor, that of Apia on the island of Upolu, steamed a small armada of British, American, and German warships. It seemed that in 1889 interimperialist rivalries would precipitate an armed conflict at the crossroads of the Pacific Ocean. How often the outcome of military offensives has been determined by unforeseen weather! While the crews of the warships surveyed each other across the harbor, they were engulfed by a sudden and savage typhoon. All ships save one British cruiser were cast aground and put out of action.

Immediately following the disastrous Samoan confrontation, the three powers agreed, by a treaty signed in Berlin with the Samoan king, to guarantee the autonomy of the islands. To this end, they were to provide a chief justice and an administrator for the district around Apia. The so-called condominium lasted for nine years, until 1898, when the chief justice provoked civil war once more by his decision to appoint a successor to the throne that had just been vacated by the king's death. The situation became so complex that, by 1900, the great powers had canceled the Berlin treaty, had participated in the civil war, and following a conference in Washington had divided up the islands: the United States received everything east of and including Tutuila, which contained its naval base at Pago Pago; Germany received the rest, compensating Great Britain for withdrawing from Samoa by making concessions elsewhere. The king of Sweden arbitrated the whole matter.

During the fourteen years following the partition of the islands—the period when Angenheister had first come to know the place—Samoan wealth funneled into Germany through a Hamburg trading firm, the Deutsche Handels- und Platagens-Gesellschaft der Süd-See Inseln (D.H. & P.G.). To work the land held by the Hamburg company and by independent planters came thousands of indentured Chinese laborers, whom the Germans found to be more easily controlled than were native Samoans and other Micronesians. On the eve of the First World War, the Chinese outnumbered Westerners by more than four to one. Despite various dissatisfactions radiated by a rump of independent planters, the state-sanctioned export mo-

nopoly guaranteed commercial viability. It turned Samoa into one
of Germany's few profitable colonies.

From his previous service, Angenheister knew that the Samoan
geophysical observatory, whose research programs in exact sciences
produced little of economic consequence, had emerged out of a German policy of cultural imperialism. Following their desire to legitimate German claims on the archipelago, the imperial colonial office
and the Prussian Kultusministerium funneled hundreds of thousands of marks into the institution. The observatory, indeed, came
to be known in Berlin ministerial circles as an institute. The government justified lavishing such attention on pure learning half a world
away by claiming to uphold the German name abroad. Scientists at
Göttingen sensed the drift of government policy. They apprised the
government of overtures made by Louis Agricola Bauer, at the Carnegie Institution of Washington, to take over the observatory in the
event that German sources reduced their financial support. Such arguments sat well in the council chambers at Berlin. The future of
geophysics at Apia seemed secure when in 1911 Wilhelm Solf, for a
decade the governor of Samoa and a firm friend of the Göttingen
scientists, rose to assume charge of the colonial office. Solf, who in
the 1920s paved the way for a German-Japanese *rapprochement*,
would have been the person least surprised by what happened at the
observatory after 1914.

A German Observatory in British-Occupied Samoa

When the First World War began, New Zealand and Australia decided to sweep through German possessions in the Pacific. To this
end, New Zealand raised an expeditionary force of clerks, laborers,
and schoolboys and sent them off in a convoy of Australian and British vessels; the presence at Tsingtau of the two fast battle cruisers
Gneisenau and *Scharnhörst* required such prudence. Joined by the
French cruiser *Montcalm* in New Caledonia, the expedition proceeded to Apia in six days. When on 29 August 1914 the task force
appeared off Samoa, the Germans at first thought that it consisted
of their own vessels and began to prepare a welcoming celebration.
Their plans were interrupted when one of the British ships signaled
to the Apia radio station to cease broadcasting. Presented with a

demand for capitulation, Governor Schultz's secretary replied that the Germans would not surrender, but that they would offer no resistance to a landing. While more than a thousand New Zealand troops were being ferried to the beach, Schultz paid off his officials, native and European, in hard currency. At the same time, the government radio operator destroyed some equipment and, soldiers later claimed, rigged a booby trap for the rest. The soldiers said that they defused his device and captured the damaged plant of steel and concrete with its great transmission mast. For their troubles Schultz and his minions, together with their families, were immediately taken to New Zealand, where they passed the war in internment.[7]

Angenheister had realized as early as 1912 that, in the event of war, Samoa would fall to the British, and the observatory would be seized as the property of a hostile government. Within two weeks of the declaration of hostilities, he made plans to secure the survival of his institution. He obtained a letter from Governor Schultz, clarifying that the observatory was the property of the private Göttingen Scientific Society, that the observers were not paid by any branch of the imperial German government, and that according to articles 46 and 53 of the Hague convention, the observatory could not be seized as if it had comprised enemy assets.[8] When the New Zealand expeditionary force landed, ten soldiers occupied the observatory, located on a strategic peninsula at one end of the harbor. The New Zealand military administrator, Lieutenant-Colonel Robert Logan, soon visited the observatory and promised that the observatory would be left alone if its work were carried out as it had been before the war.

For the first two years of occupation Angenheister managed relatively well. At the outbreak of the war the new seismographic house of concrete and wood was only half finished. Angenheister completed it. In October 1914 he began to construct a new cooking house, and in 1915 he undertook to build a new bathhouse, with water closet and sewer, and to add two verandahs of concrete and wood to the main residence. He had been able to buy enough wood for the western verandah when the New Zealanders liquidated the assets of the D.H. & P.G., including its lumberyard, at a fraction of the property's real value. Angenheister added a hut for atmospheric electricity as well as a meteorological tower. By the end of 1915 all the buildings had been made secure against mosquitoes. These improvements cost M 20,000. Because of scarcity of materials and labor, some of the construction had to be left unfinished and so exposed to the ravages of tropical weather.[9]

One of Angenheister's greatest problems turned around obtaining competent help for his many research tasks. His postdoctoral assistant, Ludwig Geiger, a Swiss citizen, departed for home in 1915 because his contract had expired with the preceding year; he traveled on private funds. Josef Passi and Ignatz Auva'a, native Samoans, had been working as assistants at the observatory at the time of Angenheister's arrival; Auva'a left in October 1914. Until 1916 Angenheister was hoping for a quick end to the war, and for this reason he initially marked much data for reduction in Göttingen. When his hopes faded, he proceeded to organize an office at Apia. He hired former customs officer Max Christoph as an assistant. Angenheister's wife worked as a calculatrix until early 1917, when Auva'a reentered the service of the observatory. When Christoph died suddenly of cerebral apoplexy in March 1917, in his place came Ernst Demandt, a former engineer who before the war had been assistant to the government biologist. Auva'a returned as a mechanic, joined in succession by two Chinese, Ah-Nau and Lo-Chin, and for a brief time a half-blooded black, Mingo. Local talent in the form of other Samoans and Chinese cleaned and tended the observatory grounds. Angenheister's supporting staff remained better than he had any right to expect under the circumstances.[10]

What was the fiscal balance sheet of a German geophysicist in wartime British territory? After May 1915 only English currency was legal tender in Samoa, with a conversion rate of 1 shilling for 1.03 marks. Angenheister negotiated in pounds, shillings, and pence, but kept books in marks and pfennigs.[11] During the course of four years, expenditures came to nearly M 100,000 or £5000. Angenheister calculated them in the following way.[12]

Director	M 38,465.92
Assistants	9,920.96
Mechanics	8,380.90
Orderlies	5,894.40
Computers	6,788.56
Travel	2,000.00
Construction	14,580.42
Furnishings	1,424.56

Instruments	612.92
Library	1,077.38
Matériel	5,184.84
Postage	708.88
Other	1,053.92
Total	M 96,093.66

Consider typical expenditures early in 1917. Angenheister paid himself half a normal salary at M 554 a month. His first assistant, Christoph, received M 268 a month. The two junior Samoan assistants received M 103 and M 41 respectively; the two Chinese assistants received M 176 and M 37 and, because they were former indentured workers, free medical care.[13]

Where did Angenheister's money originate? He maintained three separate accounts. One was for income, one for petty cash, and one for the M 300 a year earned from coconut palms on the observatory grounds. Just before New Zealand troops landed, Governor Schultz cleaned out his government coffers. Angenheister received M 23,000. In August 1915 Angenheister received an interest-free loan of M 20,000 from Karl Hanssen, director of the D.H. & P.G. Then, in December 1916 and March 1917 the New Zealand administration liquidated Hanssen's firm. Angenheister was cut off later that year. "This is to confirm my verbal advice to you that as many of the officials of the German Government have not kept their agreements with the above Company," the liquidator Mathison wrote to Angenheister, "I have decided to discontinue making advances to any person other than officer[s] of the Company."[14] From September 1917 through 1918 Angenheister no longer had even the D.H & P.G. orts to finance his operation. Instead, he managed on private loans from Germans in Samoa, totaling nearly M 30,000. This method of financing continued well into the postwar period. The generous compatriots included Dr. O. Thieme, Father W. Müller, and especially Karl Hanssen. On 16 October 1919 Angenheister was forced to sell the piano left in Apia by prewar observatory mechanic Paul Liebrecht, then back in Göttingen. It netted £30.[15]

According to Angenheister's books, at the outbreak of the war around a quarter of the money that he subsequently used was present in the government treasury. Three quarters came later from loans, either through the D.H. & P.G. and its British liquidator or from private sources in Samoa. To June 1918 the balance sheet recorded:[16]

Income from the government treasury	M 23,000.00
Income from the D.H. & P.G.	45,846.08
Other loans	29,870.00
Interest on loans	659.20
Other funds	14,545.78
Total	M 113,921.06

The personal loans—to be repaid by the Göttingen Scientific Society after the war—amounted to a small fortune. Under usual circumstances German colonists would not have shown such a pronounced interest in pure learning. The New Zealand occupation, however, imposed regulations that made it financially profitable to invest in the observatory. Private, German assets were frozen, and the threat of liquidation hung over all German businesses and plantations. When the D.H. & P.G. went into liquidation, the New Zealand administration sold its assets at only a fraction of their true value. In lending money to the observatory, Germans were confident that their capital would be guaranteed at face value by a German institution. That after 1915 the occupying army allowed such indentures to be negotiated is a measure of the value that it placed on German learning.

Directing a wartime observatory, Angenheister had to put up with inventory shrinkage and shortages. Two telescopes, a Heyde passage instrument with two lenses and a small Hildebrandt theodolite, disappeared in 1915.[17] Just before the war the observatory's magnetic theodolite had gone with astronomer Bruno Meyermann from Göttingen to Tsingtau and from there was supposed to have been delivered to Samoa by the East-Asiatic cruiser squadron. It never arrived.[18] Obtaining regular supplies of photographic paper was a serious matter, for emulsions did not keep long in the tropics. During the early years of the war, and then again between the time of the armistice and his repatriation, Angenheister received photographic supplies and other material from American geophysicist Louis Bauer in Washington.[19]

Until 1918 Angenheister was able to correspond with a number of colleagues in neutral and allied-controlled lands. He sent Bauer a short report in German on magnetic observations made at Apia during the solar eclipse of August 1914, and Bauer published it in English in his journal devoted to terrestrial magnetism. In May 1915 Angenheister and W. C. Parkinson, Pacific observer with the Department of Terrestrial Magnetism, spent several days together standardizing their magnetic instruments.[20] Mail from the United States seems not to have been obstructed. Angenheister corresponded with the Johns Hopkins University Press, for example, over a subscription to the journal *Terrestrial Magnetism.*[21] Angenheister wrote to Dutch geophysicist Willem van Bemmelen on Java in 1915 to report an error in the times that Van Bemmelen attributed to registration of magnetic storms of 1912. Van Bemmelen replied that there was indeed the error that Angenheister computed.[22] As part of the same project, Angenheister borrowed magnetograms from the Department of Terrestrial Magnetism in Washington.[23] Angenheister corresponded, as well, with his former patron Karl Hanssen, who had been transported to New Zealand. At his internment camp, Hanssen and several of his fellow prisoners had, so far as their "layman's knowledge" permitted, taken up the study of astrophysics and celestial mechanics using books provided by Angenheister.[24]

During the war life probably remained easier for Angenheister at Apia than for his colleagues at Göttingen. His observatory comprised two residences for Europeans; a Samoan house; a workshop and Chinese quarters; an office, library, and seismograph building; a hut for the magnetic instruments; a hut for the instruments that tested and calibrated the magnetic recorders; a transit hut for astronomical observations; a meteorological instrument hut; an atmospheric electricity hut; and a laboratory building. The complement of instruments at Apia was equally impressive: an astronomical mean-time clock with a Riefler pendulum and mercury compensator; chronometers; a spectrograph with photographic camera; terrestrial-magnetism and atmospheric-electricity instruments; the Wiechert horizontal and vertical seismographs; and balloons, kites, and meteorological recording devices.[25] Before it was curtailed, his recompense turned out to be competitive with that given full professors in Germany.[26] The observatory, its grounds, and the instruments were worth well over £8000 (M 160,000), and its operating budget was twice that of the physics institute at the University of Berlin.[27] Angenheister pursued big science.

Angenheister's activity at first scandalized the Australians. In December 1916 the prime minister of Australia was shocked to find from a newspaper report that Angenheister had been carrying on "scientific work without interference." The New Zealand administrator at Apia confirmed that Angenheister was indeed proceeding unimpeded, adding that he was forwarding his reports to George Hogben, the Wellington seismologist who was also secretary of the seismological committee of the Australasian Association for the Advancement of Science.[28] Angenheister never hesitated about continuing his work under the nose of and in tacit collaboration with the enemy. He wrote in 1921: "Different scientists of foreign countries interested in our work suggested to me during the war to avoid any friction and to do all in my power to continue the work."[29] In a letter that he sent to Wiechert in 1919, Angenheister outlined his wartime activity. He had followed Bauer's request that Apia observe the solar eclipses of 1914, 1918, and 1919. In 1916 he forwarded all his magnetic observations to Bauer. He had continued to publish meteorological reports in the newspaper run by the New Zealanders, the *Samoa Times*, and he provided the Apia pilot station with accurate astronomical time and with timetables for the tides. He had undertaken expansion of the observatory's physical plant, bringing it up to ten buildings.[30] In another letter to Wiechert, Angenheister argued that the "scientific character of the institute" always remained foremost in his mind. He had acted "above all to facilitate the unbroken continuity of its tasks. Only an unbroken continuity has value and increases the importance of work that has been in preparation for so long, work that has just recently begun to bear its best fruit."[31] Angenheister's activity had to be seen as an adventure of the intellect, he wrote to Wiechert: "Our work serves few practical aims, but quite entirely pure scientific research." When foreign savants recognized his work he would feel rewarded: "One must expect understanding to come earliest from scientific corporations of an international stamp. . . . This cannot be emphasized often enough."[32]

Angenheister received strikingly liberal treatment from the occupying New Zealand army. Yet it would be false to think that he had been willing to sacrifice his country for the pursuit of science and for the greater glory of his observatory. He in fact walked a narrow line. As a professor he remained a spokesman for the German nationals in Samoa. Early in the occupation he participated in a delegation to the New Zealand authorities, asking without success that German dependents be sent to a neutral port. When the occupying army requisitioned the German casino in Apia, both Angenheister and Geiger protested vigorously to the military governor.[33] The impression is given that his first protests were those of one gentleman to another. Angenheister's cordial agreement with the New Zealand army ended in April 1916, the time when the authorities confiscated his revolver and cartridges.[34] At the end of the war Angenheister was by no means above suspicion. As a result of an unknown infraction he ran afoul of the authorities in August 1918 and was placed under house arrest.[35]

Barbarians Loose in a Place of Learning:
The German View

Back in Göttingen at the beginning of the war, Wiechert was no doubt alarmed over Angenheister's future, for he knew that members of the German eclipse expedition of 1914 had been interned in the Crimea and that scientists in recently captured Tsingtau had been sent off to camps in Japan. By March 1915 he learned from Geiger, who had returned home to Switzerland, that the Angenheisters were well. As a consequence of receiving this information, he tried mightily to restore contact with Apia. First he thought to send Angenheister money by way of Angenheister's mother-in-law, wife of Göttingen chemist Gustav Tammann. Then, late in 1915, he wrote to the foreign office in Berlin, asking that Angenheister be sent M 3,000 by way of the United States ambassador, who had agreed to act as intermediary. Wiechert emphasized that in the transaction the ambassador was to write to *Doctor*, not *Professor*, Angenheister and refer to the Göttingen Scientific Society without its royal prefix.[36] Despite these precautions, the transfer of funds never occurred. Wiechert received a few reports on Angenheister from people who had left Samoa. The wife of a German doctor returned to Germany in 1916 and indicated that all was well; in December 1917 the observatory's prewar observer, Franz Defregger, wrote to Wiechert from a detention camp in Australia.[37] About other activity Wiechert remained ignorant until well after the war.

By 1919 peace had long since come, but Angenheister was not yet permitted by the New Zealand authorities to write to Wiechert. Throughout most of 1919 he labored to reestablish communication with Göttingen. He wrote to Alfred G. Mayor of the Marine Biological Laboratory at Princeton, which was financed by the Carnegie Institution of Washington. Mayor had originally planned to visit Apia, but an influenza quarantine forced him to travel to Fiji instead. The American promised to get in touch with Louis Bauer and Arthur L. Day, director of the geophysical laboratory of the Carnegie Institution and son-in-law of one of the Kohlrausches.[38] Desperate to establish himself before scientific colleagues, Angenheister then wrote to Day in English. He described research on the propagation of seismic waves through the earth: "Since I had the opportunity to see the change of physical behaviour under high pressure in the laboratory of my father-in-law G. Tammann in Göttingen [,] I extended the seismological working programme of the Samoa Observatory in these lines." Angenheister observed earthquake tides in the earth's crust, and he determined the depth of the crust from the length of the period of the main seismic waves. According to his calculations, the crust was about fifty kilometers thick. If the elasticity of materials under high pressures, as measured in the laboratory were used, it would be possible to calculate the density of different layers inside the earth. Angenheister concluded that the crust had a higher density under the oceans than under the continents.[39]

Beginning in 1919 Angenheister tried to communicate with Göttingen through the peace office at Berne. In May 1919 he spoke with the Samoan administrator, who told him that the governor general of New Zealand had telegraphed about his financial situation.[40] Angenheister badly needed to reestablish contact with his staff of trained assistants at Göttingen, who could work up the data that he had accumulated. His instruments were wandering, he wrote to Wiechert in September 1919, and they needed to be calibrated and repaired. The roofs leaked. He appended a long list of necessary supplies.[41]

The Samoan administrator forwarded this letter of Angenheister's to the Department of External Affairs in Wellington with the recommendation that Göttingen be informed immediately. Cited was a communication from Reginald Aldworth Daly, a Harvard geologist who had just visited Apia. Daly had urged that Göttingen be approached about financing and maintaining the observatory and assuming its skyrocketing debt. In Daly's view, Louis Bauer at the Carnegie Institution of Washington "might help in financing this quite unique and highly important observatory."[42] Angenheister also wrote directly to Bauer, describing how he had proceeded during the war by borrowing funds. Ever a friend of German science, Bauer relayed the letters to Göttingen and also asked Angenheister for a list of his indebtedness.[43] Angenheister finally succeeded in wiring Wiechert through Swedish physical chemist Svante Arrhenius, in English: "Observatory annual expenses 28400 all borrowed cable instructions wanted from Wiechert." Göttingen received the cable early in October 1919 and soon thereafter Angenheister's list of indebtedness.[44]

Even before he had heard from Samoa, Wiechert figured out what would surely happen, and he confided in Wilhelm Solf. Wiechert knew that Germany would lose control of Samoa. The unanalyzed data that had been accumulating since 1913, however, were another matter. They had to be seen as German property. Angenheister was to bring them home. If he could not, then Wiechert was prepared to send as a replacement a Dr. Almstadt, an observer in Brunswick who had been prepared for service in Samoa.[45] Wiechert, together with his colleague applied mathematican Carl Runge, wrote to the colonial office, relaying information from Ada Osbahr, who had just returned from Apia, and from the D.H. & P.G. In Osbahr's words, Angenheister had enjoyed "no support from the enemy during the whole period." His activity was "a beautiful consequence of German energy and German learning." The D.H. & P.G. reported that their liquidated property was used to pay Angenheister and that during the last two years of the war he had borrowed M 50 a month from the company for living expenses. In their letter, Runge and Wiechert urged that the colonial office forward a large sum to Angenheister for his current expenses and another for expenses incurred during 1915–18. They added that the Carnegie Institution might be interested in running the observatory, and that, if so, Angenheister could be placed in the employ of this American organization.[46] But the colonial office, or what was left of it, expressed no interest in the matter. The observatory fell under the jurisdiction of the new Prussian ministry for learning, art, and public instruction, and this ministry would support the observatory only if the institution remained German property.[47]

Wiechert, who had so easily used nationalist rhetoric to aid his projects before the war, persisted in constructing old arguments for a newly unresponsive bureaucracy. To one government office he expressed the view that the institution had to remain German: "For a more valuable contribution to the future of the German name, it is evident that the observatory [must] remain in our possession as a fulcrum of German character and German science."[48] Wiechert emphasized to the Kultusministerium how this "outpost of German character" had to be supported, especially in view of Angenheister's heroic efforts. "Under the most difficult circumstances, under the enervating sun of the tropics, surrounded by hateful enemies, Professor Angenheister has persisted at his post and kept the observatory in full operation."[49] Runge telegraphed to request an audience in Berlin.[50] According to a ministry memorandum, Runge and Hugo Andres Krüss of the Kultusministerium met early in October 1919 and discussed Angenheister's case in detail. Wiechert and his colleagues corresponded with the administration of the University of Göttingen in an effort to pry money out of state or federal governments.[51]

At this very time Angenheister had on his own initiated correspondence with A. W. Donald, the Swiss consul in Auckland, through the mediation of a certain S. Helg. Donald was willing to transfer funds to Angenheister. Helg quoted Donald: "In the event that [Angenheister] and the family have no[t] been made Prisoner of war or otherwise [are] in distress . . . the British Government have stated that there is no objection to such allowance being made."[52] The German government refused to authorize funds for an institution that was governed by the New Zealand parliament, but Wiechert cabled to Angenheister that credit could indeed be drawn from the Swiss consul, Donald. Wiechert urged extreme prudence in the matter. By January 1921 the unfavorable exchange rate of nine marks for one Swiss franc was rapidly deteriorating, and the cost to the German government of even minor loans negotiated through Auckland would have been astronomical. Angenheister was forced to continue operating the observatory by assuming its debts personally.[53] This was the only way that he could pay his Samoan and Chinese assistants and thus maintain an unbroken series of observations.[54]

In February 1920 Angenheister finally received written permission from Sir James Allen, the minister of external affairs at Wellington, to communicate freely with Göttingen on observatory matters. It was a welcome change, for according to Wiechert's records only about two thirds of Angenheister's communications had until that time found their way to Germany either directly or by way of Washington, Stockholm, and Berne. In the first letter of his new wave of missives, Angenheister described how, as an outcome of the visit of New Zealand astronomer Charles Edward Adams to Apia, New Zealand had decided to assume control of the observatory by instituting a new governing board, one that would assume the observatory's deficit.[55] Angenheister then began to receive regular shipments of periodicals and supplies from Göttingen. Just as lines of communication were reestablished, the situation threatened to disintegrate. In May 1920, Angenheister reported that several months hence the Samoan Germans would be sent home on a steamer, whether they wanted to go or not. He was uncertain whether he would be on board, for his successor had not appeared, and it would take half a year to train him. The New Zealand government had decided to apply the value of the observatory against war-incurred reparations. It would not pay for the observatory in cash.[56]

Angenheister and his family decided not to abandon the observatory. They remained in Samoa when most of the German settlers shipped out. The tropical geophysicist felt "personally answerable" for the future of the observatory, he wrote to Wiechert in September 1920, even though he did not know how far Germany would go to support his activity.[57] Angenheister's was a wise decision. He realized from a letter of Wiechert's that, however isolated, he was better off in Samoa than in Germany. Wiechert had emphasized that, upon his return, Angenheister would have trouble finding a university chair or its equivalent, especially because the state of Prussia wanted to convert university positions into jobs at mass-oriented *Volks-Hochschulen*.[58]

Confusion about the Goals of Pure Science:
The New Zealand View

New Zealand, as did other countries, developed a special vision of the interaction between knowledge, education, and practical activity. By war's end the national university, consisting of colleges at Dunedin, Christchurch, Wellington, and Auckland, had produced

Nobel laureate Ernest Rutherford and had employed a future president of the Massachusetts Institute of Technology, Richard Cockburn Maclaurin. Although academic tradition and circumstances permitted occasional forays into original research, such as Rutherford's wireless experiments and Maclaurin's masterful mathematical treatise on light, systematic experiment continued to be little valued in the new dominion. Pure learning was regarded with suspicion by the New Zealand educational authorities.[59]

The desideratum of receiving a sound commercial return on investments dominated intellectual activity in New Zealand. One philanthropist, Thomas Cawthron, for a time considered giving £30,000 for a solar observatory in the provincial town of Nelson. He died before the money was signed over, and the bulk of his estate, some £250,000, went to found an agricultural research institute. Writing to a former assistant in 1917, Wellington astronomer Charles Edward Adams lamented: "It seems that the Cawthron Observatory will not be formed, the Trustees are busy with other sciences, those that are alleged to give a so-called direct 'bread and butter' return, and Astronomy seems doomed, because it is too 'academic' and has no practical applications!!!"[60] Around 1920 Yale University offered to equip and staff a large photographic refractor on the South Island if New Zealand would agree to cover certain minimal costs. A local newspaper saw great value in the plan, "if only as a corrective to the idea, which must now be pretty general in the outside world, that this country's only interest lies in keeping up the prices of wool and mutton."[61] Yale's overtures shared the same fate as Cawthron's plans; because of New Zealand parsimony the observatory fell stillborn. It is typical of the situation that around 1920 by far the largest telescope in New Zealand was a nine-and-a-half-inch-objective instrument manned by J. T. Ward, an enthusiastic amateur and Fellow of the Royal Astronomical Society, in the small town of Wanganui.[62] The climate against scientific research in New Zealand would not have been widely known in Germany. For this reason, Angenheister and his colleagues hardly suspected the complex maneuvering that had taken place in New Zealand and throughout the English-speaking world over continuing the Samoan observatory and its program of pure research.

During the years after the armistice, geophysics at Apia frequently occupied the attention of leading statesmen at Wellington and London. The observatory first received notice in August 1919, when the New Zealand minister of defense and finance, Sir James Allen, asked John Rushworth, Viscount Jellicoe of Scapa and admiral of the British fleet, to look into the matter, for Jellicoe would soon visit Samoa.[63] Allen indicated that ownership of the observatory remained obscure: Angenheister looked to Göttingen for authority, and a title to the university appeared "to have been rushed through by the German Administration on the eve of the war," but Allen suspected that the title could be contested. Allen cited a telegram from Angenheister to Wiechert, where Angenheister calculated that to maintain the observatory would cost £1,500, or 28,400 prewar marks, annually. The Samoan administration had already urged Allen to make contact with Göttingen in the hope that the Germans would continue to finance the observatory. Allen refused, arguing that "it does not appear to be good policy to have a German in control of an observatory in Samoa." He felt, moreover, that £1,500 a year was exorbitant.[64] To assist Jellicoe in surveying Angenheister's operation, Allen sent him C. E. Adams.

Adams was the government's natural choice for reporting on the Samoan observatory. In 1911 he had become the first official government astronomer of New Zealand, although to bring his annual salary up to a reasonable figure of £480, for more than a decade afterward he was obliged to carry on as chief computer in the Lands and Survey Department.[65] Adams was descended from an old New Zealand family, his father having been a government surveyor on the South Island and a scientific assistant at Wellington. Early in the 1890s, as a student at Christchurch, Adams became a founding member, with Ernest Rutherford, of the short-lived Canterbury College Science Society. Before the society Adams read papers on electromagnetism and electrical engineering, to the evident pleasure of fellow student Rutherford.[66] Having been subsequently trained at Cambridge in tidal theory, Adams returned home to a government career in mathematical computation and astronomy. He commanded the £1,500, one-story, brick Hector Observatory, erected high above Wellington harbor in 1907.

In his report to Jellicoe, Adams gave high marks to Angenheister, "a highly trained and capable scientific officer, who, under exceptional circumstances, has maintained the observatory working at its full capacity." Adams noted that Angenheister suspected that he

would be asked to leave, and in this case "he has generously offered
to train any officer sent to take charge of the observatory." Angenheister also asked, through Adams, for permission to take with him
the observatory records. The Carnegie Institution would be a fine
choice to become patron of the observatory, Angenheister offered,
but Adams felt that "the work of the observatory is so valuable and
important to the British Empire that the control of it should be British."[67] He saw that the Apia observatory was more impressive than
his own institution. Lord Jellicoe agreed with Adams's evaluation.
Writing to the Earl of Liverpool, then governor-general and com-
mander-in-chief of New Zealand, Jellicoe pointed out why the observatory was valuable to the navy. Its meteorological observatory
would be useful in forecasting hurricanes and in guiding aircraft; its
tidal observations would help in navigation and in construction of
harbor facilities; its seismological observatory could help indicate
safe routes for laying submarine telegraphic cables. Jellicoe had
praise for Angenheister. If it were not possible to replace him with a
British subject "of sufficient distinction to undertake this important
work . . . then the alternative might be considered of Dr. Angenheister, who, I understand, is a real scientific enthusiast of special at-
tainments, being asked to continue the work in peace time, a British
Assistant being provided for him." Jellicoe, who had commanded
the British fleet when it engaged the Germans at Jutland, did not
hesitate for a moment in supporting a German scientist at Apia. He
strongly urged that both Canberra and London contribute to the cost
of the observatory.[68]

Adams supported Jellicoe's recommendations with a detailed
memorandum to Sir James Allen, having in the meantime become
minister of external affairs. The astronomer insisted that because of
the location and past achievements of the Samoan observatory, "No
other existing observatory could take its place." Adams cited a draft
letter from Angenheister to Wiechert, in which Angenheister explained how until 1917 the observatory's operating expenses had
been covered by loans from the D.H. & P.G. in liquidation, and afterward, when this source was cut off, from private credit. Through July
1919 Angenheister, as the delegate of the Göttingen Society, owed
£7,000, and the amount was growing. "It is obvious," Adams insisted, "that Dr. Angenheister ought to be relieved of this responsibility, but it is difficult for me to say by whom; possibly these liabil-

ities might be liquidated out of the War Indemnity New Zealand is to receive; in any case this matter seems to require a legal opinion." Adams wanted to bring Angenheister's observatory under his own jurisdiction. He optimistically seconded Jellicoe's proposal that other countries in the British empire be asked to fund it.[69]

One of the possibilities that Adams raised, and rejected, consisted in turning over the observatory to Bauer's Department of Terrestrial Magnetism. Bauer had in 1919 equipped a geomagnetic observatory at Watheroo, Western Australia, under his long-time observer, New Zealander Edward Kidson.[70] Bauer would obviously have had an interest in assuming operations at Apia. His position having been intimated in correspondence with Reginald Daly late in the fall of 1919, Bauer wrote directly to Clinton Coleridge Farr, professor of physics at Christchurch. The American relayed that he had received several letters from Angenheister and that he was upset that New Zealand was not moving to cover Angenheister's £7,000 debt.[71] Bauer also lamented to Adams about New Zealand's reticence to provide support to the Samoan operation:

It would seem a very great pity, and a great loss to science, indeed, if the Government of New Zealand does not make some provision for the continuation of the Samoa Observatory. If your Government does not make such provision, I fear that the Germans can justly contend that science is not so favoured and advanced under the administration of British countries as under German auspices.

Accordingly, in behalf of the Section of Terrestrial Magnetism and Electricity of the International Geodetic and Geophysical Union, permit me to urge that every possible representation be made to the New Zealand Government that the necessary funds and personnel be provided to assure the continuation of the Samoa Observatory, at least as well as had been the case under the German Administration.[72]

In support of Bauer's arguments, Farr wrote to the New Zealand Institute about how "the World has become indebted to [Angenheister] to the extent of about £8000." Farr continued, "It is a debt of honour to Dr. Angenheister and for the good name of New Zealand which has accepted the [Samoan] Mandate, [it] must be promptly paid."[73]

As astronomer at the Hector Observatory, Adams was paid by the

Department of Internal Affairs. Armed with letters from Bauer, he assailed the Department of External Affairs on behalf of the Apia observatory. Adams led a deputation of New Zealand scientists to the Ministry of Defense. As part of the remarks of the envoys, Thomas Hill Easterfield, recipient in 1894 of a Ph.D. in chemistry from the University of Würzburg and in 1919 appointed director of the Cawthron Institute at Nelson, emphasized that if New Zealand did not carry on the work of the Germans, "the enemy [could] say we are an exceedingly unscientific people." At any rate, Angenheister "should be paid his out of pocket expenses, particularly since this work had been carried out with the authorization of the military." Ernest Marsden, professor of physics at Victoria University College, noted "that it was pure love of science that led the Germans to put the observatory in Samoa." Marsden then stated what had been suspected for some time: "What [the German scientific mission] . . . proposed to do in Samoa was infinitely more than what we are doing in New Zealand." James Allen seemed friendly enough.[74] The prime minister informed the governor general about the government's commitment to maintenance of the observatory and its decision to order from Göttingen the supplies that Angenheister required.[75] In private, however, Allen was more circumspect. To his secretary he wrote that Angenheister could not be allowed to ask for money from Germany, that Angenheister in any case could not be retained permanently, and that if the courts allowed it the observatory would wind up under New Zealand's control as a prize of war.[76] Honor seems to have been a rare commodity among elected representatives at Wellington.

Throughout the early part of 1920 Angenheister corresponded with New Zealand scientists about his research. "Hardly any one, except a real expert in magnetic observatory work," he wrote in English to Farr, "may realize the very high amount of care and experience necessary for a trustworthy computation of base lines for horizontal intensity and especially vertical intensity on a tropical station." It was difficult to distinguish thermal variations from magnetic and to take seasonal changes into account. The magnetic balance constantly fluctuated because of the friction of steel points on agate-stone plates. For these reasons there were hardly any reliable magnetic measurements in the Southern Hemisphere. More southern measurements were necessary to verify, among other theories, that of Arthur Schuster, who claimed that the forces causing diurnal variation possessed a potential.[77]

Angenheister's elaborate arguments had produced no results, the German told Marsden in June. Furthermore, even though the foreign minister, Sir James Allen, guaranteed noninterference in the observatory's mail, Angenheister had not received two of Wiechert's letters, and two registered letters of his own had not arrived in Göttingen. Angenheister thereupon resigned.[78] From having considered Angenheister's wartime activity and his correspondence with Wiechert, we know that his resignation was a gesture of despair. He was uncertain where to go, for he realized that geophysics had a clouded future in Germany. He asked Bauer if any positions were available in the United States.[79] Bauer replied that it would be best if Angenheister stayed on at Apia as long as possible. The North American job market was impossible: Willem van Bemmelen had recently been unsuccessful in trying to find a post in the United States upon his retirement from Batavia.[80]

After he learned of Angenheister's resolve, Marsden was supportive: "I am very disturbed at the circumstances which have caused you to resign from your position although I fully sympathize with your motives and feel that your objections are fully justified."[81] Fearing that Angenheister might take along most of the data that he had collected, Adams wrote to Allen's secretary, J. D. Gray, urging the government to require that Angenheister hand over his observations and calculations.[82] Adams's concern stemmed from Angenheister's practice of sending notes and short articles to Göttingen by way of either America or the Netherlands. Six communications of his were read before the Göttingen Scientific Society in 1920 and subsequently appeared in the Society's *Nachrichten.* The longest and most significant of Angenheister's papers concerned the structure of the crust of the earth under the Pacific Ocean and continental land masses, as revealed by analyses of earthquakes throughout the preceding fifteen years.[83] To set Adams at rest Angenheister agreed to publish some of his results in a New Zealand journal.[84]

From Christchurch, Coleridge Farr offered Angenheister "what little influence" he had to promote the Samoan observatory. Farr also urged Angenheister not to resign, for the records of the observatory had "at the eleventh hour" to remain unbroken.[85] While encouraging Angenheister, Farr revealed his worst fears to English physicist Arthur Schuster:

The whole financial burden of the Observatory is a little too large for N.Z. (a country of unscientific people—mostly farmers) to bear alone, and we are hoping to get the work done there financed by contributions from Canada, Australia and England as well as New Zealand. If each contributed £100 p.a. it would mean that New Zealand's contribution per capita would be about five times that of Australia and forty-five times that of England, and it would moreover put the Observatory on a very permanent and satisfactory footing. If such a plan fails we shall have to try and persuade the Government of New Zealand to undertake the lot, but I am very doubtful whether we shall succeed in such a proposal.[86]

Farr wrote to Bauer in September 1920 about hopes of having the observatory funded by New Zealand, Australia, Canada, and England. An announcement had gone out in New Zealand for an assistant, at £400 plus fringe benefits. A director would be chosen by a committee in England consisting of Charles Chree, Arthur Schuster, Gilbert W. Walker, and Ernest Rutherford.[87]

Central authority vested in England for a scientific institution administered by the federal government at Wellington is, one thinks today, a shameful abrogation of power. Germany would not have tolerated an analogous relationship with the United States; Australia, India, and Canada refused to contribute to financing the Samoan observatory because the institution would be in the hands of New Zealanders.[88] Yet, in 1920, New Zealand was a minor nation. It had allowed an armed enemy alien to direct an observatory at Apia, adjacent to thousands of New Zealand troops. Wellington actively solicited foreign entrepreneurs to run the Samoan geophysical operation. New Zealanders felt themselves, at any rate, bound to the seat of empire in London, and in exchange for filial obedience they expected a financial allowance. They were not disappointed. London was willing to provide £800, and the Department of External Affairs in Wellington subsequently voted £5,000 to keep the observatory running on a temporary basis.[89]

In November 1920, the minister of external affairs, E. P. Lee, sent Ernest Marsden and C. E. Adams's assistant Charles J. Westland, accompanied by Marsden's wife and young daughter, to relieve Angenheister of his charge. Marsden received free passage and about £180 for his trouble.[90] As New Zealand's brightest young scientific prospect, the Wellington physicist was glad to do the government's bidding. He had studied under Ernest Rutherford at Manchester and in 1911 had collaborated with Hans Geiger on the scattering of alpha particles from a thin gold foil; Marsden and his German collaborator verified Rutherford's hypothesis of a positively charged atomic nucleus. In 1915 Rutherford persuaded him to accept the chair of physics at Victoria University College in Wellington, whither he emigrated. Between 1916 and 1919 Marsden served in Europe. Undoubtedly because of his close, prewar friendship with Geiger, he harbored no ill feelings toward German scientists.[91] He and Angenheister were compatible researchers, and they got along well together.

Once on the scene, Marsden wrote to the Samoan administration to propose an increase in Angenheister's salary until his departure late in June 1921, first-class transport to Germany, ten weeks' severance pay, and, finally, assumption of Angenheister's debts. In December 1920 Angenheister received from the Samoan administration a six-month appointment as director of the observatory. His salary would be £750 a year; his first-class passage would be paid to Germany; his debts would be taken over. He had only to train his successor, Westland—according to Marsden "a very nice quiet reserved man"—and prepare his wartime observations for publication. The latter task would net him £500.[92]

Even before Marsden's visit with Angenheister, Secretary Gray had been pessimistic about the long-term future of Samoan science. He did not anticipate, however, that during the first half of 1921 his government would completely reverse its position on the operation. With a departure date fixed for July 1921, Angenheister learned that New Zealand had decided not to repay his loans in cash, even though the English government was in favor of such a transaction.[93] In March 1921 Prime Minister William Ferguson Massey telegraphed Sir James Allen, by then his high commissioner in London, that the "Government of New Zealand have decided to close" the observatory.[94] The cost was too great for New Zealand to bear. Given "the ridicule which will fall to the lot of New Zealand in the event of the permanent closing," Thomas Easterfield of the New Zealand

Institute urged the minister of external affairs to support a less ambitious observatory scheme.[95] Marsden was nevertheless ordered to
send Angenheister out on the next available steamer and pack up
the instruments. The observatory began to close down. Equipment
that in some cases had stood for nearly twenty years was studiously
dismantled. Wellington received books, records, and many pieces of
apparatus, including the astronomical clocks and a spectograph. To
the Christchurch observatory went Angenheister's instruments for
measuring atmospheric electricity. To the Apia harbor master went
some meteorological and hydrographical instruments; to the Crown
went the contents of the well-equipped workshop and darkroom, a
boat, and a bicycle. The seismographs and magnetic recorders were
boxed awaiting possible future activation.[96]

   Angenheister was the last to know that these events would take
place. "I can scarcely write because of the dreadful news," Marsden
informed him late in April 1921. "Our Prime Minister is a farmer &
because sheep now sell at 1/6 each he thinks the world has come
to an end." The New Zealand "Cabinet had a sudden attack of financial nervousness and began to curtail every possible expenditure.
Unfortunately at this particular moment the final arrangements for
Samoa were put before them for ratification. They decided to close
the observatory forthwith." Marsden had persuaded the prime minister, however, to keep the observatory going on a reduced budget.
Only seismological and magnetic measurements would be made,
and Westland would take charge. There was a bright side: "We are
at present negotiating with Bauer to take over for two years & then
NZ will continue."[97]

   A deux ex machina descended to reverse the government's verdict.
Louis Agricola Bauer was pulling its strings. He telegraphed his support for a smaller observatory limited to seismology, terrestrial magnetism, and related meteorological measurements. Prime Minister
Massey, at Marsden's request, authorized £800 for each of the next
two years, and he felt sure that the British admiralty would renew
its offer of £800 a year.[98] Bauer's yacht Carnegie arrived in Apia during June 1921. It carried Harry M. W. Edmonds and his assistant
Donald G. Coleman. Seasoned in carrying out magnetic surveys of
Asia and the Pacific Islands, Edmonds stationed himself at Apia
while Coleman traveled to nearby islands. Both were under the

nominal direction of Westland, chief of what was left of the observatory.[99]

With Angenheister and his family receding into the background, the New Zealand government flatly refused to honor the German geophysicist's debts. By this time Angenheister had no illusions about the promises made by the former enemy, but he never slackened in his attempt to have New Zealand assume the cost of his loans. In May 1921 Angenheister queried Tate again on the debt question.[100] His final response, received just before he steamed away from Samoa, was clipped and uncompromising: "As the property has been taken over in accordance with the terms of the Treaty of Peace, the German Nationals should have no trouble in obtaining repayment from the German Government."[101] This reversal of policy was greeted with shock by the administrator of Western Samoa. He telegraphed Gray:

Samoan Observatory[:] Do you realize that repudiation of maintenance debts breach of faith with Angenheister under arrangements made by Marsden 14th December which Angenheister accepted and thereby made complete contract maintenance debts not mortgages[?] Angenheister has already partly performed his part of contract stop All lenders resident Germany. Is not settlement possible with German Government through Clearing House against reparation[?] Angenheister leaves 2nd July.

Gray replied four days later:

Samoan Observatory[:] Minister will treat observatory exactly same as other German properties accounting for value through clearing office showing all encumbrances whether mortgages or debts[,] and leaving German Government to repay its nationals whether owners mortgages or creditors of properties retained under Article No. 297 Peace Treaty.[102]

Angenheister took up pen and wrote a final, long, and eloquent letter to the minister of external affairs in Wellington. He tempered his indignation at bureaucratic deception with the right amount of humility. Angenheister insisted that his research and observations were *wanted as well by the N.Z. as the Samoan Government, and that the results of the observations are delivered for use to the benefit of the Samoan and the N.Z. Government and N.Z. scientists. . . . It seems therefore that not only a moral obligation for the N. Zealand and Samoan Government exists to pay for work, they wanted to be done. . . . I hope I have honestly done my part; but I fear that [the international scientific community] will be as much disappointed as I must be, if the affaires of the Observatory should not be settled in fair and broad minded manner and if in addition to the hard work I have done continually during 7 years in tropics, I should be personally responsible for payment of debts incurred in continuation of scientific work done in other people's interest.*[103]

On his way home to Germany, Angenheister stopped by Washington, and he urged Bauer to intercede with New Zealand to recover his debts. Bauer obliged.[104] The American assured Angenheister that the New Zealand government would do the right thing about the observatory debts. One wonders what he meant.[105]

Attempting to Carry the Torch

The future of the observatory emerged more clearly in the wake of Angenheister's departure. In the words of conservative Prime Minister Massey, "We have made up our minds to keep the observatory going."[106] Henceforth the observatory was to have an annual subsidy of £800 from the British admiralty and a resident geophysicist, Harry Edmonds, paid by Bauer's Department of Terrestrial Magnetism. Edmonds handled magnetic and electrical observations, his specialty. Westland became director at £700 a year, and the New Zealand government provided around £800 for annual expenses. Governing the observatory was an honorary board of advice, consisting of New Zealand's most distinguished physical scientists and several administrators, including Farr, Adams, the distinguished Wellington geometer Duncan McLaren Young Sommerville, and Marsden as chairman. Supreme authority came from the New Zealand Department of External Affairs, then in control of captured German territory.[107]

By virtue of his long-standing interest in the observatory, Bauer emerged as a foreign authority regularly consulted by the governing board. In 1922, returning from the Rome meeting of the international union of geophysics, Bauer stopped by Wellington. There he met with the board and aired his thoughts on a research strategy. Bauer supported Westland's inclinations to reject all undertakings except those that pertained to the extraction of data from natural phenomena in the immediate vicinity of Samoa. In his annual report of 1921, Westland had justified his decision not to record distant earth tremors: "I do not, however, consider these distant shocks to be the province of this observatory, because there are plenty of observatories on the continents not far from these epicenters to deal with them."[108] Among his remarks before the board, Bauer elaborated dozens of empirical tasks for the observatory, none of which was directly related to ongoing theoretical work.[109] In a world of increasing mathematical and theoretical specialization, Bauer was a Baconian fact-gatherer. He fathered a worldwide network of stations that recorded information on terrestrial magnetism and atmospheric electricity, spending millions of dollars to assemble data that finally served little purpose. Theoretical inclinations were absent among those who sat on the board of advice, and Bauer's directives seemed to them to indicate precisely the way that the observatory should function. No one realized, or cared, that the observatory had been reduced from Wiechert's laboratory for testing theoretical seismology and theoretical astrophysics to a mere source of data for the most part irrelevant to broader concerns in earth sciences.

The observatory was ill equipped to carry out time-consuming geophysical routines. Edmonds fell sick and was replaced by Andrew Thomson, a man described by his employer, Bauer, as skilled principally in measuring atmospheric electricity.[110] Thomson, a Canadian from Ontario, had in 1916 received an M.A. in physics from the University of Toronto. The following year he spent as a Townsend Scholar at Harvard. Although he served honorably in the United States Army during the First World War, he had retained a British passport.[111] Thomson and Westland were each in command of a portion of the observatory's resources, and each continually trod on the other's toes. To make matters worse, neither was enthusiastic about taking and reporting meteorological readings, the sole matter of concern to the Royal Navy, which underwrote a large part of the observatory's costs. Westland was not having an easy time of it: "I feel pretty well run down and cheap in the evenings," he confided to his patron Adams, "Samoa has a tireing [sic] climate."[112]

The situation deteriorated so quickly that Marsden went to Apia in October 1922. He laid the blame squarely on Westland, whom he characterized to Secretary Gray in the Department of External Affairs as a "queer 'bird'" and "a ghastly failure socially." Westland, in Marsden's estimation, "had worked very hard indeed on magnetic work but had neglected meteorology & wouldnt even let Thomson do it although our agreement with Carnegie Inst. was that they would do meteorological work. The result is that although Carnegie people have spent much money we have not had the benefit of it as it has been devoted to special purposes."[113] By the end of his tour, Marsden wanted Westland removed. He "has proved himself not fit to be in charge from a technical point of view. I am of opinion that he be deposed by Thomson & offered assistantship at a lower salary. He will probably refuse, if so we should advertise for assistant from NZ at 450–500 net & send him over to work under Thomson."[114]

It is not often that a scientist allows himself to be demoted for incompetence, but with no prospects of a better position in New Zealand, Westland agreed to become Thomson's subordinate at a lower salary. His new position was that of seismologist. Thomson advanced to acting director, even though he was an officer of the Carnegie Institution.[115] Bauer then moved to place Thomson on a New Zealand payroll; the Samoa investment was becoming too great for him to justify.[116] He wrote to Gray pointing out that Thomson's salary was not commensurate with his new tasks. He intimated that he would favor Thomson's resignation from the Department of Terrestrial Magnetism to accept the position of director at Apia. Gray, of course, resisted Bauer's plan and countered that Thomson had "stated quite definitely that he does not desire to sever his connection with the Carnegie Institution, but prefers to remain as your representative."[117] Everything was solved amicably. The Carnegie Institution continued to finance research in atmospheric electricity; Thomson left Bauer's payroll and became permanent director at Apia in 1924.[118]

Thomson's research interests wandered far from seismology, the area in which the observatory had previously made an important contribution. He undertook projects such as measuring solar radiation and studying the effect of the moon on barometric pressure.[119] Citing a central achievement of the thirty years of meteorological records at Apia, he pointed with childlike wonder to the peculiar observation of Sir Gilbert Walker "that whenever the Nile flood is above average the temperature at Samoa . . . for the following six months will be below the average, and vice versa."[120] For a time Thomson had the means to indulge his scientific fantasies. There was a staff physicist, paid at the rate of a third of his own salary.[121] Thomson's budget ran to £2,500, or about 50,000 prewar German marks—approximately the equal of that under Angenheister's regime.

Realizing that it would get nothing of great value from the observatory, the admiralty in 1925 cut its annual stipend to £400.[122] In a final move to rid itself of what had become a weighty albatross, the Department of External Affairs in 1926 unloaded responsibility for the observatory on the Department of Scientific and Industrial Research. Marsden, as newly appointed permanent secretary of that department, had thence forward to finance the observatory as well as run it.[123] He managed on reduced funds from the British admiralty and from Bauer's Department of Terrestrial Magnetism. In 1929 he provided half of Thomson's salary through the budget of the New Zealand meteorological office.[124]

Thomson became discouraged at the neglect of his operation. During the entire decade of the 1920s Wellington spent only around £300 for renovation, essential expenditures when dealing with wooden buildings in the tropics. Real estate crumbled for want of major repairs.[125] Returning after an absence in March 1930, Thomson wrote to Marsden:

At Apia the buildings have fallen into a worse condition than I had anticipated. Dilapidated and run down in appearance, doors falling to pieces, reefs collapsing, the observatory needs a thorough overhaul. My house was in a particular shocking state. Rats had chewed the backs of one cupboard of books. Probably £35 worth of books are damaged beyond hope of rebinding. My gramophone which cost over £35 had been smashed by people who had had a party in my house. My mattress and bedding had been stolen.[126]

Westland and Sanderson were both thin. Sanderson complained that the heat made it impossible to study for his university degree.

Thomson defended the meteorological instruments of the observatory from navy charges of inaccuracy, although he did not dispute that his calculations were more than a year in arrears.[127] Thomson began to look elsewhere for funds. While touring North America in 1927 he interested the United States Navy in his research on wind currents. His "old friend" Herbert Hoover, then secretary of commerce and later president of the United States, conveyed Thomson's research to the United States hydrographical office; the admiral in charge there was "very enthusiastic over what we had already done at Samoa and promised speedy aid."[128] More sanguine about maritime rivalries in the South Pacific, Marsden admonished Gray: "If the United States people care to contribute, let them do so; but we cannot go cap in hand to them officially: Thomson has done it unofficially, and that probably is sufficient."[129] The United States never came through with support, although the Carnegie Institution did continue to provide a grant for terrestrial magnetism.[130]

Bauer's iron-free magnetic survey vessel *Carnegie* docked at Apia for the last time in November 1929. While taking on provisions, it exploded and sank, a few kilometers from the geophysical observatory. The captain and cabin boy perished, and several crewmen were seriously burned.[131] The fate of the *Carnegie* symbolized sinking fortunes at the observatory. In 1930 both Westland and Thomson resolved to leave Samoa for better parts. The two had never gotten along well together, and in parting they remained nasty toward each other.[132] The new director, J. Wadsworth, was even less interested than his predecessors had been in seismology. His budget, along with all others in the government, suffered from the depression. This time Marsden went, cap in hand, to the Rockefeller Foundation, asking for $15,000 over three years to carry the observatory through hard times.[133] Marsden received $10,000 for the years 1931–33.[134]

His luck had, however, run out. Inactive and in failing mental health since 1928, Marsden's old patron Bauer committed suicide in 1932. Bauer's successor, Alexander Fleming, was less sympathetic to the cause of Samoan science. The Rockefeller Foundation would not renew its grant, which had been seen as a temporary measure. The New Zealand government could provide only £280 a year. In the absence of other support, Marsden was compelled to abandon seismology.[135] Money for the observatory arrived in small amounts, each expenditure requiring the special attention of the prime minister. In 1935 the observatory grounds were overgrown and strewn with garbage, leading to a serious mosquito problem; the showers were cited as a menace to public health:

The shower in the Observatory grounds needed by the Samoan staff and by the coolie has been found in such disrepair and of such design as to contravene the requirements of the local Health Ordinance.

. . . Many tins, bottles and an accumulation of coconut husks and undergrowth have been removed. The job is hardly complete yet, but the reduction in mosquito numbers has been very considerable. The work has been done by prison labour and should cost the Observatory nothing.[136]

Notes from Marsden to the prime minister reveal the extreme difficulty with which funding was arranged. Marsden asked the leader of the government to approve an expenditure of £80 for rebuilding the verandah on the director's residence, then Wadsworth's home.[137] Three years later Marsden wrote to the minister of scientific and industrial research, pleading for £161 to carry out repairs on the lavatory and verandah.[138] In February 1938 Marsden asked his minister once more for a new toilet at Apia.[139] In July 1937 the cabinet approved £3,275, £1,850 of which was for meteorological buildings, but the lavatories had to wait until 1940.[140] With a new war the observatory went under the control of the New Zealand air department.[141] The main building, finished on the eve of the First World War and rapidly deteriorating, somehow served in the Second.[142]

The Extramural Professor Returns Home

We leave the raw New Zealand geophysicists with their substantial scientific and financial woes and return to Gustav Angenheister. When in 1921 he arrived in inflation-racked Germany, he had by no means left off dealing with the Apia observatory. By May 1922 the Commission of Crown Estates was still trying to get out of honoring the New Zealand commitment to repay Göttingen and Angenheister. The commission repeated the questionable logic that since the observatory was "not actually taken over by the New Zealand Government until the 1st January, 1920," Angenheister's debts were incurred under the administration of the Göttingen Scientific Society![143] The argument did not convince the representative of the German clearing office in London, A. von Friedberg, who claimed a credit to the German government of £15,000.[144] Time has obscured the exact number of ounces of gold saved from crossing the English Channel as a result of New Zealand's seizure of one of Germany's finest geophysical observatories.

Realizing that his larger claims for reimbursement would never be granted, Angenheister proceeded to work up the data that he had accumulated at Apia. The New Zealand authorities had indicated that they would underwrite his publication costs. Angenheister especially wanted £200 promised for having turned over his observations for the years 1912–15; he proposed to send the manuscript to and receive payment from the high commissioner for New Zealand in London.[145] Before any money changed hands, however, Angenheister's calculations had to be certified properly by a British geophysicist. An authority emerged in Charles Chree, meteorologist at Kew Observatory outside London. Lacking means to pay an assistant and without great desire to work up the results himself, Angenheister decided to proceed in stages; payment for portions of his results would go toward reducing the remainder of the data.

When Angenheister approached Chree, the British meteorologist had only recently become apprised of the arrangement: "I had not heard anything of the agreement between you and the New Zealand Government until a few weeks ago, when I had a letter from Dr. Marsden asking me to report on the first consignment of the work, with a view to certifying that the first payment was due." Chree noticed that the requirements imposed on the results by New Zealand and the program outlined by Angenheister did not rally [sic], and Chree undertook to approach the New Zealand high commission about payment.[146] Angenheister urged that the matter be given priority. He had had to stop work because his funds were exhausted, and to continue he needed partial payment.[147] Eight weeks later Angenheister sent Chree results that he himself had computed. Chree was not satisfied, and proposed alterations.[148] The pattern of partial payments began when, sometime late in 1923 or early 1924, Angenheister was sent £175 for his efforts.

The arrangement proved cumbersome and inefficient. Angenheister had continually to beg for payment, and the New Zealand high commissioner had always to obtain instructions from home before sending the money to Germany.[149] Angenheister complained in English to the high commissioner: "Instead of acknowledgement for the difficult work so carefully done, you have only complaints for me." Without money Angenheister could not finish his task, and soon he would take up a new post as director of geophysics at the Potsdam observatory.[150] To Chree, Angenheister wrote that the high commissioner was "not aware of the amounts of careful work necessary to reduce the records of the magnetic balance. You, of course, know what it means, and that it is better not to hurry the work and to make it as carefully as possible."[151] Chree was sympathetic. He asked the high commissioner to meet Angenheister's views.[152] When in 1926 Angenheister completed as much of his program as he could, he was several years behind schedule. He described how difficult it had been to obtain the data: "I had to overcome during this work many political and economical difficulties in Samoa and here in Germany too."[153] He sent the results to the New Zealand high commissioner, asking for the remaining £75 on the £500 originally authorized him. Angenheister hoped to receive official acknowledgment for his labors![154] One wonders how the New Zealand government could have officially thanked a German national for having contributed to the advancement of science in wartime territory occupied by New Zealand troops, or, indeed, what Angenheister would have done with such thanks had he received them.

At the time that he was writing up his Samoan observations and looking for a better position, Angenheister was saddled with legal claims advanced by those from whom he had borrowed money in Samoa. The widow of Dr. O. Thieme, a half-Samoan woman whose property had not been confiscated by the New Zealanders in 1920, brought a suit against him for M 2,500.[155] Angenheister claimed that he could do nothing about collecting the loans, and he reminded the initiators of the suit that he had been asked to speak at Thieme's funeral.[156] Karl Hanssen also claimed his money from Angenheister. To the D.H. & P.G. Angenheister explained how, under military liquidation, company money that could not leave Samoa had come his way. Writing from Buitenzorg on Java, Hanssen argued that his seized assets had not been used to pay the observatory.[157]

The simplest resolution was the one that the observatory curators chose. As the German inflation of 1922 began its inexorable course, the curators repaid the loans. In February 1922 Paul Liebrecht received M 720 for his piano, sold in 1919 by Angenheister for £30; the repayment was about a quarter of its preinflation value. The D.H. & P.G. was likewise repaid. The company fumed at having been so treated.[158] The affair dragged on for years. Hanssen claimed that the Samoa curators owed him pounds sterling, and as late as 1927 Angenheister was summoned from Potsdam to appear before the Göttingen Scientific Society to testify on the matter.[159]

The Samoan tangle consumed the best part of Angenheister's working life. He published little from his post at Potsdam. As a reward for disciplinary felicity and, it may be suspected, for heroism in the face of the enemy, in 1928 he succeeded Emil Wiechert in the chair of geophysics at the University of Göttingen.

Conclusion

Through the early 1920s, Gustav Angenheister remained in direct and friendly contact with his colleagues in New Zealand, Australia, the United States, and the United Kingdom. Could he have known about his special fate in 1915, as he continued all manner of observation, calculation, correspondence, and construction at his observatory? What thoughts crossed his mind as he sat in his finely appointed office only a few kilometers away from two thousand New Zealand troops, a loaded revolver in his desk drawer? He did sense, after the armistice, how much better he had been able to provide for his wife and young children than if he had remained at Emil Wiechert's side in Göttingen. Angenheister knew, too, how to turn difficult circumstances to his advantage.

Having spent more than three years at Apia before the war, Angenheister realized that the existence of the observatory hung on a policy of German cultural imperialism. After the war, he and his English-speaking colleagues sought to maintain the observatory's prewar functions. It was a conspiracy of kindred spirits. The scientists joined forces to pressure civil servants and politicians into supporting scientific research; for the former, science transcended nationalist sentiment, while for the latter, science served national goals. Although knowledge of weather patterns, magnetic declinations, and like phenomena is vital for military operations, in the minds of the scientist conspirators, geophysics had never been at war. Angenheister, indeed, issued weather reports and carried out tidal surveys for the use of the New Zealand armed forces, and he did so without feeling that he was in any way collaborating with the enemy. Scientists at Wellington, Washington, and London never questioned the veracity of his reports.

Allied authorities had their own programs of cultural imperialism, however, and for them Samoa simply did not merit a major institute of pure research. Only reluctantly were they pressured— were they shamed—into matching the German munificence that had catapulted the Samoan observatory to the top rank of institutions of its kind. Original research stagnated and then ceased as the observatory came under the direction of uninspired geophysicists. Funding dried up as the New Zealand government, the British admiralty, and the Carnegie Institution of Washington sought to divest themselves of a marginal acquisition. In view of the strategic umbrella maintained by the United States naval base at Pago Pago in Eastern Samoa, the Commonwealth sensed little urgency in supporting even a meteorological office at Apia.

Angenheister's experiences in Samoa revise the picture that has come down to us of German researchers ostracized, after 1914 and through much of the 1920s, by their French-speaking and English-speaking colleagues. The Pacific basin, indeed, formed the backdrop for a number of related dramas. The Japanese gladly dismantled and carted off the German Institute of Technology and the observatory at Tsingtau; although the French harassed the German Institute of Technology in Shanghai, however, the federal Chinese government protected it well after China had officially joined with Britain and France. Throughout the war, the Australian government directly financed the work of anthropologist Bronislaw Malinowski, even though he was technically an Austrian subject; but for a rigid attitude on his part, physicist Peter Pringheim, interned while attending the 1914 Melbourne meeting of the British Association for the Advancement of Science, might have gone free. And while New Zealand removed the German national George von Zedlitz from his university chair at Wellington, the philologist remained a free man.[160]

The farther away one looks from Burlington House and the Quai de Conti, then, the weaker seems the intended Allied boycott of German science.

In turning to the international dimensions of science and science policy during the early decades of this century, political commentators have taken tedious pronouncements by one or another aging scientist Tartuffe as an accurate reflection of scientific activity.[161] Such commentators have as a rule overlooked the circumstance that, just as researchers half a world away from the trenches remained collegial during the war, so, beyond the meeting rooms of Berne and The Hague, practicing scientists made their own peace. Administrator whips in the parliament of science could not force all their backbenchers to toe the party line. Gross oversimplifications have been the result of the view that pure science remained outside the international political sphere before the early 1920s. The fine structure of geophysics at Samoa reveals deeper levels of meaning that have escaped writers who, in pronouncing on scientific ideology, have avoided consideration of scientific practice.

This essay is drawn from part of one chapter of *Cultural Imperialism and Exact Sciences: German Expansion Overseas, 1900–1930* (New York: Peter Lang Publishing, 1985) and it is reprinted here with the generous consent of the publisher.

1. An excellent discussion of the rise of nineteenth-century geophysics is provided in Walter Kertz, "Die Entwicklung der Geophysik zur eigenständigen Wissenschaft," Gauss-Gesellschaft (Göttingen), *Mitteilungen* 16 (1979): 41–54.

2. A descriptive account, based entirely on published sources, is given by Gustav Georg Angenheister in "Geschichte des Samoa-Observatoriums von 1902 bis 1921," in *Zur Geschichte der Geophysik*, ed. H. Birett, K. Helbig, W. Kertz, and U. Schmucker (Berlin: Springer Verlag, 1974), pp. 43–66.

3. F. Burmeister, "Angenheister, Gustav Heinrich," *Neue Deutsche Biographie* 1 (1953): 291–92. Angenheister's contract, dated 18 May 1914, is preserved in the file "Samoa Observatorium, Verträge," located in the archives of the Institut für Geophysik, Göttingen (hereafter cited as IG, preceded by file title).

4. Hermann Wagner, "Vorgeschichte und bisherige Entwickelung des Samoa-Observatoriums," in *Ergebnisse der Arbeiten des Samoa-Observatoriums der Königlichen Gesellschaft der Wissenschaften zu Göttingen: I. Das Samoa-Observatorium*, ed. Hermann Wagner (Berlin, 1908), pp. 7–26, published as part of volume 7 of the *Abhandlungen* of the Mathematical-Physical Class of the Göttingen Society. Wiechert's group published most of their work in the *Nachrichten* of the Mathematical-Physical Class of the Göttingen Society. See James Bernard Macelwane, *Introduction to Theoretical Seismology, I: Geodynamics* (New York, 1936).

5. A complete picture is provided in *The Cyclopedia of Samoa, Tonga, Tahiti, and the Cook Islands* (Sydney, 1907).

6. The political and economic domination of Samoa by foreign powers has been chronicled exhaustively. General treatments in English include: Felix M. Keesing,

Modern Samoa (Stanford, 1934); James W. Davidson, *Samoa mo Samoa: The Emergence of the Independent State of Western Samoa* (Melbourne, 1967); Richard P. Gilson, *Samoa 1830 to 1900: The Politics of a Multi-Cultural Community* (Melbourne, 1970). Delightful journalism is found in the treatment by Western Samoa's most famous resident writer, Robert Louis Stevenson, *A Footnote to History: Eight Years of Trouble in Samoa* (New York, 1892). Essays by Stewart P. Firth, John A. Moses, Peter J. Hempenstall, and Paul M. Kennedy grace the volume edited by Moses and Kennedy, *Germany in the Pacific and Far East, 1870–1914* (St. Lucia: University of Queensland Press, 1977). The last volume contains an excellent bibliography. Two useful government handbooks are Government of New Zealand, Department of External Affairs, *Handbook of Western Samoa* (Wellington, 1925), and United States Department of the Navy, *American Samoa: A General Report by the Governor* (Washington, D.C., 1927).

7. "Capture of Samoa," Sydney Morning *Herald*, 22 September 1914; "Prisoners of War," Auckland *Star*, 25 September 1914. The New Zealand detention camps are described in "Unsere Leidengenossen in Neuseeland," *Der Kamp Spiegel: Wochenschrift der deutschen Kriegsgefangenen* (Liverpool, Australia), 17 February 1918. See also Ernest Scott, *Australia during the War* (Sydney, 1936), pp. 119–37. The occupation of Samoa is treated in Edmund Dane, *British Campaigns in Africa and the Pacific, 1914–1918* (London, 1919), p. 196; S. J. Smith, "The Seizure and Occupation of Samoa," in *The War Effort of New Zealand*, ed. H. T. B. Drew (Auckland, 1923), pp. 22–41; L. P. Leary, *New Zealanders in Samoa* (London, 1918); S. J. Smith, *The Samoa (N.Z.) Expeditionary Force, 1914–1915* (Wellington, 1924); Otto Riedel, *Der Kampf um Deutsch-Samoa: Erinnerungen eines Hamburger Kaufmanns* (Berlin, 1938), pp. 228–43; Frieda Zieschank, *Ein Jahrzehnt in Samoa (1906–1916)* (Leipzig, 1918), pp. 117–52. The sources do not agree on details.

8. Governor's Office to Angenheister, 9 August 1914, "Korrespondenz Ang.-Wiechert Obs.-Grundstück," IG.

9. Angenheister, "Kassenbericht (Copie)," 1918, "Abrechn. Bericht. Verträge in Samoa," IG.

10. "Kassenbericht," ibid.

11. Ibid.

12. "Uebersicht über die Ausgaben Juli 1914–Juli 1918," ibid.

13. Original receipts from Samoa, ibid.

14. Receiver of the D.H. & P.G. in Liquidation to Angenheister, 3 September 1917, "Schuldschein d. S. Obs. Korr. Amer. u. Schweizer Consul," IG.

15. J. D. Gray to Allen, 14 April 1920, plus appendixes, DEA, 18/1/Pt. 2, National Archives, Wellington, New Zealand (hereafter cited as NA); "Report on Accounts of Samoa Observatory, by H. C. Tennent," 24 January 1921, Appendix A, DEA, 18/1/Pt. 3, NA.

16. Angenheister, "Uebersicht über die Guthaben & Schulden der Observatoriumskasse in Apia, Samoa," "Abrechn. Bericht. Verträge in Samoa," IG.

17. Angenheister, "Inventar II des Samoa Observatoriums," pp. 66, 190, "Korrespondenz Ang.-Wiechert Obs.-Grundstück," IG.

18. Wiechert to Angenheister, 12 January 1920; Angenheister to Wiechert, 14 March 1920; Compare Geiger to Wiechert, 17 November 1913, "Samoa Observatorium 1913," IG, where Geiger reports Meyermann's plans.

19. Angenheister to Bauer, 26 February 1919 and 1 October 1919, "Correspondenz Marsden, Bauer, Farr, etc.," IG.

20. "Report on Results of Comparisons of the Samoa Observatory Magnetic Instru-

ments with Standards, May, 1915," typescript dated Department of Terrestrial Magnetism, 30 March 1916, [Untitled folder], IG.

21. E. W. Ditters to Angenheister, 27 May 1917, [Original receipts from occupied Samoa], IG.

22. Angenheister to Van Bemmelen, 26 August 1915 (draft); Van Bemmelen to Angenheister, 9 October 1915; [Angenheister's notebook of magnetic recordings, 1912], IG.

23. E. Lester Jones to Angenheister, 21 March 1916, "Correspondenz Marsden, Bauer, Farr, etc.," IG.

24. Hanssen to Angenheister, 2 February 1917 and 27 March 1918, ibid.

25. C. E. Adams to Minister of External Affairs, 11 December 1919, DEA 18/1/Pt. 1, NA.

26. "Samoa. Observatorium. Verträge," IG.

27. J. D. Gray to Sir James Allen, 14 April 1920, plus appendices, DEA, 18/1/Pt. 2, NA, Paul Forman, John L. Heilbron, and Spencer Weart, *Physics circa 1900: Personnel, Funding, and Productivity of the Academic Establishments* (Princeton: Princeton University Press, 1975), Historical Studies in the Physical Sciences, vol. 5, p. 70.

28. Prime Minister of Australia to Acting Prime Minister of New Zealand, 12 December 1916; reply 17 January 1917; DEA, 18/1/Pt. 1, NA. Hogben to Angenheister, 17 April 1916, "Correspondenz Marsden, Bauer, Farr, etc.," IG.

29. Angenheister to New Zealand Minister of External Affairs, 29 June 1921, DEA 18/1/Pt. 3, NA.

30. Angenheister to Wiechert, 21 May 1919, English translation by the New Zealand government, DEA 18/1/Pt. 1, NA.

31. Angenheister to Wiechert, 26 December 1919 (draft), "Administrator," IG.

32. Angenheister to Wiechert, 8 December 1919, "Korrespondenz Ang.-Wiechert. Obs.-Grundstück," IG.

33. "Auszug aus dem Kriegstagbuch des Herrn K. Hanssen, Apia," entries for 30 August 1914, 19 September 1914, 5 October 1914, Reichskolonialamt 2624, pp. 186–218, Zentrales Staatsarchiv Potsdam, Democratic Republic of Germany (hereafter ZStA Potsdam).

34. Receipt for the revolver and cartridges, "Angenheister, Inventar III des Samoa Observatoriums," IG.

35. Provost Marshall to Angenheister, 7 August 1918, "Administrator," IG.

36. Wiechert to Foreign Office Secretary, 16 October 1915; Wiechert to Wilhelm Solf, 27 September 1915; "E. Wiechert, Persön. Akten," IG.

37. Wiechert, "Bericht . . . 1914/15," in Göttingen Gesellschaft der Wissenschaften, *Nachrichten: Geschäftliche Mitteilungen, 1915*, pp. 11–12; "Bericht . . . 1915/16," ibid., *1916*, p. 9; "Bericht . . . 1916/17," ibid., *1917*, p. 9; "Bericht . . . 1917/18," ibid., *1918*, p. 42.

38. A. G. Mayor to Angenheister, 14 September 1919, "Correspondenz Marsden, Bauer, Farr, etc.," IG.

39. Angenheister to A. L. Day, 6 October 1919, ibid.

40. Angenheister, "Geschäftsbericht," 1918/1919, "Abrechn. Bericht. Verträge in Samoa," IG.

41. Angenheister to Wiechert, 8 September 1919. Government English translation, DEA, 18/1/Pt. 1, NA. Compare Minister of Defense to Governor General of New Zealand, 17 November 1919, ibid.

42. "Samoan Observatory," n.d., ibid.

43. Bauer to Secretary of the Göttingen Scientific Society, 2 November 1910; Bauer

to Angenheister, 3 November 1919 and 4 November 1919; "Correspondenz Marsden, Bauer, Farr, etc.," IG.

44. Protocol, Samoa Curators, 8 October 1919, "E. Wiechert, Persön. Akten," IG.

45. Wiechert to Solf, 19 June 1919, Reichskolonialamt 6214, ZStA Potsdam.

46. Runge and Wiechert to Colonial Office, 13 August 1919; D.H. & P.G. to Samoa curators in Göttingen, 5 August 1919 (copy); ibid.

47. Memorandum from the Kultusministerium, 19 September 1919, ibid.

48. Untitled draft, 1919, "E. Wiechert, Persön. Akten," IG.

49. Samoa curators to Kultusministerium, 11 August 1919 (draft), ibid.

50. Runge and Wiechert to Colonial Secretary, 29 September 1919; Runge to Colonial Office, 30 September 1919; Reichskolonialamt 6214, ZStA Potsdam.

51. Undated memorandum; Samoa curators to the curator of the University of Göttingen, 17 November 1919; H. A. Krüss of the Kultusministerium, internal memorandum, 23 September 1920.

52. S. Helg to Angenheister, 30 August 1919 and 23 October 1919, "Schuldschein d. S. Obs. Korr. Amer. u. Schweizer Consul," IG.

53. Wiechert to Angenheister, 13 January 1920 and 16 February 1920, "Korrespondenz Ang.-Wiechert. Obs.-Grundstück," IG.

54. Angenheister to Wiechert, 9 August 1920, ibid.

55. Angenheister to Wiechert, 14 March [1920], ibid.

56. Angenheister to Wiechert, 20 May 1920, ibid.

57. Angenheister to Wiechert, 23 September 1920, ibid.

58. Wiechert to Angenheister, 27 August 1920, ibid.

59. The configuration of science in New Zealand is recorded in Sidney H. Jenkinson, *New Zealanders and Science* (Wellington, 1940); John C. Beaglehole, *Victoria University College* (Wellington, 1949); F. W. Furkert, *Early New Zealand Engineers*, ed. and rev. W. L. Newnham (Wellington, 1953); Peggy Burton, *The New Zealand Geological Survey: 1865–1965* (Wellington, 1965); William P. Morrell, *The University of Otago: A Centennial History* (Dunedin, 1969); J. D. Atkinson, *DSIR's First Fifty Years* (Wellington, 1976); Hugh Parton, *The University of New Zealand* (Auckland and Oxford, 1979); and W. J. Gardner, E. T. Beardsley, T. E. Carter, *A History of the University of Canterbury, 1873–1973* (Christchurch, 1973).

60. Adams to Charles J. Westland, 28 June 1917, Seismological Observatory Archives, Geophysics Division, Department of Scientific and Industrial Research, Wellington, New Zealand (hereafter cited as SO).

61. Lyttleton *Times*, 12 January 1921; C. E. Adams to F. Schlesinger, 19 January 1921, SO; "Studies of the Sky: National Observatory," New Zealand *Times*, 29 January 1921.

62. C. E. Adams to C. J. Westland, 21 December 1916, SO.

63. Reginald H. Bacon, *The Life of John Rushworth Earl Jellicoe* (London, 1936), p. 413.

64. Allen to Jellicoe, 21 August 1919; memorandum, Allen to Cabinet, 18 August 1919; DEA, 18/1/Pt. 1, NA.

65. George A. Eiby, "The New Zealand Government Time-Service: An Informal History," *Southern Stars* 27 (1977): 15–34.

66. Minute Book of the Canterbury College Science Society, entries for 6 August 1892, 16 September 1893, and 12 May 1894, Canterbury University Archives, Christchurch, New Zealand.

67. Adams, "Notes on Samoan Observatory," 14 October 1919, DEA, 18/1/Pt. 1, NA.

68. Jellicoe to Liverpool, 15 October 1919; Jellicoe to Ferguson, 20 October 1919; Jellicoe to Admiralty, 20 October 1919; ibid.

69. Adams to Minister of External Affairs, 11 December 1919, ibid.

70. Isabel M. Kidson, *Edward Kidson* (Christchurch, [1941]); G. R. Wait, "The Work and Equipment of the Watheroo Magnetic Observatory," in *Proceedings of the Pan Pacific Congress*, ed. Gerald Lightfoot, vol. 1 (Melbourne, [1923]), pp. 505–9.

71. Bauer to Farr, 18 November 1919 (copy), DEA, 18/1/Pt. 1, NA.

72. Bauer to Adams, 9 December 1919 (copy), ibid.

73. Farr to New Zealand Institute, 24 January 1920, ibid.

74. "A Deputation to the Hon. Sir James Allen from the New Zealand Institute, Thursday, 12th February, 1920;" Adams to Allen, 11 February 1920; ibid.

75. H.O. 2000/1, February 1920, ibid.

76. Allen to Gray, 13 February 1920, ibid.

77. Angenheister to Farr, 8 April 1920 (copy). "Correspondenz Marsden, Bauer, Farr, etc.," IG.

78. Angenheister to Marsden, 7 June 1920 (copy), ibid.

79. Angenheister to Bauer, 11 August 1920 (copy), ibid.

80. Bauer to Angenheister, 16 September 1920, ibid.

81. Marsden to Angenheister, 25 August 1920, "Administrator," ibid.

82. Adams to Gray, 16 July 1920; Adams to Angenheister, 26 August 1920 (copy); DEA, 18/1/Pt. 2, NA.

83. Gustav Angenheister, "Beobachtungen an pazifischen Beben: Ein Beitrag zum Studium der obersten Erdkruste," Göttingen Gesellschaft der Wissenschaften, Mathematisch-physikalische Klasse, *Nachrichten, 1921*, pp. 113–46.

84. Angenheister to Adams, 5 August 1920, DEA, 18/1/Pt. 2, NA. The publication appeared as G. Angenheister, *A Summary of the Meteorological Observations of the Samoa Observatory (1890–1920)*, ed. E. Marsden and D.M.Y. Sommerville (Wellington, 1924).

85. Farr to Angenheister, 3 June 1920, "Administrator," IG.

86. Farr to Schuster, 23 August 1920 (copy), DEA, 18/1/Pt. 2, NA.

87. Farr to Bauer, 18 September 1920, DEA, 18/1/Pt. 1, N.A.

88. New Zealand High Commissioner in London to New Zealand Prime Minister, 4 September 1920, DEA, 18/1/Pt. 2; H. J. Read to the New Zealand High Commissioner in London, 10 December 1920, DEA, 18/1/Pt. 3, NA.

89. Gray to Lee, 8 December 1920, DEA 18/1/Pt. 2, NA.

90. Gray to Lee, 10 November 1920; Gray to the Administrator of Western Samoa, 29 October 1920; ibid.

91. C. A. Fleming, "Ernest Marsden, 1889–1970," *Biographical Memoirs of the Fellows of the Royal Society* (London), vol. 17 (1971), pp. 463–96.

92. Marsden to Angenheister, n.d., "Administrator"; Angenheister to Wiechert, 20 December 1920, "Correspondenz Ang.-Wiechert Obs.-Grundstück"; Administration of Western Samoa to Angenheister, 14 December 1920 (copy), [Untitled folder], IG.

93. Angenheister to Wiechert, 31 January 1921, "Korrespondenz Ang.-Wiechert Obs.-Grundstück," IG.

94. Allen to Massey, 2 March 1921, DEA 18/1/Pt. 3, NA.

95. T. H. Easterfield to Minister of External Affairs, 9 March 1921, ibid.

96. Marsden to Department of External Affairs, received 31 March 1921, ibid.

97. Marsden to Angenheister, two letters of 20 April 1921, "Correspondenz Marsden, Bauer, Farr, etc.," IG.

98. Marsden to Gray, 1 April 1921; Marsden to Minister of External Affairs, 10 May 1921; DEA, 18/1/Pt. 3, NA.

99. Bauer to Administrative Officer, Apia, 25 May 1921 (Adams' copy); Gray to Bauer, 27 May 1921 (copy); ibid.

100. Angenheister to Tate, 18 May 1921, "Administrator," IG.

101. Secretary of the Government of Western Samoa to Angenheister, 30 June 1921 (copy), ibid.

102. Both telegrams cited in a memorandum from B. J. Smith, Acting Secretary of External Affairs, to the Secretary of the Samoan Administration, 28 June 1921, DEA, 18/1/Pt. 3, NA.

103. Angenheister to Minister of External Affairs, 29 June 1921, DEA 18/1/Pt. 3, NA.

104. Bauer to Gray, 16 November 1921, ibid.

105. Marsden to Bauer, 12 January 1922 (copy), ibid.

106. Extract from "Discussion on Cook Islands Estimates, 29 Sept. 1922," ibid.

107. "Honorary Board of Advice, Minutes," 22 November 1921, DEA 18/1/Pt. 2, N.A.

108. Westland, "Annual Report," 27 November 1921, ibid.

109. Bauer, cited in "A Committee, consisting of Dr. L. A. Bauer . . . on the 4th July, 1922 . . . in Regard to Magnetic Observations," DEA 18/1/Pt. 6, N.A.

110. Bauer, ibid.

111. Biographical file on Thomson, Department of Terrestrial Magnetism, Carnegie Institution of Washington, Washington, D.C.

112. Westland to C. E. Adams, 6 July 1921, SO.

113. Marsden to Gray, n.d. and 20 October 1922, DEA 18/1/Pt. 3, NA.

114. Marsden to Gray, 10 December 1922, ibid.

115. "Board of Advice, Minutes," 19 December 1922, DEA 18/1/Pt. 2, N.A.

116. "Board of Advice, Minutes," 18 May 1923, DEA 18/1/Pt. 6, N.A.

117. Bauer to Gray, 12 February 1923; Gray to Bauer, 22 May 1923 (copy); DEA 18/1/Pt. 3, N.A.

118. "Board of Advice, Agenda Paper," 25 January 1924, DEA 18/1/Pt. 6, N.A.

119. Thomson to Col. J. W. Hutchen of the Samoan Administration, 31 May 1927, DEA 18/1/Pt. 3, N.A.

120. Thomson, quoted in "The Apia Observatory, Described by the Director," *The British Australian and New Zealander*, 1 September 1927, p. 7.

121. "Apia Observatory, Estimated Expenses," 25 November 1924 and 2 December 1926, DEA 18/5, NA.

122. Joseph Gordon Coates (Prime Minister of New Zealand) to Sir C. Ferguson (Governor General), 15 September 1925, ibid.

123. Public Service Commissioner to Gray, 1 September 1926, DEA 18/1/Pt. 4, N.A.

124. "Apia Observatory, Quarterly Report, January–March 1931," DEA 18/1/Pt. 6, N.A.

125. Gray to Secretary of External Affairs, 19 September 1929, SIR 49/15/8, NA.

126. Thomson to Marsden, 7 March 1930, ibid.

127. Thomson to Marsden, 17 March 1930, 2 April 1930, ibid.

128. Thomson to J. D. Gray, 27 July 1927, DEA 18/5, NA.

129. Marsden to Gray, 2 September 1927, ibid.

130. F. R. Callaghan to Minister of the Department of Scientific and Industrial Research, 27 June 1939, SIR 49/15/8, NA.

131. J. Harland Paul, *The Last Cruise of the Carnegie* (Baltimore, 1932), pp. 328–31.

132. Westland to Secretary of the Administration of Western Samoa, 29 September 1930, SIR 49/15/8, NA.

133. "Apia Observatory. Quarterly Report, January-March 1931"; "Council of Scientific and Industrial Research, Minutes . . . Observatory Committee," 19 May 1931; DEA 18/1/Pt. 6, NA.

134. Department of Scientific and Industrial Research, "Memorandum for Members of the Observatory Committee," 21 September 1931, ibid.

135. "Unconfirmed Minutes of Meeting of Observatory Committee," 21 September 1931, ibid.

136. H. F. David to Secretary of the Department of Scientific and Industrial Research, 18 September 1935, DSIR 49/15/8, N.A.

137. Marsden to the Prime Minister, 18 April 1934, ibid.

138. Marsden to Minister of the Department of Scientific and Industrial Research, 16 February 1937, ibid.

139. Marsden to Minister of the Department of Scientific and Industrial Research, 11 February 1938, ibid.

140. Marsden to Minister of the Department of Scientific and Industrial Research, 7 July 1938; H. Bruce Sapsford to Marsden, n.d. (received 7 March 1939); ibid.

141. H. G. Lawrence to Air Department, 14 March 1941, ibid.

142. H. Bruce Sapsford to Director, Meteorological Office, Wellington, 21 November 1940, ibid. The economic decline of Western Samoa is considered in Maria Holtsch, *Die ehemaligen deutschen Südseekolonien im Wandel seit dem Weltkrieg* (dissertation, University of Marburg, 1934).

143. J. Collins to Secretary of the Administration, Apia, 9 May 1922, DEA 18/1/Pt. 3, NA.

144. Friedberg to G. E. Robinson, 20 September 1922, ibid.

145. Angenheister to Marsden, 1 November 1922 (copy); C. Coleridge Farr to Marsden, 28 May 1923 (copy); ibid.

146. Chree to Angenheister, 17 April 1923, "Correspondenz über Verarbeitung, Chree," IG.

147. Angenheister to Chree, 22 April 1923, ibid.

148. Angenheister to Chree, 18 June 1923; Chree to Angenheister, 23 June 1923; ibid.

149. J. K. Campbell (Assistant Secretary to the New Zealand High Commission in London) to Angenheister, 25 September 1924 and 1 September 1924, ibid.

150. Angenheister to High Commissioner, 8 September 1925, ibid.

151. Angenheister to Chree, n.d. [August 1925] (draft), ibid.

152. Angenheister to Chree, n.d. [August 1925] (draft); Chree to Angenheister, 14 September 1925; ibid.

153. Angenheister to Chree, 12 February 1926 (draft), ibid.

154. Angenheister to New Zealand High Commissioner, 12 February 1926 (draft), ibid.

155. Angenheister to Wiechert, 20 December 1920, "Korrespondenz Ang.-Wiechert Obs.-Grundstück"; Notary Dr. jur. Mehring to Angenheister, 17 March 1924, "Schuldschein d. S. Obs. Korr. Amer. u. Schweizer Consul"; ibid.

156. Angenheister to Mehring, 17 January [1924]; "Schuldschein . . ." ibid.

157. Angenheister to D.H. & P.G., 9 November 1921; Hanssen to Angenheister, 16 February 1922; ibid.

158. Angenheister to Liebrecht, 16 February 1922; D.H. & P.G. to Samoa curators, 30 March 1922; ibid. On the inflation, see Gordon A. Craig, *Germany, 1866–1945* (New York: Oxford University Press, 1978), pp. 450 ff.

159. Angenheister to Göttingen University curator, 17 June 1926; Tiersch of the Göttingen Scientific Society to Angenheister, 30 March 1927; "Schuldschein . . ." IG.

160. These points are considered in Pyenson, Cultural Imperialism and Exact Sciences.

161. Simplistic and unconvincing statements are the order of the day in the surveys constituting the third part of Science, Technology, and Society: A Cross-Disciplinary Perspective, ed. Ina Spiegel-Rösing and Derek de Solla Price (London: Sage, 1977); for example, Sanford A. Lakoff: "The tradition of scientists' neutrality and internationalism was rudely shaken, if not totally destroyed, by the outbreak of World War I," (p. 358).

Period and Process in Colonial and National Science

DAVID WADE CHAMBERS

A great deal of scholarly time and energy have been invested in the division of the past into chronological periods. The boundaries of these periods are defined by reference to significant turning points and to emergent and declining social and intellectual forces. Unfortunately, many of these studies lose sight of how artificial, how contrived to special purpose, and how finally arbitrary such chronological periods may be. The conventional triumvirate of ancient, medieval, and modern has been subdivided and recombined into a complex and contradictory array of periods, epochs, ages, and eras, predicated on such factors as political change, intellectual development, geographical boundaries, technological revolution, and institutional transformation—all, of course, seen in ideological and cultural perspective.

Historians of science and technology are in a position to look at the whole question of periodization with a critical eye, since the well-known periods imposed by European political history seem largely irrelevant for many, if not most, of their purposes. Historians of science have made lesser use of periods defined by political considerations than of those defined by intellectual movements, such as the Renaissance, the Scientific Revolution, and the Enlightenment. And even these intellectual periods have been adopted with a degree of reflection and restraint, as when the study of the medieval-to-modern transition gave rise to useful and informative debate. Indeed, it might be said that the discipline first came of age in the critical discussion of the Burckhardtian invention of the Renaissance.

Another milestone in the field, the internalist-externalist split, also has substantial implications for the periodization problem. As noted, in a close look at European science, periods are found that are largely defined by intellectual, or internalist, criteria. Colonial scientific periods, on the other hand, are far more likely to be defined by contextual, or externalist, criteria. Why should this be true? Historical explanations that incorporate an astute balance of internal and external factors might well produce a need for different periodization schemes in both the European center and the colonial periphery. Under the domination of internalist thinking, the belief was widespread that the initial generation of scientific ideas was a process not subject to social conditioning. In a colony, however, ideas were not "generated" but were "imported" or "received" and thus were considered subject to social conditioning. This is truly nonsense. The generation and construction of ideas and the importation and reception of ideas are all processes that occur constantly in all countries.

It is more productive to think of ideas as existing in a process of continuous construction, reconstruction, and transformation, always on the basis of both social and intellectual parameters and variables. The concept of reception often implies wholesale and passive transplantation of certain significant ideas, which then displace, or overcome, local, outdated ones. On the other hand, if reception is regarded as a more complex, interactive process, then study of it will involve the selection and interpretation of foreign ideas, whose subsequent redefinition occurs as these exotic threads are mixed into the native intellectual fabric.

The imposition of European intellectual periods onto colonial settings may have an unfortunate influence on the history that is written. How does the use of the scientific revolution, for example, that most sacrosanct internalist period, affect the analysis of colonial scientific development? Principally in one way: by concentrating attention on the question of how quickly, and in what manner, knowledge and ideas were diffused from center to periphery. While this is a legitimate question, it is not a balanced focus for the writing of, say, Asian or Latin American history. Even in the face of powerful cultural and economic influences from abroad, the focus of colonial history must be the colony itself. The scientific revolution did not take place in Upper Canada, in New England or New Spain, in Manchester, or even—and we often wonder why not—in China. Intellectualist periodization, based as it is on certain European centers, has engendered neglect of important contextual questions and has also made the study of provincial science itself a second-rate enterprise, no matter what approach is taken.

In the colonial setting, for example, the vital question of modernization is directly related to what is often seen as a causal connection between scientific revolution and industrial revolution. Because internalist periodization focused this discussion on the European experience—and on the Whiggish notion of modern science coming into being in one place, at one time, for all time—historians of science have had little to say to third world aspirations to modernize, apart from emphasizing the importance of emulating Europe or, more to the point, the importance of joining the European scientific enterprise, albeit as inferior partners. The implicit message to the non-European world that science advances in linear progression through predictable stages seems to allow no alternative patterns of development. Thus, description becomes prescription.

While the Scientific Revolution, considered as a chronological period, may have only limited value for the colonial historian, it might be supposed that the period of the Enlightenment, on the other hand, by its very nature and essence, would prove more relevant to the colonial setting. Certainly a great many Latin Americanists have found the Enlightenment helpful in their work. In this essay I shall attempt to show that that sort of periodization is often misleading and, in the hands of some authors, actually procrustean. In spite of a small band of Peraltas, Alzates, and Franklins, the colonial New World exhibits European Enlightenment only in traces and in part.

Moving on to the nineteenth century, it has been suggested that the organizational and professional changes in science which occurred in this period are so important that they deserve to be called the Second Scientific Revolution.[1] In my view one scientific revolution is more than enough. Nevertheless, the point is well taken that the traditional periodization framework underemphasized such nineteenth-century developments as professionalization, specialization, and institutionalization, whose significance has been ably demonstrated in recent years by a number of the participants in this conference.

If scholars have made it clear that the institutional revolution of nineteenth-century science is as consequential as the theoretical revolution of seventeenth-century physics, they have also shown that contextual aspects of scientific activity constitute more than a colorful backdrop to scientific change. For one thing, treating the social and institutional framework of science as problematic—and not fixed or merely "given"—enables a more constructively nationalist approach to the problem of scientific development and modernization, especially in those former colonial outposts that now constitute both the second and the third worlds.[2]

In this essay I shall tell a remarkable story of a serious eighteenth-century attempt to establish an institutional base for European science in one of those so-called colonial outposts. I shall then consider three possible frameworks, or periodization schemes, that might be applied: the scholasticism-enlightenment model, the three-phase diffusion-of-science model, and the colonial-national model.

Mexican Science before 1867

The history of Mexico City does not, of course, begin with the arrival of Europeans. At the time of the conquest, which can be arbitrarily dated 1520, the population of Central Mexico was probably as large as that of present-day Australia, around 15 million, and Cook and Borah have estimated it to be as large as 30 million. Within a hundred years the indigenous population had been reduced to a small fraction of that number, but by the time of Alexander von Humboldt's arrival in 1803 the population of the Central region is known to have been around 6 million, at a time when New York, Massachusetts, and Pennsylvania together would have totaled less than 3 million. Mexico City was, as it is today, the largest city in all the Americas, North or South, the great majority of its inhabitants either Indian or Mestizo. Racially, culturally, and intellectually, indigenous culture has had a vast influence on modern Mexico—as of course has European culture. Some perspective on the maturity of

European civilization in Mexico can be gained by recalling that, in the year of Humboldt's visit, 1803–4, the University of Mexico was almost exactly twice as old as the University of Melbourne is today. The University of Mexico was founded nearly a century before Harvard.

For two hundred years in the colony of New Spain science was leisurely, aristocratic and eclectic: the serious concern of a few isolated scholars located at the University, in the *colegios,* and in the monasteries.[3] First indications of the penetration of modern European science appeared in the eighteenth century in these institutions, where science served principally as a subject for rational discourse rather than empirical investigation.[4]

Then, shortly before the end of the eighteenth century and thirty years before political independence was achieved, a period of greatly heightened scientific activity got under way, marked by the formation of a small but lively scientific community and the earliest stages of scientific professionalization. In institutional terms, I am referring to the appearance of the *Gazeta de Literatura,* edited by José Antonio Alzate, in January, 1788, to the founding of the Botanical Garden and its course of lectures in May 1788, and, above all, to the opening of the School of Mines on January 1st, 1792.[5] The focus in this paper will be on the School of Mines.

In curriculum, faculty, and organization, the Mexican School of Mines represented a greater departure from Latin American educational tradition than did l'École Polytechnique from European tradition. But not only that, the Mexican School of Mines, founded in 1792, included most of the pedagogical innovations introduced in Paris two years later.

To be sure, the Mexican reforms were less comprehensive, less influential in history, and probably less consciously revolutionary; yet many of the educational innovations for which l'École is often given credit were in place in Mexico before the Parisian school opened its doors. Furthermore, it appears that principally German and Swedish, not French, technical schools were studied by the Mexican educators.[6] This may seem surprising in the light of Mexico's general intellectual orientation to France.

Without any doubt Alexander von Humboldt was right to suggest in 1804 that Mexico City was a scientific center to compare favorably with Philadelphia or Boston.[7] Not only did the organizational plan of the School of Mines parallel that of the best technical schools in Europe, to my knowledge no other school anywhere in the Americas offered a wider range of scientific subjects: mathematics, including calculus; theoretical chemistry combined with practical laboratory work (quantitative and qualitative analysis); experimental physics, including mechanics, electricity, optics, and astronomy, in addition to hydrostatics and hydrodynamics; and of course mineralogy, paleontology, and geology.[8] The latest scientific discoveries, moreover, were being incorporated into the undergraduate curriculum with very little delay. At the commencement ceremonies of 1797, for example, two ranking seniors addressed themselves to the general principles of chemistry; treating these issues "according to the new theory of Señor Lavoisier, adopted by the principal chemists of the day."[9] The students had learned Lavoisier's principles from his *Traité elementaire* in a Spanish translation made expressly for use at the Mexican School of Mines and published within eight years of the original French edition.[10]

All lectures and textbooks at the mining school were in Spanish, unlike those at the University and *colegios*, where lectures were given in Latin. A course in Spanish grammar was part of the first year of study, and French was also taught, though irregularly. The school charged no tuition fees and maintained an exceptionally large number of scholarship students. At all times the majority of degree candidates were receiving full scholarships. Thus, financial support to students, coupled with the gradual relaxation of caste restrictions on admission, promised that greater use of human manpower resources would be made.

From the time of its founding the mining school was almost entirely secularized. Admission was based on previous training and on competitive examinations, with consideration given to geographic distribution across the country.[11] Teaching and research were brought together, as indeed were theoretical and practical concerns. After 1801 advanced students were used as teaching assistants (*ayudantes de clases*) with the responsibility of assisting professors, both in laboratory exercises and in actual research.

During the early years money was available in some abundance. Mineralogical collections, demonstration equipment, scientific apparatus, and books were being purchased locally and in Europe. I. B. Cohen has noted that the school's instrument collection "greatly resembled that of Harvard in both its scope and the nature of the instruments."[12] Furthermore, great pains were taken to assemble an outstanding faculty. The first professorships at the school were filled by men educated in Europe, but as vacancies developed native talent was put to use.

Thus, the School of Mines—and to a lesser extent the Botanical Garden—financed by government and industry, provided jobs, facilities and tools, professional training, intellectual stimulation, channels of communication to Europe and within Mexico, popular prestige, and the beginnings of professional authority for the fledgling scientific community.

In spite of such achievements and bright prospects, these institutions never fulfilled their potential. In 1811, at the start of the war for independence, they fell into a decline from which they did not begin to recover for fifty years. The extent of this abruptly downward plunge may be demonstrated by comparing the absolute number of teachers, students, and journals of science during the twenty-year period of activity with the same figures taken from the years of decline that accompanied and followed the war of independence.

In 1795 perhaps a dozen men earned their living in scientific occupations of teaching, research, or journalism; thirty years later it is not certain that even one received a regular salary for his scientific work alone. Between 1788 and 1795 had been published the *Gazeta de Literatura*, a journal containing both popular and technical science; yet the *Gazeta* was not equaled in quality until 1832, or in regularity of publication until the 1850s.[13] Furthermore, during the period between 1788 and 1811 well over a hundred persons received extended professional scientific training in Mexico City—a number that includes only graduates of the School of Mines and the Botanical Garden—whereas during the first years of the republic science courses, when offered, were ill attended.

Why did this happen? Why should Mexican scientific progress suddenly cease altogether for several generations after the war for independence? Great cultural advancement is not ordinarily expected in a small, economically exploited poverty-stricken, war-torn country during its first years of independence. Nevertheless, the seemingly complete break in the continuity of scientific institutions established in reasonably auspicious circumstances needs further analysis.

It might at first be supposed that these institutions were simply a premature attempt by Europeans to give Mexico science, that the institutions were a purely European activity imposed on a primitive Mexican landscape. The reality is far more complex. After all, in the founding of the schools, Mexicans were intimately involved in all aspects of institutional planning and finance. Although in the beginning the schools were staffed by European settlers in Mexico, as early as 1805 the majority of the science professors were Mexican-born and Mexican-educated.[14] These young Mexican teachers had been trained at the School of Mines.[15] Furthermore, the numerous attempts recorded in the annals of Mexican legislation to revitalize the scientific schools throughout the nineteenth century indicate that their decay was not indicative of a lack of Mexican concern.[16]

The first and most important reason for this decline was economic. Industrialization and capitalization were almost entirely lacking in the Mexican economy; a number of governments were overthrown because of bankruptcy. Production in the mines was at an all-time low. Mexico was an economically exploited country, held in feudalism by the depredations, first of the Spanish, then of the British, and finally of the United States.

Another explanation for the decline of scientific institutions is the pattern of severe civil disorder and social upheaval that prevailed throughout the first half of the century. War is not necessarily destructive of institutions. In modern times war—and the threat of war—have marched hand in hand with scientific and technological advance. Schools and research centers are sometimes mobilized into feverish activity in support of the military effort.[17] On the other hand, violent class warfare accompanied by sustained economic instability and political chaos may have quite another effect. In Mexico nearly fifty years of class warfare, revolution, corruption, foreign exploitation, and military invasion devastated the country's economic and cultural institutions.[18] Apart from bringing financial ruin, the civil disorders took a certain toll in the lives of men with scientific training. Several of the most promising graduates of the School of Mines were killed in the war with Spain, and scientific personnel were not spared in the decades of civil war that followed.[19] Considering the economic and political circumstances of the first half of the nineteenth century, it is surprising to find scientific establishments surviving at all.

An upturn in science finally came around 1870. By the end of the century Mexico was in the iron grip of the dictator Porfirio Díaz and for a time under the control of a group of men whose allegiance to Compte and Spencer led them to be called "los científicos."

A General Framework for Colonial Science

The foregoing pages provide a simplified chronological account of a brief episode in the scientific history of one country. In recent years, many such historical cases, representing great cultural diversity, have begun to emerge far from the European centers of scientific activity. If these cases, taken together and taken separately, are to be understood, the attempt must be made to formulate a general developmental model or comparative periodization scheme that will accord with accepted theoretical perspectives.

Of course, the division of Mexican history into discrete periods useful to the historian of science and technology is not a simple matter of discerning a factual or necessary pattern of development and giving names to successive stages of that development. The validity of the periods chosen will depend on the sorts of question historians wish to pursue. If, for example, they hope to facilitate cross-cultural comparisons or to explicate the diffusion and reception of European ideas, they should either stay with European periodization or else attempt to find a general model for colonial development. If on the other hand their aim is to see a science that is distinctively Mexican or to understand Mexico on its own terms, then historical analysis might be better served by a strictly contextual periodization scheme. In either case, finding a useful historical framework for colonial science is a problem shared by all those who study Australia, the United States, French and English Canada, or Latin America, and even by those who study colonial influences in the great cultures of Asia.

In sum, valid reasons for constructing a general analytical framework include the following closely related aims: one, to facilitate cross-cultural comparison of the many and varied social and intellectual environments in which "modern science" has come into being; two, to help understand the processes by which institutional accommodation has enabled modern science to prosper or has al-

lowed it to fail; three, to clarify the nature of modern science itself—
is it unitary and invariant or is it multifold and culturally derived;
four, to identify the contribution to human intellectual advancement of particular cultural traditions; five, to understand the process of diffusion of ideas from one cultural setting to another.

In the remainder of this essay I shall explore three possible analytical schemes as they are related to the Mexican case and shall
conclude with some general observations.

The Scholasticism Enlightenment Model

The European periodization scheme most widely applied to Latin
American history—Scholasticism-Enlightenment—was elaborated
primarily to discredit the old belief that the Spanish Empire remained isolated from the Enlightenment in Europe. Evidence was
produced to document a revolutionary intellectual ferment that
reached its high point in the reign of the Spanish King Charles III
(1757–88).[20] The scholarly task of assessing the influence of Enlightenment thought on Latin America was one undertaken by a whole
generation of historians, and their work has enriched our understanding—indeed transformed our perspective—of the intellectual
history of that southern continent. They have convinced us that
even in the far reaches of the Spanish colonial empire, literate individuals who cared to read were aware of the intellectual transfor-
mations occurring in Europe. Indeed, many intellectuals showed
themselves to be well versed in the writings of French philosophers,
and many cultural institutions in the New World were unquestion-
able products of European Enlightenment.[21] Yet it is still open to
question whether the Enlightenment is an appropriate explanatory
device for any period of any country in the Americas.

Those historians who advocate the importance of the Enlightenment in Latin America rest their case heavily on the widespread
dissemination of European books.[22]

*The full force of what is commonly referred to as "the French En-
lightenment" reached a zenith in Mexico at the end of the eighteenth century. Evidence is voluminous. In 1795 no less than three
sets of the* Encyclopedie *were revealed at one time to the Holy Tri-
bunal, all owned by prominent* criollos.[23]

In these accounts much significance is attached to the discovery of private libraries, the size and quality of which would have been admired even in Paris.[24]

Without doubt, the extensive book trade in various Latin America countries belies the solemn reputation of the Index as a great obstacle to the transmission of ideas.[25] The Index was ineffective for several reasons. Often the primary impact of a book preceded its formal prohibition. Frequently, detailed discussions of proscribed writings were readily available. Father Fejoo's popular discussions of Cartesianism, for example, were highly influential. The bureaucratic inefficiency of the Inquisition, moreover, has been well documented by historians.[26] In spite of intensified Holy Office censorship of French writings after the storming of the Bastille, knowledge of French authors was widespread among intellectuals in the capital, as Humboldt pointed out.[27] Although the "liberal" Viceroy Revillagigedo expressed confidence in the careful measures he had taken to prevent the discussion of French political ideas, he made no such attempt to discourage French *scientific* writings.[28]

The majority of foreign scientific works found in the scholarly libraries of New Spain were French and English, sometimes in translation, sometimes in the original editions.[29] Members of the scientific community, with only a few exceptions, could read French. José Antonio Alzate, who in 1771 was named corresponding member of the French Academy of Sciences,[30] demonstrated in his *Gazeta de Literatura* an acquaintance with English and German scientific work, in addition to French. José Mariano Mocino began his study of European science by concentrating on the scientific writings of France and only later went on to those of England. This seems to have been a typical progression in the scientific self-education of those Mexican scholars before the founding of the scientific schools.[31]

A second type of evidence used to demonstrate the penetration of the Enlightenment involves detailed analysis of the lives and the works of certain Latin American thinkers whose cardinal intellectual commitments were European and enlightened. No better examples of this need be cited than the personnel of the new scientific schools who formed a community of scholars with a set of common loyalties quite unlike any previously seen in Mexico or anywhere in Latin America. The community was composed of several dozen men in a population of perhaps 6 million. They were optimistic in the face of occasional hostility and continual financial distress. They were progressive in the most tradition-bound culture of the West. Their intellectual orientation was French in a land where the word *European* was almost synonymous with *Spanish*.[32] Their familiarity with contemporary European thought cannot be doubted.

Finally, one of the strongest arguments for the influence of the Enlightenment is the School of Mines itself: boldly secular, fully committed to the need for experimental investigation, and wholly devoted to the promotion of useful knowledge.

These considerations are convincing evidence that Latin America was, at that time, no replica of medieval Europe; in an attempt to set straight this record, however, some historians have devoted themselves to searching out Enlightenment hues in the richly complex mosaic of Latin American history.[33] In this they have corrected one imbalance, but they have created another. To claim that the "Enlightenment in Spanish America, as in Europe, supplanted scholastic speculation by experimental investigation" is to misconstrue the evidence at hand seriously.[34] The statement that the Enlightenment "was well underway in America a half century before the wars of independence broke out" must be carefully qualified if misunderstanding is to be avoided.[35]

In some of these historical studies, a tone of hyperbole, an almost defensive posture, can be sensed. This tone can perhaps be explained by the writers' awareness that, after all is said and done, the Latin American colonies were not notable exemplars of Enlightenment society. Mexico produced as few liberated philosophers as it did noble savages—which is to say, the Enlightenment was no more true than was the Black Legend. Among the tiny minority of Mexicans who cultivated book learning, many were aware of what was being said in Europe, but only a handful fully identified with the new philosophy. Indeed, the attitudes that characterize Enlightenment thought were scarcely relevant to the political and economic realities that prevailed in Mexico at the time. Finally, these interpretations usually do not take indigenous cultural influences into account at all, a grievous deficiency in any examination of the history of Mexican thought.

It is now generally acknowledged that political aspects of the Enlightenment lagged behind scientific aspects. Analysis of the revolution against Spain demonstrates the weakness, not the strength, of native agents of the Enlightenment.[36] Independence came to Mexico in 1821, only after ten years of civil war and a generation of political intrigue operating in a maze of ill-defined cross-purposes. At this time the liberals were virtually powerless before the forces of reaction—the clergy, the military, and the propertied classes—who were the real mainstay of the revolution against Spain. Frightened of the prospect of social change instigated by the enlightened Spanish regime of 1820, Mexican reactionaries joined with Mexican liberals to bring about the existence of a separate Mexican nation. But Mexico had no bourgeoisie; Mexican liberals had no political or economic base of support and no authoritative leadership until the coming of Benito Juárez thirty-six years after the war of independence.

In vivid contrast to the United States, which after 1776 was much freer of feudal institutions than either Great Britain or France, Mexico in the nineteenth century had military and clerical establishments as firmly entrenched as those of France and centralist and elitist factions as strong as those of England. Following the failure of the liberal thrust to power in 1833, under which major reforms were attempted by Vice President Gomez Farias, liberal sentiment grew steadily throughout the country, with the mestizos developing as an opposition force to the creole aristocrats. Finally after 1840, especially after the war with the United States, the conservative coalition began to weaken, leading up to the Reform of 1857 and eventually to the first noteworthy support of cultural and scientific institutions since the Spanish Viceroys Bucareli and Revillagigedo.

This lack of any political or economic foundation for liberal thought before Juárez has led Latin American historians to emphasize scientific, rather than political, aspects of the Enlightenment, especially with regard to "useful knowledge."[37] It is true that "useful knowledge" was often offered up by politically conservative groups for the purpose of strengthening the viability of the old regime. Indeed, it is precisely because of the many traces of scientific Enlightenment coupled with the absence of enduring commitment to scientific reform that it is finally necessary to conclude that the term *Enlightenment* is not entirely appropriate for this period.

Scientific periodicals were short-lived and of no fixed periodicity.[38] Some Mexicans did possess extensive collections of scientific treatises, but needless to say, the presence of scientific books in a private library does not always indicate scientific sophistication on the part of the owner. In any case, the best of the Mexican scientists would surely have made a more substantial contribution to European science had they lived in an environment that authentically reflected Enlightenment ideals.

The point here is not to belittle the efforts of those hardworking gentlemen who identified progress with the importation of European science, but rather to highlight the difficulties that they faced. These same difficulties also plague any historian who describes this period as "the Enlightenment." Transplanting intellectual periods holus-bolus from one continent to another, and from one century to another, doesn't really work.

The complexities of Latin American culture defy the attempt to apply such fundamentally European concepts as the Enlightenment. The shoe doesn't fit.

Some historians seem to consider the Enlightenment a sort of moral standard against which the intellectual progress of any eighteenth-century country or even a nineteenth-century country can be measured. In my own work the question whether a Mexican Enlightenment took place has not seemed so important as have questions about the ends that such European ideas have been made to serve.

The Diffusion Phase Model

Clearly, individual ideas and sets of ideas do make their way around the world. Are patterns to be found in this "diffusion" process that provide a framework for understanding colonial science? A framework that will expedite the comparative analysis of science in different colonial contexts? Fifteen years ago George Basalla proposed such a model, which he felt was valid for describing the introduction of Western science into nonscientific countries, specifically, "Eastern Europe, North and South America, India, Australia, China, Japan and Africa."[39] While the model received mixed reviews on publication, during the last five years it has been uncritically applied in a number of studies. The model describes three overlapping stages. In chronological sequence these stages are phase one, in which the nonscientific society provides a source for European science through the medium of the scientific expedition; phase two, in which the colonial science begins to develop but is still "based primarily upon institutions and traditions of a nation with an established scientific culture"; and phase three, in which an independent and national science is established in the former colony.

Although the model purports to consider "the introduction of modern science into any non-European nation," during phase one scientific information is transmitted only in the opposite direction—from the New World back to Europe. Therefore phase one is utterly irrelevant to the diffusion vector that Basalla hopes to explain. Certainly the Mexican experience provides little justification for considering European expeditions as the first stage of its scientific growth or even as a separate phase at all.[40]

European scientists did travel to Latin America during the sixteenth and early seventeenth centuries, but, as Basalla notes, the effect of these expeditions was limited almost entirely to Europe. No evidence has been produced to show that they had more than a negligible effect on the stimulation of scientific growth in the New World, and they therefore cannot be considered the first phase of that growth. During the seventeenth and early eighteenth centuries scientific expeditions did begin to function in a limited manner to encourage local scientific activity. Carlos Siguenza y Góngora, for example, born in Mexico City in 1645, one of Mexico's first scientists of talent, established an international reputation by meeting all the distinguished scholars who came to Mexico, often corresponding with them after their departure.[41] Later, during the first half of the eighteenth century, European expeditions to Latin America—especially that of La Condamine—brought about the election of seven Latin American scientists to the French Academy of Sciences.[42]

Not until late in the eighteenth century did expeditions help the "spread of western science" by fostering the foundations of New World institutions. In Mexico, the Botanical Garden was established in 1788 by the Royal Botanical Expedition, and the School of Mines (1792) was part of a larger project, sponsored by Charles III of Spain, which included a mission of German mining experts.[43] Other Spanish expeditions functioned, in the same way and at the same time, to foster scientific growth elsewhere in Latin America.[44] During the nineteenth and even during the twentieth century, scientific expeditions and travelers continued to encourage scientific activity, though in a less substantial manner than in the late eighteenth century.[45] Thus, scientific explorations and expeditions, the signal for Basalla's phase one, have influenced the growth of Mexican science in *all* periods of its history *except* the earliest.

In Mexico, the first phase of Basalla's model best describes the last two decades of the eighteenth century, precisely the years of the founding of the first Mexican scientific institutions—the hallmark of his phase two. Furthermore, this same period, 1788–1811, also seems to mark the transition to national science, or to phase three.

Thus, in Mexico, the three phases of the Basalla model are seen to be so much intermingled as to be of little value in analyzing scientific growth. Other problems with the model remain. Natural science, for instance, does not completely dominate the "colonial" period in Mexico. The studies of early Mexican scientists, like those of seventeenth-century Europeans, tended to be eclectic, showing lively interest in astronomy, mathematics, engineering, and botany. The tendency for colonial scientists to favor the natural sciences over the experimental is to some extent a reflection of similar European predispositions during much of the eighteenth century. Since the Western Hemisphere was, in a sense, the world's laboratory for natural sciences, the scientists living in its midst might be expected to confine themselves almost exclusively to problems of taxonomy and natural description. As a matter of fact they did no such thing. Early Mexican scientists drained swamps and observed astronomical phenomena with as great enthusiasm as they named new plants. By the time the founding of the School of Mines was under way, the focus became social utility, whether in the natural or the physical sciences.

The diffusion-phase model makes dubious assumptions about the nature of science, history, and culture. It is a strictly linear analysis of extremely varied and complex cultural and scientific matrixes. Science is everywhere the same, an unproblematic given, a positivist conception that fails to make accurate distinctions among sciences and technologies. The model is presumed to have universal application, relating similarly to cultures as varied as Mexico and India, Australia and Japan, Africa and Quebec. The model is blind to time, with the same phase occurring hundreds of years apart in different locations. The so-called phases of the model make sense only when seen as interactive processes, not separate periods, as reflexive patterns, not chronological stages.

Finally, on close examination Basalla's diffusion model is found to be simply a special case of the colonial-national model. Phases one and two are clearly seen as stages of colonial development, which overlap with phase three, denoted by increasing national independence.

The Colonial-National Model

The conventional periodization of Latin American political history is, for the most part, unrelated to the history of science in those countries. These divisions were drawn by political historians with exclusive regard to political hegemony; thus we are told that the indigenous period ends in 1492, while the colonial phase lasts until the inauguration of the national period in 1824. Since the only two transition points in this model are the making and breaking of Spanish power in the New World, it is not surprising that the periods defined correspond reasonably well to European history and are roughly applicable to all Spanish American countries. Although such simplistic periodizing provides mnemonic aid to students of political development, it may actually obscure the analysis of social, economic, and intellectual history. If the concept of colonialism is considered to include economic criteria as well as political, for example, it becomes necessary to extend the colonial period into the twentieth century, or at least until late in the nineteenth, when the forces of nationalism began to develop.[46] In other words, colonialism is peculiar to no one historical period but has been a dominant force in Latin America since the consolidation of Spanish power in the sixteenth century.[47] Furthermore, the majority of terms that have been employed to designate narrowly defined periods—Spanish, colonial, patrimonial, national, independent—have been vaguely and inconsistently applied.

If the historical perspective is broadened so as to include the periods of colonial science histories in other parts of the world, the terms *Federalist, Republican, Jeffersonian, Age of Jackson, Banksian, Baconian, emergent, antebellum,* and so on are encountered.

Needless to say, cross-cultural comparisons of diverse provincial traditions are made more difficult by such a bewildering array of periods. In other words, without a more general framework, we sink into a sea of local histories.

It was suggested earlier that distinguishing colonial and national periods of a scientific tradition is often of doubtful validity. Not only is there usually a considerable overlap between the supposedly colonial institution and the supposedly national institution, but also colonial science is seldom in any significant sense transformed into national science. (The United States, the Soviet Union, and Japan are exceptions that show how it is done.) Today, European and American journals are the world repositories of scientific information, no less than were the museums of Europe during the eighteenth century. Colonial science remains colonial even in politically independent nations.

Colonial institutions remain on the periphery of international science, which is often focused on problems of great interest to the dominant scientific powers but of much less interest to economically developing countries. The relations of center to periphery are those of status and access, of power and control.[48] Even when internationally recognized institutions emerge in a geographically provincial setting, they usually remain subject to most of the restraints imposed by their peripheral location, and they certainly remain Eurocentric in orientation, at least in the absence of an effective wedding of the production of knowledge with the production of goods and services in the peripheral country.

This Eurocentric orientation in fact becomes vitally important in the examination of certain questions concerning the uses of science in the colonial setting. It may even be important for the understanding of scientific developments in the center. Is it possible, for example, that the history of continental drift theory would have developed very differently had not much of the supporting evidence been located in the periphery and supporters of the theory dispersed in such places as Tasmania and Canada? On the other hand, when internationally recognized scientific institutions appear in a former colony, they may be said to illustrate the growing ability of the center to appropriate knowledge—as well as primary resources and staple commodities—from the economically dependent nation.

A colonial scientific institution becomes national not just when it is financed by the national treasury and staffed by its own citizens, but rather when it is economically and politically integrated into the national interest. The United States, the Soviet Union, and Japan may be cases in point. Most Latin American countries are not.

It might be suggested that the Mexican scientific schools of the eighteenth century simply provide a contrast between two varieties of Spanish colonialism: the university symbolizing Hapsburg rule and the School of Mines symbolizing the "enlightened" Bourbon approach. The intellectual, economic, and sociostructural foundations of the scientific institutions, however, complicate the case.

Never a purely colonial establishment, the School of Mines drew not upon the Spanish scientific tradition, but rather upon educational precedents in northern Europe. This fact, combined with substantial Mexican contributions, both financial and intellectual, produced an institutional order more closely aligned with the later national models than with the medievalism of the original Spanish institutions. And yet even in a country as steadfastly independent as modern Mexico, scientific institutions are often marginal in the setting of the Mexican political economy. Such institutions may be said to serve as colonial outposts for dominant American and European interests.

Conclusion

In summary, the three models considered have several common characteristics that severely limit their explanatory power. First, they are Eurocentric in conception. While to some degree inevitable, this has the effect of minimizing local contributions, trivializing distinctive aspects of local development, and focusing the discussion entirely on science and technology to the neglect of other social values and cultural products. Furthermore, science itself is seen as inevitably and monolithically European in format, a dubious assumption, especially for the long run.

Second, the models are all chronologically linear and progressive; that is, they pass through a series of stages which move not only from then to now, but also from darkness to light, from error to fact, from poverty to wealth, from wrong to right. Such naive assumptions ignore modern understandings of the nature of science and the nature of culture, which are both complex and interactive matrixes, not one-dimensional unidirectional strands. In short, the models use positivist theories of science and imperialist theories of development.

Third, the models pretend to be universal in their applicability. This feature seems highly desirable, if it can be achieved. Not surprisingly, pressure for universality often stems from social and economic aspirations in the former colonies. What does the third world want from case studies of scientific development? A guide to modernization. Why do they seek to modernize? Put in the simplest terms, they seek greater wealth—or higher standards of living—and they seek positions of greater power and independence in the world, but not at the cost of cultural integrity.

Historians of science and technology wishing to serve such aspirations must attempt to understand the development of scientific and technological institutions in a rich diversity of cultural settings. They must seek neither to force this diversity into a single acultural mode or to deny the possibility of patterns of regularity.

It is still too early to delineate the fine structure of such patterns of regularity. But it is clear that a satisfactory approach to the problem must derive from, and successfully account for, a multitude of case studies, most of which remain undone. Such an approach will highlight whatever natural groupings and patterns emerge in these case studies and will also include consideration and disposition of the differences that emerge from diverse cultures, diverse geographic locales, and diverse economic structures. Cultures will be seen as interactive, not merely receptive. Science and technology will be seen as a complex of ideas and processes, not just staged development.

Thus, the final significance of the Mexican case considered here must await detailed research in other colonial settings. Mexico, may after all, appear to be atypical and unique; or it may manifest a pattern most common in the Spanish New World; or it may represent one in a series of possible patterns of cultural and economic response to the explosion of ideas and developments that is often called Western science.

Even in its current limited form, the Mexican case reveals that any account of colonial science which sees the generation of scientific culture purely as the delayed importation of European institutions and ideas—or which sees it as an earlier, more primitive form of what was later to turn into European science—cannot be presumed to hold any general validity because it fails to accord specifically with what is already known about Mexico.

1. I do not know who originated this term, but I have seen it used by Everett Mendelssohn, in "The Emergence of Science as a Profession in Nineteenth-Century Europe," in *The Management of Scientists*, ed. Karl Hill (Boston: Beacon Press, 1964), p. 4, and ten years later by Arnold Thackray, in "Natural Knowledge in Cultural Context: The Manchester Model," *American Historical Review* 79 (June 1974): 674.

2. For a detailed treatment of this question with reference to China, see Peter Buck, "Science, Revolution, and Imperialism: Current Chinese and Western Views of Scientific Development," *Proceedings of the XIV International Congress of the History of Science* (Tokyo, 1974); see also D. W. Chambers, *Red and Expert: A Case Study of Chinese Science in the Cultural Revolution* (Geelong, Victoria: Deakin University Press, 1979).

3. For a useful survey of Mexican science, see Eli de Gortari, *La ciencia en la historia de México* (Mexico, D.F.: Fondo de Cultura Economica, 1963). For a brief review of science in Mexican education before 1800, see D. W. Chambers, "Two Stages in the Development of Science in Mexico" (Ph.D dissertation, Harvard University, 1969); J. T. Lanning, *Academic Culture in the Spanish Colonies* (New York: Oxford University Press, 1940); E. Beltran, ed., *Memorias del primero coloquio mexicano de historia de las ciencias*, 2 vols. (Mexico, D.F.: Sociedad Mexicana de Historia de Ciencia y la Tecnología, 1964).

4. Bernabé Navarro, *La introducción de la filosofía moderna en México* (Mexico, D.F.: El Colegio de México, 1948); for comparison, see J. T. Lanning, *The University in the Kingdom of Guatemala* (Ithaca, N.Y.: Cornell University Press, 1955) and *The Eighteenth Century Enlightenment in the University of San Carlos de Guatemala* (Ithaca, N.Y.: Cornell University Press, 1956). The latter suggests that "students . . . droned out [Newton's] laws in the Schoolman's singsong Latin."

5. For a general introduction to Alzate's work, see Rafael Moreno, "Alzate, educador, ilustrador," *Historia mexicana* 7 (January–March 1953): 371–89. On the founding of the Botanical Garden, see H. W. Rickett, "The Royal Botanical Expedition to New Spain," *Chronica Botanica* 11 (Summer 1947): 1–86. On the opening of the School of Mines, see Chambers, "Two Stages in the Development of Science in Mexico." Walter Howe, *The Mining Guild of New Spain* (Cambridge: Harvard University Press, 1949), gives the best general account of mining education, emphasizing matters of administration. J. J. IzQuierdo, *La primera casa de las ciencias en México* (México, D.F.: Ediciones Ciencia, 1958), gives the best treatment of curriculum, facilities, and students. The summary description of the Mining School that follows is based on these three works and on my dissertation, "Two Stages."

6. In their *Representación* to the Spanish king, Velazquez de Leon and Lassaga, the founders of the school, point to German and Swedish mining schools as exemplary. Elyuhar and del Rio received their technical training at Freiburg, later touring other parts of Germany, the Scandinavian countries, and England. The writings of Duhamel, professor at l'École des Mines and later at l'École Polytechnique, are occasionally cited in Mexican scientific treatises; no reference is made, however, to the format of French technical education. Humboldt had visited l'École Polytechnique before his

arrival in Mexico in 1803, where he was in close touch with Elyuhar and the faculty of the mining school, but by this time the Mexican school had been in operation for eleven years.

7. Alexander von Humboldt, *Political Essay on the Kingdom of New Spain*, 2d ed., trans. John Black (London: Longman, 1822), vol. 1, p. 22. This work, originally published in 1811, received a mixed critical response. In particular, several doubted the validity of Humboldt's assessment of the scientific institutions he visited. This matter will be treated in detail in D. W. Chambers, "Centre Looks at Periphery: Foreign Accounts of Mexican Science in the Nineteenth Century," MS submitted for publication; an earlier version was presented to the Midwest Junto of the History of Science Society in 1977.

8. My knowledge of European technical schools is based on such monographs as that of F. B. Artz, *The Development of Technical Education in France, 1500–1850* (Cambridge: MIT Press, 1966), and on discussions with Alexander Ospovat of Oklahoma State University.

9. *Gazeta de México* 8 (29 November 1797).

10. Antoine Lavoisier, *Tratado elemental de chimica* (México, D.F., 1797).

11. For further discussion of the admissions policy, especially with regard to the relaxation of race and caste restrictions, see Chambers, "Two Stages," pp. 81–89.

12. I. Bernard Cohen, *Some Early Tools of American Science* (Cambridge: Harvard University Press, 1950), p. 63. The fact that the collection included apparatus—such as orrery, reflecting telescope, camera obscura, and Nollet apparatus—of little use in mining education or research indicates the depth of the commitment to the establishment of a comprehensive instructional base in the physical sciences. The original list of instruments ordered is given in full in Howe, *Mining Guild*, pp. 501–8.

13. This journal, edited by José Antonio Alzate, was published at least once a month for more than seven years. It and the *Gazeta de México* are the most important sources of information about certain scientific controversies that stirred the scientific community during these years. *Registro Trimestre*, published 1832–33, was associated with the name of the scientist and politician Pablo de la Llave, who contributed more than half the papers. *Boletín de la Sociedad Mexicana de Geografía y Estadística*, published in 1839 and from 1853 to the present, is the oldest continuously published scientific journal in Mexico.

14. Most of the early records of the school have disappeared; useful primary materials, however, are collected in Santiago Ramírez, *Datos para la historia del Colegio de Minería* (México, D.F., 1890), including a list of professors, pp. 187–91.

15. None of the founding generation of distinguished Mexican savants—Bartolache, Alzate, Leon y Gama, Montaña—was offered a professorial position in any of the new institutions; young graduates of the technical schools, once the schools were under way, were given preferential treatment, illustrating the professional character of the training. An analysis of the social origins, education, and scientific contributions of the twenty persons who constituted this first scientific community is given in Chambers, "Two Stages," pp. 31–49.

16. Among the political groupings there was no firm ideological base for opposition to the School of Mines. Anticlerical liberals would not wish to destroy the most important school that was free of the authority of the church. The royalists would presumably not wish to do away with that which they had created. The aim of the third important group, the Creole moderates, was not to destroy the institutions of the country but to wrest control of them from the Spaniards. All political parties, moreover, recognized the usefulness of trained engineers in an independent republic. For further discussion of the politics of Mexican science during this period, see Chambers, "Two Stages."

17. See, for example, the profound influence of World War I on every phase of U.S. science, as described by A. Hunter Dupree in *Science in the Federal Government* (New York: Harper & Row, 1957).

18. By comparison the U.S. war for independence was almost gentlemanly. In *Some Early Tools*, pp. 51–53, Cohen retells the charming story of the expedition of the Harvard Hollis Professor of Mathematics and Natural Philosophy into British-occupied territory to observe an eclipse of the sun. The good professor felt that the period of safe conduct given to him by the British was "wholly inadequate." In Mexico no such incidents seem to have occurred. No such expeditions were organized, and it is not likely that they would have been respected either by the plundering guerrilla chieftains or by such vicious Spanish generals as Calleja.

19. Ramírez, *Datos*, pp. 217–19.

20. See especially Jean Sarrailh, *L'Espagne éclairée de la seconde moitié du XVIII siècle* (Paris: Imprimerie Nationale, 1954); Richard Herr, *The Eighteenth Century Revolution in Spain* (Princeton: Princeton University Press, 1958); and Arthur Whitaker, ed., *Latin America and the Enlightenment*, 2d ed. (Ithaca, N.Y.: Cornell University Press, 1961).

21. Though now dated, the best historiographical discussion of the Enlightenment in Latin America is Arthur P. Whitaker's "The Intellectual History of Eighteenth Century Latin America," reprinted in *Latin American History*, ed. Howard Cline (Austin: University of Texas Press, 1967), vol. 2, pp. 723–32. Whitaker believes that the Enlightenment "was the central theme of the most significant thought of [the eighteenth century] both in and about Spanish America." He makes an excellent case for the proposition that "changing attitudes towards [the Enlightenment] have been a major factor in the interpretation of the intellectual history of Latin America." I question, however, his notion that the Enlightenment forms a natural—or even a useful—periodization of Latin American development. Whitaker himself states that "there is some reason to suspect that the importance of the [Enlightenment] may be found to diminish rather than increase as the problem is subjected to the more systematic and discriminating study which it still sorely needs."

22. See the various essays in *Latin America and the Enlightenment*, ed. Whitaker.

23. Hugh M. Hamill, "The Mexican Criollos and the Hidalgo Revolt of 1810" (Ph.D. dissertation, Harvard University, 1956); p. 15.

24. See, for example, Lawrence S. Thompson, "The Colonial Libraries of Latin America," in *History of Latin American Civilization*, ed. Lewis Hanke (Boston: Little, Brown and Company, 1967), pp. 36–46.

25. See Jefferson R. Spell, *Rousseau in the Spanish World before 1833* (Austin: University of Texas Press, 1938), p. 40, for the frequent evasion of the Index. José Torre Revello, *El libro, la imprenta y el periodismo en America durante la dominación española* (Buenos Aires, 1940) indicates the near impotence of the censors. Irving A. Leonard, *Books of the Brave* (Cambridge: Harvard University Press, 1949), and Monelisa Perez-Merchand, *Dos etapas ideologicas del siglo XVIII en México* (México, D.F., 1945) offer interesting accounts of the book trade.

26. Richard E. Greenleaf, "The Mexican Inquisition and the Enlightenment, 1763–1805," *New Mexico Historical Review* 41 (July 1966): 181–91.

27. Humboldt, *Political Essay*, p. 210.

28. Revillagigedo to Valdes, 14 January 1790, Archivo General de la Nación, Mexico City, p. 111.

29. See, for example, the libraries catalogued in Navarro, *La introducción de la filosofía moderna en México*, pp. 305–8. In these and other colonial libraries the names most frequently encountered are Bacon, Boerhaave, Newton, Nollet, Buffon, Descartes, Duhamel, Fontenelle, Gassendi, and Pascal. Indeed, these are precisely the

names most frequently cited in scientific treatises and journals in eighteenth-century Mexico.

30. Institut de France, *Index biographique des membres et correspondants de l'Academie des Sciences de 1666 à 1939* (Paris: Gauthier-Villars, 1939), p. 8.

31. *Gazeta de literatura* (Mexico City, 1792), vol. 1, p. 285.

32. Humboldt speaks of meeting Mexican citizens in the "remote provinces" who were shocked to learn that some Europeans did not speak Spanish; such ignorance of their language the Mexicans presumed "to be a mark of low extraction." Humboldt, *Political Essay*, p. 210.

33. Arthur Whitaker, ed., *Latin America and the Enlightenment* (New York: Appleton, 1941), offers the best example of this one-dimensional approach to the eighteenth century. No one can deny the heuristic value of the book nor its solid grounding in primary research; its publication stimulated such a vast quantity of scholarly research emphasizing the strength of Latin American Enlightenment, however, that the proper proportions have been lost. In Mexico City in 1963, for example, thirty-two papers were presented at a symposium on the general topic "La ilustración en la America Latina." Twenty-six of these were subsequently published in *Memorias del primer coloquio mexicano de historia de la ciencia*, ed. Beltran. Though many of the investigations presented are extremely useful, when taken together the essays paint a distorted picture of Mexican reality.

34. Harry Bernstein, "Some Inter-American Aspects of the Enlightenment," in *Latin America and the Enlightenment*, ed. Whitaker, 2d ed., p. 53.

35. J. T. Lanning, "Reception of the Enlightenment," ibid., p. 76.

36. Charles Griffith, ibid., pp. 119–43, questions the simplistic notion that the European Enlightenment was, in some manner, causally related to the wars for independence. Historians now reject that idea.

37. See, for example, the commentaries on the paper by Eli de Gortari, "La ilustración y la introducción de la ciencia moderna en México," in *Memorias*, ed. Beltran, vol. 2, pp. 48–49.

38. *Mercurio volante* (1772–73), edited by José Bartolache, is of some interest. Bartolache held that physics, medicine, and logic were all alike when considered as sciences. *Gazeta de literatura* (1778–95), edited by Antonio Alzate, is the most interesting scientific journal of the period.

39. George Basalla, "The Spread of Western Science," *Science* 156 (5 May 1967): 611–22.

40. The first natural history expedition to Mexico was commissioned by Philip II in 1570. Francisco Hernández had close contact with many Mexican doctors, but this was in his capacity as a medical examiner, not as a botanist. See his *Obras completas*, ed. Germán Somolinos d'Ardois, 3 vols. (México, D.F.: Universidad Nacional de México, 1959–60). The first volume contains a comprehensive biographical study.

41. For biographical information, see Irving A. Leonard, *Don Carlos de Siguenza y Góngora* (Berkeley: University of California Press, 1929).

42. V. W. von Hagen, *South America Called Them* (New York: Alfred A. Knopf, 1945), pp. 1–85; Institut de France, *Index biographique*.

43. The definitive study of the garden and the expedition is Rickett's "Royal Botanical Expedition to New Spain." A good account of the mission is to be found in Clement Motten, *Mexican Silver and the Enlightenment* (Philadelphia: University of Pennsylvania Press, 1950).

44. See Frederico A. Gredilla, *Biografia de J. C. Mutis* (Madrid, 1911), and A. R. Steele, *Flowers for the King* (Durham: Duke University Press, 1964). Steele's delightful account of the Ruiz and Pavon expedition to Peru includes an excellent discussion of the interactions of scientific travelers and Latin American scientists, pp. 12–25.

45. In this respect, the Humboldt expedition in 1804 was the most successful. After Humboldt, Darwin, Spruce, and other scientific travelers seem remarkably indifferent to the prospect of communicating with Latin American scientists. In 1882 Alexander Agassiz, for example, who had traveled at least twice to invest in the exploitation of Mexican mineral resources, found it "appalling what barbarians the Mexicans still are, at least a hundred years behind the age." After being generously entertained by a former student at Harvard, Agassiz wrote to his mother, "How anybody who has spent four years in the United States and subsequently studied eight years more in France, can have gone back to this semi-barbaric state passes my comprehension." Utterly ignorant of the archeological investigation being conducted by Mexicans, Agassiz thought it a pity that the "ruins are not in a civilized country where they could be studied and preserved." Unfortunately, Agassiz's attitude seems to have been fairly representative, both of European attitudes and those of "los norte americanos" who traveled to Latin America in the nineteenth century. See G. R. Agassiz, ed., *Letters and Recollections of Alexander Agassiz* (Boston: Houghton Mifflin Company, 1913), pp. 200–201.

46. Richard M. Morse, "The Heritage of Latin America," in *The Founding of New Societies*, ed. Louis Hartz (New York: Harcourt Brace, 1964), p. 165, suggests that the "colonial period" should extend from 1760 to 1920.

47. See Mario Góngora, *El estado en el dereche indiano, época de fundación, 1492–1570* (Santiago, 1951). Góngora presents a convincing case for extending the end of the indigenous period to 1520.

48. Access to funds, to journals, and to the best laboratories, the best teachers, and the best students, and so on. Interesting studies of these relations include those of Ben-David, Shils, Schon, and von Gizycki.

Differential National Development and Science in the Nineteenth Century: The Problems of Quebec and Ireland

RICHARD A. JARRELL

The problem of the way science passed from Western Europe to new societies or to societies undergoing some process of Westernization has been typified as the problem of "colonial science," derived from a diffusionist model enunciated by Basalla.[1] More recently, a model of the interaction of science in developed and developing countries based upon the concept of center and periphery has gained adherents.[2] These models are not specific to the history of science but are rather drawn from more general sociological considerations. Historians of science tend to shy away from theorizing, at least explicitly, but all historians employ social theory implicitly. The social history of science, since it lies on the border between sociology and history, cannot escape the need to provide a theoretical basis for its selection and analysis of data. As Reingold has noted recently, this is rarely done with much clarity.[3]

A discussion of scientific change, which may be seen as one aspect of social change, is necessary if we are to account for the spread and growth of science in colonial societies, both historical and contemporary. By scientific change I do not mean the conflicts or shifts in the methodological prescriptions or epistemological foundations of a particular science at a particular time, but rather the social structure and relations between scientists and their institutions. It can be argued that the *content* of science is a function of the social relations of science—and vice versa—in any given society; I will not deal with it here. History is an interpretation of the past and as such must assume something about the nature of social activities and their changes. Faced, in my own researches, with the growth of science in nineteenth-century Quebec and Ireland, I have found that they are not easily reconciled with the "colonial science" model. Likewise, the center-periphery model runs into difficulties under closer analysis. These two models and their consequences will be discussed later.

In the historical, as opposed to the political or sociological, analysis of the spread of Western science, examples are typically drawn from among those societies that have successfully integrated science into their cultures.[4] These societies are also those that became industrialized; the other former colonial societies tend be lumped as the third world or developing countries and have offered little scope to historians. Our available examples, therefore, are of winners as opposed to losers in the development stakes, and we may fall prey to the dangers of a onesided view. The problematic societies that I shall deal with later, Ireland and Quebec, fall somewhere between the extremes.[5] Neither society was on the margins of European civilization, yet neither adopted science as an integral part of its culture during the nineteenth century. We require, therefore, some notions that are general enough to account for differential national development and the concomitant growth of science. I will sketch some of these notions before turning to brief critiques of current theories.

Differential National Development in a Colonial Setting

For science to develop in a colonial setting, several criteria have to be met. The following list is not exhaustive, but I suggest that these criteria must be fulfilled, both for the industrial development of a country and the associated development of science. Let us assume that no country can be said to have a developed scientific base—in institutions, manpower, and production—unless it also has a significant industrial base. Exactly what the nature of the science–industrialization linkage is can be ignored for the present purpose. What must be explained is why some countries develop both industrially and scientifically more rapidly than others. We cannot stop there, however, for we must also explain why some countries that *ought* to have developed have failed to do so: this is the main theme of this essay. It is therefore necessary not only to know the factors that are sufficient for development but also to have some sense of those that are necessary. The peculiar mixture of these factors for each emerging colonial country provides the basis for understanding the differential development.

1. The immigrant base must be sufficiently large and steady to ensure the exploitation of the natural resources of the colony. The American colonies that became the United States provide the best example. The influx of peoples was large and on the whole steady over a long period, ensuring an early move to industrialization and the cultivation of science. Canada and Australia, slowly and thinly settled, required far longer to reach a level approaching that of the United States.

2. The social and political structure of the colony must remain stable throughout long periods. Social stability rests upon many factors, not least upon the particular mixture of the population, while political stability is intimately related to social stability as well as to colonial policies. This factor is important in the study of Spanish and Portuguese colonies in the Americas. Mexico, Brazil, and Argentina, for example, were scarcely resource-poor and they were not bereft of adequate inflows of population. Nonetheless, neither industrialization nor science developed to a great extent until quite recently. Social and political instability loom as large forces in the slow development of these societies.

3. Sufficiently large urban centers and a middle class must emerge before industrialization can take off. This factor is related to the first two. If too few immigrate, cities cannot form as focal points for industry, commerce, education, and the other institutional apparatus necessary for rapid industrialization and scientific development. Industrialization feeds upon itself for growth, so there must be a ready stock of entrepreneurs, workers, and consumers. British industrialization is equally the history of urbanization, and in Canada, Australia, and South Africa, it is even more strikingly so. The middle class, however narrowly or broadly it is defined, is the class

of industrialization and science. The lack of political and social stability militates against the emergence of a large and active middle
class.

4. The natural resources of the colony must be sufficiently diverse for several modes of exploitation. I do not believe the absolute
amount of exploitable resources to be quite so important as the variety. A colony with fish, furs, agricultural land, forests, and minerals is a far more likely site for industrialization than one in which
only limited agriculture can be developed. This helps account for
the lack of rapid industrialization of colonies such as Cuba and Ja-
maica, where immigrants exploited agriculture primarily and,
worse, developed near monocultures. European countries have developed industrially and scientifically with very little in the way of
natural resources, as the experience of the Netherlands shows. The
Dutch entrepreneurial outlook enabled them to exploit resources
from elsewhere for their own development. On the other hand, both
Canada and Australia had an embarrasment of resources, but both
exploited relatively few of them until later in the nineteenth century.

5. The immigrant group must transfer a significant amount of
the diversity of the mother country to the colony. This, I suggest, is
the most potent factor of all and one that is commonly overlooked.
By diversity I mean the complex of goals, behaviors, outlooks, ideas,
institutional forms, and technologies of a society.[6] This diversity is
transferred to the colony, though not perfectly or wholly, of course.
A society that is richly diverse will spawn colonies that are richly
diverse; one that is poor in diversity will, consequently, produce colonies with narrow outlooks and fewer possibilities for development
from the outset. Britain was the most diverse country from the late
seventeenth century to the mid nineteenth, so it is no surprise that
its colonies—which also met the other criteria—were the most successful in industrialization and the cultivation of science. Spain, one
of the least diverse colonizers of Europe, produced pale images of
itself. A certain degree of diversity appeared in Spanish colonies because of interactions with their physical environments and with the
native populations already occupying the land when they arrived.[7]

Diversity, as a concept describing the nature of a particular social
group is superior to such psychologists' terms as "group conscious-

ness" and "the American way of life," because with diversity specific items such as institutions, practices, technical methods, and political procedures can be enumerated. This diversity evolves in time and is affected by subsequent arrivals of immigrants who transfer further diversity from the homeland, which has also changed. There is, of course, a limit to diversity, and societies have means to control its growth and spread; otherwise, the disintegration of society would ensue.

All five factors are obviously interrelated. The differential growth of the United States and Mexico reflect their interaction. The Americans were the most successful colonists of all in the entire history of European colonial expansion. They had a large, steady flow of immigrants, stability, urban centers, a large middle class, and diverse natural resources that were exploited early, and the majority of immigrants sprang from the most diverse country on earth. Mexico, on the other hand, developed much more slowly. It had a sizable inflow of population, but its social and political instability prevented early growth of more than one large urban center and produced a tiny middle class, slow exploitation of resources, and an inheritance of poor diversity. Countries that begin to take off at a later date, as Mexico has, have to import diversity from outside, a difficult task at best.

Up to this point, I have attempted to describe the factors involved in national development of colonies, but it must be noted here that there are, in general, two classes of colony: those in which the colonial power moves into territory that is either unpopulated—almost a null class—and those in which it moves into territory inhabited by an indigenous population that can be assimilated, destroyed, or pushed to a marginal position. Early European colonies were of this type. The aboriginal populations of the Americas and Australasia were disposed of by these means. A second type of colony was created, that in which the indigenous population was neither assimilated nor destroyed and which could not be ignored. Most of the African and Asian colonies of Europe were of this type. The industrially and scientifically successful colonies, such as the United States, Canada, Australia, New Zealand, and South Africa, were of the first type. The less diverse colonies of the first class and the colonies of the second class now constitute the third world. Large and organized indigenous populations obviously curbed the factors elsewhere that made for success in the so-called white colonies of Britain.

That the indigenous populations made few strides themselves toward industrialization and science during their colonial periods was due, in large part, to the stifling (or crushing) of their native diversity. Further, the nature of the diversity of such societies before the European conquests was so different in kind from that imported by the immigrants and colonial administrators that science and industry would have been unlikely to emerge in the now familiar pattern.

Alternative Views of Science and Differential National Development

Among alternative ideas on scientific development, pride of place goes to Basalla's 1967 model, which, although ostensibly dealing only with science and not industrialization, must cover both simultaneously. This model of the transmission of Western science to other countries or cultures assumes three distinctive phases. In the first, science is practiced in the new land by men from the colonizing or exploring country for the uses of that country. This scientific activity tends to be centered on exploratory or classificatory sciences such as natural history, geology, and geography. The flow of information goes only one way, to the initiating nation. After some time, when science begins to get a foothold in the newly settled land, the amount and intensity of this early activity wanes.

Concurrently, a new phase of activity—colonial science—begins to grow. This phase is characterized by a dependency relation with the initiating country such that institutional forms, the means of education, hardware, books, theories of science, and a significant share of the manpower come from Europe. Colonial scientists, few in number, relegated to the tasks of collecting data to enrich the treasury of science elsewhere, must look to the mother country for recognition, honors, stimulation, and media for publication. The level of this activity rises to greater heights than the earlier activity. Even as this phase is passing its peak, a final phase, the movement toward the establishment of autonomous national science, begins.

Once this third phase has been entered, the territory is familiar.

The scientific community of the new country is largely educated at home by native teachers, while stimulation of the development of new areas, recognition, the exchange of information, and, generally, the obtaining of legitimacy, are all generated internally. On one hand, the foundations for this enterprise must be laid by the social system at large: encouragement and government support for science and an adequate technological base. On the other hand, internal developments in science, such as the establishment of scientific education, respectable journals, and home-grown organizations, must take place.

I have no quarrel with Basalla's description, since it does reflect the stages through which new scientific communities seem to pass.[8] What requires discussion is the fact that Basalla's model is not, strictly speaking, a model at all. There are no mechanisms to explain how these phases transform into one another. In fact, the "model" reduces to a diagram showing the rise and fall of the levels of scientific activity, and it implies three distinct processes.

The first requirement is that the new nation—which initially is not a nation at all—must be a developing social system. To limit the discussion, I will refer only to three colonial nations founded by Britain—the United States, Canada, and Australia—rather than those which already had complex social systems in place but were "Westernized," such as Japan, China, and Russia, or those in which imperialist policies assumed major proportions in the life of the colony, as they did in India and the African colonies, which are more complicated. In the cases of the North American and Australasian colonies, nations had to be built up from virtually no bases, which is a simpler task than transforming and assimilating the social system of a large and organized indigenous population. All three nations are examples of developmental social systems that eventually moved into exponential growth. The family resemblances that these societies share have as their source the fact that a considerable proportion of the diversity of British society was transferred to them holus-bolus. That all these nations differ from one another and from Britain arises from the fact that the three were founded at different times—the transferred diversity differed—environmental constraints differed, the influx and mix of populations differed.

The developmental demands on the colonial nations differed from the home society also in the sense that vast empty areas needed to be explored and populated and vast stores of resources tapped. The differentiation of populations and the particular forms of institutions and other social groups were necessarily different. The resultant economies, too, differed from that of Britain, but more striking was the different evolution of social relations, which tended to be more open and democratic and even, to European observers, nearly anarchical.

What of science? It is often argued that science is international in scope and not constrained by particularistic national development. Many would deny that "national science" makes sense. Nonetheless, in contemporary societies, science is an integral part of national social systems. Not only are its practitioners full members of many nonscientific social groups, but the scientific enterprise is intimately linked to a number of systems such as the political and economic. To the extent that a national social system becomes highly developed through industrialization, so science in that system will be more highly developed. This will mean greater differentiation: more disciplines cultivated, more manpower, more institutions, organizations, and media for the communication and storage of information—in short, all of Basalla's criteria for a mature national science. It seems to be true that science is the same everywhere in the sense that methodology and epistemology are shared, but differences remain when educational patterns, choice of research, goals, government, and institutional support and interaction are analyzed. Were there no national differences in science, the formulation of science policy would be straightforward.

The growth of science in any one country is a unitary process, not three distinct processes, and the three phases described by Basalla must be accounted for by the dynamics of development. When a social system is on the whole undeveloped, as were those of the United States in the eighteenth century and of Canada and Australia in the nineteenth, its population is small and less differentiated, and the level of diversity is lower. The need for science in such a society is slight, and societal differentiation has not progressed sufficiently far to allow for more than a handful of scientists. Both local and external scientists would typically be naturalists, given the unexplored physical environment and undeveloped social milieu. Such societies are too little developed for sciences such as chemistry and physics. As the social system develops, so does science; in one way, science is carried along by national development, but innovations from science continually feed back into the rest of the social system, helping to drive development to higher levels. In time, the complexity reaches a point that the historian in retrospect can recognize the so-called struggle for nationhood. The struggle for national, autonomous science is not far behind. The three curves of Basalla's model are not, then, measuring three separate systems but only one. Three curves are misleading; if it were possible to define an empirical measure of the development of science, a single undulating line would instead be found tending to rise with time and approaching an exponential curve.

More general than Basalla's model is the center-periphery analysis. The concept of center-periphery relations can be seen in two senses, one general, the other particular. In the general sense, the concept is new to neither historians nor sociologists as an explanation of large-scale differences in the actions of nations or substantial social groups.[9] The use of these terms assumes a dominant-subordinate relationship and, when applied to studies of colonialism, refers to the metropolis and the colony. There is little agreement among those who have employed the center-periphery terminology. The distinction can be seen geographically—Europe as central, the Americas as peripheral—or, more commonly, as a means of distinguishing social roles. Social rank is posited as a major determinant of whether a person or institution is central or peripheral, and since this terminology is found most often in political sociological studies, the center is often identified with decision-makers. Those who are peripheral in a social system can be designated as those who have little or no access to the decisionmaking process. Langholm associates accessibility to decisionmakers with similarities between those near the center—or part of the center—and the decisionmakers; since, in his view, similarity is the basis for social integration, those at or near the center are better integrated.[10]

The more specific use of the concept applies to relations among individuals or groups in central and peripheral regions. In the history of science, the relations between scientists in the metropolis and scientists in the colony can be characterized. In this sense, the historian assumes bilateral relations between scientists that are asymmetrical—that is, scientists at the periphery may supply specimens, data, and manpower to the center, while scientists in the center provide theory, problems, professional norms, institutional patterns, recognition, and so on. This more specific version of the center-periphery argument has been applied almost entirely to the history of modern science, since the gulf between the level of scientific development of highly industrialized countries and that of most third world countries is so obvious.[11] Studies of such center-periphery differences in nineteenth-century science are fewer, although aspects of both American and Canadian science have been interpreted with this model.

There is certainly ample evidence to argue that many asymmetrical bilateral relations, both individual and institutional, have existed during the last three centuries. And it is not difficult to multiply examples of scientists in Britain, for example, deprecating science in America, Canada, or Australia; likewise, scientists in the developing countries can be found decrying their own inferior positions or playing the role of the inferior.[12] Unfortunately, this kind of evidence, while useful rhetorically, is so scattered and particularistic as to be suspect as a true indicator of actual long-term dominant-subordinate relationships. To take the case of Canada, it might be expected that a center-periphery relationship between British and Canadian scientists would characterize the nineteenth century and between American and Canadian scientists the twentieth. This expectation would be fueled by the general notion that the Canadian economy was dominated successively by Britain and the United States. More specific to the sciences, it could be noted that imported science professors of the nineteenth century tended to be British while those of the twentieth century tended to be American. This is merely impressionistic, however, since no one has attempted a substantial survey of scientific relationships of Canadians for either century. A detailed analysis of one portion of the correspondence of Sir William Dawson, perhaps the Canadian scientist of the last century with the highest-profile, suggests a shift from British to American relationships.[13] This analysis does not elucidate, however, the precise nature of these relationships. The real question is, how typical was Sir William Dawson? He may well have been very typical and it might be discovered that a shift from British to American contacts did occur for Canadian scientists. It would then be legitimate to use the center-periphery model. To do this, the following would have to be established:

• Canadian scientists' relationships—especially important relationships—were forged with foreign scientists;
• a definite shift from British to American relationships took place and is demonstrable by statistics; and
• content analysis of letters, or citation analysis, would reveal asymmetrical relationships (foreigners dominant, Canadians subordinate).[14]

This approach is, of course, impossible in any comprehensive way, given the loss of most of the scientific correspondence, to say nothing of the sheer magnitude of the task. It would be possible, however, to select representative scientists for whom data could be collected and hope to infer more general trends. This has been attempted by Duchesne for three Québec scientists of the nineteenth century, with interesting results.[15] He found that an analysis of the correspondence links of the three—chosen for their specialties of geology, entomology, and botany—were not uniform. All three corresponded with Americans, French, British, and anglophone Canadian scientists, and there was no obvious chronological pattern to the relationships apart from those of happenstance—contacts made on trips abroad, and the like. In some instances, the relationships placed the Canadians in inferior positions, in some superior, and in some equal. The only obvious pattern discernible by Duchesne was the tendency to be on the receiving end early in one's career, which is neither surprising nor enlightening with respect to center-periphery situations. Scientists in the center presumably had to begin at the bottom, too.

This impression is strengthened by my own studies of the Canadian astronomical community from just after the turn of the century until the Second World War. Most of the demonstrable linkages were internal, and those with foreign astronomers, primarily American, were typically on equal terms. The leader of Canadian astrophysics until the mid 1930s, J. S. Plaskett, developed a great many contacts with such men as Hale, Pickering, Adams, Kapteyn, and Oort, and his letters reveal that he placed himself in an inferior position only when he contemplated entering a new research area. Once he had established himself, he treated his foreign colleagues as equals and was treated the same by them.[16]

To counterbalance these cases are others, of course, in which the dominant-subordinate relationship seems to obtain. Pyenson provides a diffusionist interpretation of what he styles the "incomplete transmission" of European physics to Canada and Argentina at the turn of the century.[17] His detailed and valuable reconstruction of the growth of physics at McGill University implicitly applies the center-periphery argument insofar as Canadian physics is seen as being in a subordinate position to British physics and this largely because of structural weaknesses in Canadian science and society. McGill may be an unfortunate example, since it was anomalous when compared with the rest of Canadian universities.[18] Its physics department was dominated by British-born and British-educated men who may have attempted to recreate British physics in a colonial outpost. Indeed, Pyenson's evidence bears out this picture. Any attempt to sustain a higher level of physics in an institution embedded in a social system whose level of development could not provide the requisite support system was doomed to failure. The other Canadian departments of physics were more in accord with their social and economic surroundings. As the emphasis in McGill physics seemed to drift toward metrological work rather than the latest theoretical and experimental studies, it does not necessarily offer an example of retrogression so much as an institution settling into its natural state. This does not mean that the physics at McGill, or anywhere else in Canada, was inferior science, only that it could not maintain the same diversity and high level of activity, nor could it be expected to.

The transmission of an image of European institutions, problems, and practices of science occurred because social systems can and do import externally produced models, particularly when the social systems are developing rapidly. That these forms do not always take in the new environment is not so much an incomplete transmission as it is the inability of the new social system to support them. In other instances, the particular forms may have been inappropriate. Scientific organizations and institutions in the United States, Canada, and Australia in the nineteenth century were often near carbon copies of British exemplars, and a goodly percentage of them failed. They were adopted in the first place because people with scientific interests were drawn together naturally, and the British model, born under similar circumstances, seemed viable and familiar. But early nineteenth-century Britain was a society richer in variety and diversity than the colonial nations and could sustain such organizations more readily.

The only conclusion that can be drawn on the basis of documented historical cases is that the center-periphery model is only a limited explanatory device. There is no doubt that, on some levels, scientists of one country look to those of another for guidance. The pattern of American scientists studying in Europe is well known, but even that pattern may be misleading.[19] American scientists studied in Britain and Germany but British scientists also studied in Germany, thus raising the question, where was the center, where the periphery? In a survey of nineteenth-century Canadian scientists, I found that surprisingly few studied abroad and almost none of those in Britain, the so-called mother country.[20] The fact is, there is almost no empirical basis for invoking a center-periphery model to explain the actual behavior of scientists and scientific institutions.

I would suggest the following reasons that the center-periphery analysis seems to make sense and why it has only limited value. In the first place, any national system that is more highly developed than its neighbors should support a scientific community more diverse than that of the corresponding communities elsewhere. During the nineteenth century, not only was British science larger in absolute numbers of practitioners and institutions than, say, American, Canadian, Australian, or South African science, it was also more diverse and more highly differentiated. Clearly, anyone in a colonial nation—the periphery—wishing to take up a new line of research would be much more likely to seek out someone in the more highly developed country—the center—not so much because of some feeling of inferiority to the larger scientific community elsewhere, but for the much more practical reason of a greater likelihood of finding the requisite knowledge, guidance, or encouragement in the richer community. To speak of Canadian science as being peripheral to the central position of British science is meaningless. It might be argued that Canadian entomology from 1865 to 1885 bore such a relation to American entomology, but that seems equally implausible; indeed, the little empirical research that has been done seems to show that one would be hard pressed to demonstrate even a weak statistical description of the behavior of a scientific specialty in any one country. But it would be reasonable to assume that an individual would seek out the other individual, group, or institution best suited to provide him with what he needs. As his needs changed, so would the relationships that he forged; if he succeeded in acquiring sufficient knowledge himself, others would presumably seek him out.

It is also clear that the nature of these relationships would depend on the specialty even more than on the personalities. The majority of the important geological theorists in the English-speaking world during the nineteenth century were British, but American and Canadian geologists seem to have forged many links with each other rather than with the British. The rock formations of North America respect no political boundaries, so it is not surprising that James Hall of the New York State Geological Survey and William Logan of the Canadian Survey should rely on one another much more than ›
on their British confreres. I suspect that if a very close scrutiny of the relationships of individual scientists or small groups were made, the center-periphery division would evaporate and in its stead would be a congeries of rapidly changing relationships. Since each country develops at its own rate and in its own way, its scientific community will have specific requirements and will differentiate in a unique manner. This can mean that some specialties would be quite sophisticated and attract outsiders, while others might be new, or numerically weak, and would need to seek elsewhere for their needs. Nineteenth-century Britain is an example of this: for a period, British chemistry looked to Germany, but this did not necessarily mean that British chemistry was peripheral. Both Britain and Germany were developmentally at similar stages, and bilateral relationships were much more common and more likely.

Science and Internal Colonialism in
Quebec and Ireland

The center-periphery argument can be used in a wider context by applying the terms to whole societies where the center represents the central society, which eventually assimilates, or tries to assimilate, peripheral social groups to itself. One of the clearest examples is Britain, which, beginning from a central core in south-central England, spread outward eventually to engulf the diverse cultures of the so-called Celtic fringe—that is, Scotland, Wales, and Ireland. There are two theoretical viewpoints in sociology to explain the process by which the core assimilates the periphery, diffusionism and internal colonialism. The former, in broad form, is a growing interaction between central and peripheral societies, with the result that greater commonality is fostered, leading, in turn, to national integration.[21] Industrialization hastens the process. Counter to this view is the more recently elaborated internal colonialism, which has been applied primarily to third world examples.[22] This theory, emphasizing the domination of the periphery by the center through exploitative measures, assumes not the diffusion of cultural elements until a commonality is reached, but rather the creation and maintenance of social and economic differences between members of the central and peripheral societies, so that the former will remain in the superior position.

The selective application of either of these theories does seem to suggest that the use of center and periphery as descriptive terms is legitimate, while their use in the specific context of scientific relationships is, at least, suspect. I suggest this is so because the wider sphere in which center and periphery have some descriptive value is a stage removed from the sphere of scientific activity. Either the diffusionist or internal colonialism model can be applied to either of these societies to explain their different evolutions. From the diffusionist point of view—and I oversimplify greatly—Ireland can be seen as having been slowly integrated with Britain economically, politically, and culturally through intensive contact—planted colonists, colonial administrators, and soldiers—until the differences were small. Of course, political integration was not ultimately achieved, but cultural and economic integration proceeded remarkably far, considering the history of inter-island strife. The Quebec example superficially resembles the Irish; after some 200 years of close interaction with Anglo-Canadians and Americans, Quebec has been largely integrated with the rest of Canada politically and economically, but less so culturally.

Application of the internal colonialism model produces an entirely different picture, since it relies so often on the depiction of oppressors and victims. In Ireland, the English are seen planting colonists who seize the land, control the institutions, and enforce laws that reduce the native population to near servitude. These policies have the result of creating unequal economic situations in center and periphery, which, in turn, lead to a hardening of ethnic lines that is reflected culturally and politically. The Quebec case is much less flagrant, largely, it may be supposed, because of the much greater distance of periphery from center. The net result of the historical process of internal colonialism is continuing strife and the rise of nationalist groups and separatism.

Both theoretical approaches have some merit, because evidence can be produced that supports either; the truth may lie somewhere between the extremes. Since the internal colonialism model distinguishes between ethnic groups or social systems on the basis of ethnicity—usually identifiable by linguistic, religious, and cultural differences—in the interaction between center and periphery, an alternative would be to distinguish the groups not so much along the lines of intergroup strife, but on comparative levels of development. In the case of Canada, the "French-Canadian nation" can be distinguished from the "English-Canadian nation"—stock phrases in the decades-old debate on Canadian bilingualism and biculturalism—in the developmental sense. There is much to support this approach, and the rough indicators of development can be compared using the available historical statistics.[23] The scientific part of the social system is a little more easily measured for diversity through such indicators as numbers of workers, disciplines, organizations, educational patterns, and the like, and the crude comparisons available show real structural differences.[24]

The case of Ireland is more complicated. In the early nineteenth century, two social systems are definable, divided along linguistic (English-Irish), confessional (Roman Catholic-Protestant) and geographical (urban-rural) lines, but these become blurred toward the end of the century with the near disappearance of the Irish language. In economic and political development, there is no doubt that the Anglo-Irish minority were far ahead of the so-called Celtic majority. This can be measured by the distribution of wealth, status, power, and so on. This division was, at first, effected through punitive laws and later through strong integration within the two social systems. Even so, the Anglo-Irish "nation" felt itself to be developmentally at a disadvantage to Britain and this feeling was perpetuated because

Ireland was so closely linked with England. Many of them believed they were politically and economically inhibited in their growth.[25]

To employ an undiluted version of the center-periphery argument in its diffusionist form has a little merit for Ireland but almost none for Quebec. The internal colonial variation recommends itself in that the concept can be applied to both societies; the central feature is that the chief agents of control were the local population of British stock rather than British politicians and administrators. These populations could be seen by the indigenous population as the social equivalent of the military garrisons. Certainly in the early nineteenth century the lion's share of civil, military, and legal positions in Ireland was in the hands of the Anglo-Irish, and in Quebec it was in the hands of British or American immigrants. Nonetheless, I think a strict internal colonialist view ought to be rejected, because the reality was far more complex.

Ireland and Quebec were not colonies in the ordinary sense.[26] What makes them of great importance is that each of them represents the colonization of an indigenous population that was itself European in culture and outlook. Neither the French Canadians nor the Irish were easily subdued aboriginals whose cultures differed greatly from the colonizing nations. Neither group was ignorant of the industrial revolution that was taking place, nor was either socially incapable of ready acceptance of that revolution. Yet neither society made significant strides toward the achievement of industrialization and the cultivation of science in the nineteenth century, leaving it to the colonizers of British stock.

In Ireland, the feudal basis of the economy lingered longer than in Britain, partly because of the landowners' attitudes, partly because of official policy. The diversity of Ireland, with the exception of Dublin, was simply less than that of contemporary Britain. The penetration of diversity was also hampered by the fact that, at the beginning of the nineteenth century, the vast majority of Irish spoke no English. Of course, Britain's economic restrictions on Ireland were a substantial obstacle to industrialization, but Britain had also economically restrained the American colonies. Those colonies, however, being more diverse, took matters into their own hands with well-known results. Ireland lacks significant natural resources, but human resources could have overcome the lack. Writing in 1845,

Robert Kane, the Catholic founder of the Museum of Irish Industry, proclaimed "The fault is not in the country, but in ourselves; the absence of successful enterprize is owing to the fact, that we do not know how to succeed; we do not want activity, we are not deficient in mental power, but we want special industrial knowledge."[27]

Quebec had different disabilities, for it was a double colony; it was originally a colony of the first type, but not a very successful one. Official neglect by Paris ensured that New France never attracted many immigrants, while encouraging only subsistence agriculture, fur trading, and minor local industries. The unique ruling structure of clergy, foreign administrators, and a petty baronial system militated against the emergence of a powerful middle class and, consequently, large-scale exploitation of natural resources. The coming of the British changed the form of colonialism, yet the arrival of new diversity did not spark a move to industrialization, despite the fact that the French Canadians were far less restricted than the Irish and far greater resources were available to them.

Both Canada and Ireland underwent industrialization during the nineteenth century, but the majority of those responsible were the Anglo-Canadians and Anglo-Irish, many of whom were, not surprisingly, Scottish in origin. This industrialization was also localized: in Canada industry arose predominantly in English-speaking Ontario and in Montreal, which was half English-speaking until late in the century, while in Ireland, Ulster was the center of activity, a region dominated by those of Scottish origin. The centers of scientific activity were in the same locations (Dublin excepted).

The internal colonial model identifies oppressors and oppressed; if there is a mentality of the oppressor, there is equally a mentality of the oppressed. If this view is taken seriously, then the lack of interest in industry and science on the part of the Irish and the French Canadians can be dismissed as a consequence of the mentality forced upon them by their oppressors. This view of passivity or lethargy might be invoked for specific aboriginal groups in other colonies, but applied to the Irish and French Canadian middle classes, those who would be the natural participants in science and industry, the idea collapses. In fact, the evidence suggests rather an active turning away from industrialization and scientific activity as a means of maintaining cultural identity and social cohesion. Nationalist movements in both countries, movements led by the middle classes, often extolled the virtues of the farm and generally ignored the potential benefits of industrialization. They were, in fact, quite correct; industrialization and the cultivation of science, by increasing the diversity of a society, tend to break down the more rigid traditional cultural outlooks. The liberal professions, rather than entrepreneurial activity, allowed a young man to advance without having to take on characteristics of the British. The more creative and intellectual could turn to the cultivation of literature, art, and music, thereby maintaining cultural mythology and purity. Science could offer neither untainted advancement nor cultural stasis.

One possible reason for the rejection of science and industrialization has not been mentioned: the influence of the Roman Catholic Church. Both the Irish and French Canadians were Catholic societies, both were traditional in religious matters, and both underwent virulent ultramontane upheavals during the nineteenth century. The extreme conservative element in both the Irish and Canadian churches might be invoked as a strong force against modernization. Nonetheless, I have been unable to detect any significant antiscience bias on the part of the clergy in either Canada or Ireland during the nineteenth century. Indeed, clergymen were among the most noteworthy cultivators and proponents of science among the indigenous populations. The position of the church was, I believe, that of a natural ally to the elite, which wished to maintain cultural purity and identity. Seen in that light, the church was an admirable and formidable ally, and prominent nationalists in both Ireland and Canada were clergymen.

The Cultivators of Science in Quebec and Ireland

To say that the Irish did little in science seems, at first glance, a preposterous statement, for it was the homeland of Hamilton, FitzGerald, Larmor, Lord Rosse, Stoney, and many others of international renown.

But there were two Irelands, and almost all the scientists were Anglo-Irish. Trinity College, the Royal Dublin Society, and the Royal Irish Academy were near exclusive preserves of the minority. In centers outside Dublin, especially in Cork and Belfast, the scientific leaders were equally Anglo-Irish. Were there scientists in the

"other" Ireland? There were, of course, but few in number, and the
Catholic University in Dublin, where Hennessy, Sullivan, and Casey
were professors, was their natural center. The Queen's Colleges, created to provide Catholics equal opportunity at higher education, attracted few Catholics as professors or students because of opposition
by the church to mixed education. Even the secular College of Science, a creation of the British Department of Science and Art, had
little effect on the native population. The natural leader of a scientific revival among the Irish population would have been Robert
Kane, but he, although a Catholic, was spiritually closer to the ruling minority and, as first president of Queen's College, Cork, was
staunchly opposed to the church's attitude toward secular higher
education. The Catholic colleges taught science, of course, and the
Reverend Nicholas Callan of Maynooth College was perhaps the
outstanding Irish expert on electricity of the nineteenth century.
Few others had his aptitude or interest.

Differences between the two social groups in Quebec and Ireland
are, in part, measurable. The two central provinces of Canada, Ontario and Quebec, diverged rapidly during the nineteenth century,
and a few salient features of their demography underscore the societal differences. The English-speaking minority in Quebec shared, of
course, most of the cultural and economic traits of the majority of
the inhabitants of Ontario. From 1851, French Canadians made up
75 percent of the population of Quebec—then Lower Canada—
while the anglophone minority began to shrink within the larger
cities. Roman Catholics, including Irish immigrants who were English-speaking, constituted an even larger percentage in Quebec, but
only a small minority in Ontario. Three representative indicators
show the differences in development.[28] First, in industrial workers,
Ontario had, by 1851, surpassed Quebec in absolute numbers and
retained the lead. When the figures are analyzed by region, the two
Quebec cities are found to have been on a par with Ontario cities in
both percentage and number of workers involved in industry, but in
smaller communities, by 1870, only 7.3 percent were industrial
workers in nonmetropolitan Quebec and 12.1 percent in nonmetropolitan Ontario.

Two educational indicators display the differences equally: in
1891, the rate of literacy in Ontario was 76 percent, while that in
Quebec was 52 percent, despite state-supported educational systems

that had been established simultaneously a half century earlier and despite the fact that Ontario received more immigrants from overseas throughout the period. Further, by 1901, although both provinces had similar numbers of primary schools and students per capita, the differences in secondary and higher education were substantial. Quebec had more secondary institutions and students in both numbers and percent, but more than half were anglophone. In higher education, including professional education, Ontario had one and one half times the students per capita; at the university level, the majority of Quebec's students were enrolled at McGill and Bishop's, both anglophone institutions. Education was given high priority by the government of Quebec, yet it is clear that the young francophone in the second half of the century had less opportunity in higher education in his own language, and few took advantage of anglophone education.

The sciences, and engineering, were simply not attractive alternatives to the French Canadian student. Both l'Université Laval and l'École Polytechnique offered science, and the latter offered engineering, but neither drew a significant number of students. In fact, before 1900, Laval graduated no one in science, although a science degree had been available for more than thirty years. By contrast, McGill graduated fifty-three in science and engineering in 1900 alone. This state of affairs was a reflection of the wider views of Quebec society; certain members of the elite were unhappy with this situation, and a succession of Quebec premiers, beginning with P.-J.-O. Chauveau, took initiatives to improve the availability of scientific and technical education. Chauveau's government created the Conseil des Arts et Manufactures in 1872 to open schools in the larger centers to teach drawing, design, modeling, and geometry. These schools were in obvious imitation of those created in Britain by the Department of Science and Art, knowledge of which Chauveau may have obtained on his education tour of Britain in 1866. These schools, plus significant efforts to disseminate scientific agriculture to the rural population and a special committee of the Legislative Assembly devoted to industry were some of the various government attempts to modernize the outlook of Quebec during the 1870s and 1880s. In the end, however, they had little effect, because the population at large had little enthusiasm.

Irish demographic comparisons are more difficult to reconstruct, since the Irish language, unlike French in Canada, was systematically pushed aside by the operation of the national schools. Religion must substitute as an identifying factor, and although some Anglo-Irish were Catholic, the division was generally along Catholic-Protestant lines. In addition, the rural population was heavily weighted toward the Irish as opposed to the Anglo-Irish, the group which constituted a substantial minority in the major cities of Dublin, Cork, and Waterford and a majority in Belfast. Approximately three quarters of the Irish population were Roman Catholic during the second half of the century. Ignoring the two social groups and just taking the aggregate figures, we can see the gulf between Britain and Ireland: at the 1871 census, for example, 20 percent of the Irish engaged in agriculture, but only 7 percent in Great Britain, while only 10 percent were industrial workers and 23 percent in Britain.[29] When the two social groups are reinserted the differences are even more substantial, because most of Irish industry was concentrated in Protestant-dominated Ulster and in Dublin, which was a quarter Protestant and the center of Anglo-Irish intellectual life.

Ireland had no local, provincial government during the nineteenth century as did Quebec, so educational schemes to broaden the industrial and scientific outlook of the population were imported from Britain. The main thrust was the science schools program of the Department of Science and Art in South Kensington. The department, quite separate from the Irish educational establishment to the end of the century, worked on the basis of local initiative, with central funds for prizes, materials, and government inspection. The geographic distribution of science schools was, then, an indication of local interest. The schools were established from the late 1850s onward; by 1865, seventeen of the twenty-eight Irish schools were in Ulster. The overall Irish enrollments lagged seriously behind those of Britain. By 1885, science schools in Great Britain enrolled nearly 57,000 students, while Ireland enrolled only 4,600, a ratio of more than twelve to one, although the general difference in population was about five to one. More indicative, 63 percent of the Irish science students were enrolled in Ulster—especially Belfast—schools.[30] Social class was also an indicator: that a substantial middle class is important to the cultivation of science has already been posited, and the differences in enrollment bear this out. In 1868, Dr. Sidney, an inspector for the Department of Science and

Art in charge of all Irish schools along with some English and Scottish schools, noted that in his English and Scottish schools, 64 percent of the students were from the artisan and middle classes, while only 36 percent in the Irish schools were from the same classes.[31]

The Irish institutions of higher education, only six in number and all small, catered to the middle class, primarily to the Anglo-Irish, despite the fact that during the second half of the century, Roman Catholics were free to attend all six institutions.[32] The only institution seen by the Irish hierarchy as truly theirs, the Catholic University in Dublin, failed to develop as a center for science, although its scientific staff was eminent. The Royal College of Science in Dublin, a foundation of the Department of Science and Art, became the natural center, yet it was as successful in attracting students from Britain as from its own country. During the last decade of the century, slightly more than half its graduates were non-Irish. That Ireland offered little scope in employment for graduates in science and engineering is shown by the fact that of ninety-seven graduates of the college from its inception to 1898, seventy-five took employment in Britain or the colonies, not at home.[33]

Scientific societies and institutions provide another index of activity. Scientific societies of nineteenth-century Canada and Ireland were, almost without exception, organized for amateurs as much as for the emergent professionals.[34] In Canada, science was cultivated in purely scientific organizations, such as the Natural History Societies of Montreal and of New Brunswick, in Mechanics' Institutes and similar literary institutes, and in mixed societies with multiple purposes. The same was true in Ireland. The distribution of societies and institutions was, as might be imagined, generally limited to cities and the larger industrial towns. The striking difference is that in English-speaking Canada and those parts of Ireland where the Anglo-Irish exerted substantial influence, far smaller centers also maintained institutions or organizations, which was not true in French Canada or most of Ireland.

The Montreal Natural History Society, the only large scientific organization in Quebec, was dominated by anglophones but enrolled francophones. Of the mixed societies, those that were anglophone-dominated, such as the Literary and Historical Society of Quebec, took science seriously; those that were francophone-dominated, such as the Institut Canadien or Institut Canadien-Français in several cities and towns, tended to treat science in a minor way. The Mechanics' Institutes, ideally organized to bring scientific and industrial education to the new working class, were widespread in Ontario and located in many small centres, while few were ever established in Quebec, and those by middle-class anglophones.

In Ireland, the greatest activity was in Dublin and Belfast, where the influence of the Anglo-Irish was greatest. Other towns and cities with mixed societies, institutes or mechanics' institutes were those with populations of 10,000 or more, such as Cork, Limerick, Waterford, and Carlow, except in Ulster, where societies formed in smaller centers, such as Carrickfergus, Downpatrick, and Lisburn. Outside the two chief cities, Cork was the center of the greatest activity, but it was also the largest Anglo-Irish center in the west of Ireland. A map of the institutions shows most of the west and central portions of the island devoid of anything, with the exception of Galway, a university center. Education and institutional foci were extremely important for the production of a scientific corps, and where they were deficient or nonexistent, the chances of young people being attracted to science were diminished accordingly.[35] Equally, where the social interest in industrialization and science was slight, the chances that such an educational and organizational base would emerge were very slight. Even an unsystematic survey of the most notable scientific figures of Quebec and Ireland in the nineteenth century shows that the social differences I have sketched were real. Employing a broad definition of a person interested in science—that is, one who teaches, one who works for the government in a scientific capacity, or one who publishes scientific articles, I find that of those who worked in Quebec during the nineteenth century, twenty-six were francophones from Quebec, while seventy-six were non–French Canadian.[36] The Irish list is more difficult to differentiate;[37] a similar set of criteria generates a larger list of those who were from Ireland or worked in Ireland during the century. Eighty were alumni or teachers at Trinity College and another thirty-eight were from elsewhere or educated in other schools; these 118 were almost all Anglo-Irish or Protestants from elsewhere. The remainder include eleven known Roman Catholics and six who may have been. In short, the Irish group contributed about 13 percent of the Irish scientific manpower during the century, although they represented a group that constituted about 80 percent of the population.

Concluding Remarks

Diffusionist models, such as Basalla's, provide us with a sketch of the way science evolves in colonial societies but does not identify the factors that allow that evolution in the first place. The center-periphery and internal colonial theories offer insights into the development of colonies of the first type, those involving insignificant, indigenous populations, and those of the second type, with significant populations, respectively. Yet neither explains the growth of both types sufficiently, and both fail to account for the special cases of Ireland and Quebec. The factors involved in national development that I have enumerated make it possible to see why there is differential national development in colonial societies and why science is equally differential in its growth and social structure. The identification of factors does not constitute a theory per se, but a prolegomena to a theory. These factors could, with some modification, be applied to the colonizing countries themselves. Britain, the source of the industrial revolution, had a sufficient population base, a stable social order, urban centers, a middle class, widespread exploitation of resources both home and abroad, and, most important of all, sufficient diversity to fuel the growth of industry and science. Those European colonizing nations that failed to embark on massive industrialization—and that were consequently weaker in science—were those deficient in one or more of these factors, and their respective colonies reflect them.

The colonizers of Quebec and Ireland had these factors in their favor, and they developed. Yet the colonized, too, had most of these advantages, with one essential difference: to maintain their cultures and identities, their elites consciously chose to freeze or redirect the diversity of their societies away from the prevailing attitudes of their colonizers, ensuring their social survival but also ensuring their nonparticipation in the great currents of the nineteenth century.

1. George Basalla, "The Spread of Western Science," *Science* 156 (5 May 1967): 611–22.

2. A survey of the basic variations in usage is given in Sivert Langholm, "On the Concepts of Center and Periphery," *Journal of Peace Research* 8 (1971): 273–78.

3. Nathan Reingold, "Comment," *News and Views* (Newsletter of Historians of American Science) 1 (January 1981): 3–4, in response to a guest editorial by Margaret Rossiter, ibid. (September 1980): 3–4.

4. The first important comparative work was that of Donald Fleming, "Science in Australia, Canada, and the United States: Some Comparative Remarks," *Actes du dixième Congrès Internationale d'Histoire des Sciences* (Paris, 1964), vol. 1, pp. 179–96. American science has a large and growing literature, while Canadian and Australian historians are just beginning to explore their scientific origins. Japanese and Brazilian science have a limited literature, but even the outlines of the histories of science in other former colonies lie in darkness.

5. The terms *Quebec* and *French Canada* are used interchangeably here, as they often are in Canada, although the former properly refers to the geographic entity, the latter to the social group.

6. The word *diversity* is specific to a systems theory of social change; for details, see Henry Teune and Zdravko Mlinar, *The Developmental Logic of Social Systems* (Beverly Hills, Calif.: Sage Publications, 1978).

7. The physical environment is sometimes seen as something outside human activity, but in highly developed systems, human connections with the physical environment are too intimate to be easily distinguished. It may be more of a factor in the early stages of development. See Herbert Karp and Sal Restivo, "Ecological Factors in the Emergence of Modern Science," in *Comparative Studies in Science and Society,* ed. Sal Restivo and Christopher Vanderpool (Columbus, Ohio: Charles E. Merrill Publishing Company, 1974), pp. 123–43.

8. This mode of thought is pervasive, and many have reached similar conclusions. In a paper read before the Canadian Society for History and Philosophy of Science (CSHPS) in 1976, I sketched the growth of Canadian science a la Walt Rostow. Although I had not seen Basalla's article, I later found that our views were uncannily alike.

9. Edward Shils's ideas have had a significant effect on sociological theory; see his "Center and Periphery," in *Logic of Personal Knowledge* (Glencoe, Ill.: Free Press, 1961), pp. 117–30. The first important social history of Canada, Arthur Lower's *Canadians in the Making* (Don Mills, Ont.: Longmans Canada, 1958), employs the distinction.

10. Langholm, "On the Concepts of Center and Periphery."

11. See, for example, Christopher Vanderpool, "Center and Periphery in Science: Conceptions of a Stratification of Nations and Its Consequences," in *Comparative Studies,* ed. Restivo and Vanderpool, pp. 432–42.

12. That these feelings could excite misunderstandings as late as the turn of the century is shown by G. F. Fitzgerald's comment to W. F. Barrett concerning the future of the control of the Royal College of Science of Ireland:

You *think the question of management is safe—I do not. I know the contempt London has of all provincial places and how they think they can arrange things for us so much better than we can for ourselves that I don't believe they will give up one jot or tittle of power.*

Fitzgerald, of course, was no second-rate scientist in some backwater college. Fitzgerald to Barrett, 23 November 1896, Barrett Papers 44, Royal Society of London Archives.

13. Lewis Pyenson and Susan Sheets-Pyenson, "The Formative Years of Canadian Science: John William Dawson and His Circle, 1865–1886," paper read before the CSHPS, Saskatoon, 1979.

14. Citation analysis, in the modern sense, cannot be undertaken for much of nineteenth-century science, given the lack of uniform citation practices. As an approximation, however, a crude matrix of relationships can be arrived at.

15. Raymond Duchesne, "Science et société coloniale: les naturalistes du Canada

français et leurs correspondants scientifiques (1860–1900)," *HSTC Bulletin* 5 (May 1981): 99–139.

16. R. A. Jarrell, "The Birth of Canadian Astrophysics: J. S. Plaskett and the Dominion Observatory," *Journal of the Royal Astronomical Society of Canada* 71 (1977): 221–33.

17. Lewis Pyenson, "The Incomplete Transmission of a European Image: Physics at Greater Buenos Aires and Montreal, 1890–1920," *Proceedings of the American Philosophical Society* 122 (1978): 92–114.

18. See Yves Gingras, "La Physique à McGill entre 1920 et 1940: la réception de la mécanique quantique par une communauté scientifique périphérique," *HSTC Bulletin* 5 (January 1981): 15–39; and "Le Développement du marché de la physique au Canada: 1879–1928," in *Critical Issues in the History of Canadian Science, Technology, and Medicine*, ed. R. A. Jarrell and A. E. Roos (Thornhill and Ottawa: HSTC Publications, 1983), pp. 16–30.

19. A recent example is Bruce Sinclair, "Americans Abroad: Science and Cultural Nationalism in the Early Nineteenth Century," in *The Sciences in the American Context: New Perspectives*, ed. Nathan Reingold (Washington, D.C.: Smithsonian Institution Press, 1979), pp. 35–53.

20. R. A. Jarrell, "British Scientific Institutions and Canada: The Rhetoric and the Reality," *Transactions of the Royal Society of Canada*, ser. 4, 20 (1982): 533–47.

21. There are countless diffusionist studies. A succinct review of the main tenets of diffusionism is to be found in A. L. Kroeber, "Diffusionism," in *Social Change: Sources, Patterns, and Consequences*, ed. Amitai and Eva Etzioni (New York: Basic Books, 1964), pp. 142–46.

22. Michael Hechter, *Internal Colonialism: The Celtic Fringe in British National Development, 1536–1966* (London: Routledge & Kegan Paul, 1975), is the best introduction to the theory—especially part 1.

23. Two recent works in Canadian historiography have made substantial strides in this direction: John McCallum, *Unequal Beginnings: Agricultural and Economic Development in Quebec and Ontario until 1870* (Toronto: University of Toronto Press, 1980), and Fernand Ouellet, *Lower Canada, 1791–1840* (Toronto: McClelland & Stewart, 1980).

24. For the nineteenth century, consult R. A. Jarrell, "The Rise and Decline of Science at Québec, 1824–1844," *Histoire sociale* 10 (1977): 77–91; for the twentieth century, see Raymond Duchesne, *La Science et le pouvoir au Québec, 1920–1965* (Quebec: Éditeur Officiel du Québec, 1978). A recent structuralist study of Quebec science, based on the theoretical work of Pierre Bourdieu, set forth in "La Specificité du champ scientifique et les conditions sociales du progrès de la raison," *Sociologie et sociétés* 7, no. 1 (1975): 91–118, has appeared: see L. Maheu et M. Fournier, "Le Champ scientifique québécois: structure, fonctionnement et fonctions," ibid. no. 2 (1975): 89–114.

25. It is not surprising that much of the leadership of the Irish nationalist movement during the nineteenth century was in the hands of the Anglo-Irish, who were politically more highly developed.

26. R. A. Jarrell, "Colonialism and the Truncation of Science in Ireland and French Canada during the Nineteenth Century," *HSTC Bulletin* 5 (May 1981): 140–57.

27. Robert Kane, *The Industrial Resources of Ireland* (Dublin, 1845), p. 412.

28. Statistics are drawn from various editions of the *Statistical Yearbook of Canada*, published by the Dominion Bureau of Statistics, and from McCallum, *Unequal Beginnings*, pp. 129ff. McCallum's thesis is opposite to that outlined here. He argues that Ontario's lead was a result of better agricultural lands, which in turn created the capital for industrialization. The industrialization of Quebec City and Montreal was,

therefore, less dependent on French Canadian agriculture. He rejects any sociological differences as a causative factor in the differential growth of Quebec and Ontario. He overlooks, however, the substantial differences in agricultural technique employed in the two provinces, the fact that much of Quebec industry was created and run by anglophones, and that other countries which lacked strong agricultural bases, such as Japan, have been able to acquire industrial capital in sufficient quantities.

29. *Thom's Directory of Ireland* (1883), pp. 568, 633.

30. Department of Science and Art, 32d Report, United Kingdom, Parliament, Commons, *Sessional Papers 1884–85*, vol. 28.

31. *Sessional Papers 1867–68*, vol. 27.

32. Jarrell, "Colonialism and the Truncation of Science," pp. 149ff.

33. Royal College of Science for Ireland, *Directory 1897–98* (Dublin, 1897), pp. 26–36. The list is of Associates—that is, full graduates and distinguished nongraduates who followed partial courses. I have enumerated only those whose occupations are listed.

34. See R. A. Jarrell, "The Social Functions of the Scientific Society in 19th Century Canada," in *Critical Issues,* ed. Jarrell and Roos, pp. 31–44.

35. There were, of course, notable exceptions. John Tyndall, raised by a policeman father in a small village near Carlow, comes to mind. He, like many self-made men, had an outstanding teacher but no local social or intellectual support system.

36. These data leave out some two dozen others who may have studied or worked in Quebec but about whom I have too little information. None were French Canadian.

37. The basis of the Irish list is biographical data from dictionaries such as the *Dictionary of National Biography* and from graduation lists of Trinity College. Unlike the French Canadians, the Irish cannot always be differentiated from the Anglo-Irish by name alone. Trinity graduates were almost all Protestant, as were those of Queen's Belfast, although a few Catholics did attend. Those, like the mathematician John Casey, are included among the Catholics rather than among the Trinity graduates.

Civilizing by Nature's Example: The Development of Colonial Museums of Natural History, 1850–1900

SUSAN SHEETS-PYENSON

He who views only the produce of his own country may be said to inhabit a single world; while those who see and consider the productions of other climes bring many worlds in review before them. We are but on the borderland of knowledge; much remains hidden, reserved for far-off generations, who will prosecute the examination of their Creator's works in remote countries, and make any discoveries for the pleasure and convenience of life. Posterity will see its increasing museums and the knowledge of divine wisdom flourish together; and at the same time antiquities and history, the natural sciences, the practical sciences of the manual arts will be enriched.

Linnaeus, Museum Adolphi Friderici Regis, *1754*[1]

During the late nineteenth century the natural world was feverishly and systematically sorted into bottles, drawers, and display cases in hundreds of museums across the globe. This remarkable development has largely escaped the notice of historians, who have bestowed more attention on the universities and learned societies that often nurtured and supported museums. Those interested in the internal development of natural history sciences also have neglected the importance of these institutions. Yet those classifiers, compilers, and collectors who dominated natural history during the nineteenth century were responsible for the waxing of the "museum movement" that became powerful during the decades leading up to 1900.[2]

The proliferation of natural history collections began as early as the sixteenth and seventeenth centuries. Assemblages of the curiosities of nature nonetheless may be traced back to the third century B.C. and Ptolemy's Museum at Alexandria or to the treasures hoarded by nobles and clerics since antiquity. It was, however, the great voyages of exploration that made collecting natural objects a reasonably common endeavor among the more privileged classes. But the impetus behind these early museums was usually piety or superstition, both of which attitudes esteemed the rare or peculiar. Early collectors prized horns and bones alleged to have come from unicorns, murderers, and giants, as well as mummies or less complete human remains, as they celebrated the abnormal, the bizarre, and even the imaginary. "Aesthetic" considerations tended to determine the arrangement of museums, with the result that curators aligned objects in the most incongruous fashion, such as armadillos with ostrich eggs.[3]

Gradually, though—and especially as objects were garnered from new continents—Europeans recognized that the usual order of things offered diversity sufficient to delight the observer without recourse to monstrosities or fakes. Linnaeus, for example, recorded this sense of wonder, evoked by the striking colors, extremes of size, and fascinating structures in the natural world, as he catalogued the treasures of the royal Swedish museum in 1754.[4] Soon an initial reaction of astonishment or surprise was seen as the stimulus for promoting a deeper knowledge of the kingdom of nature. At the same time, the instruction of the observer—thereby bringing about the refinement of his character, according to nineteenth-century thinkers—came to overshadow the museum's other and previously dominant function, amusement.[5]

The new educational importance attached to museums is reflected in the fact that the eleventh edition of the *Encyclopedia Britannica* counted no fewer than 2,000 scientific museums in existence by 1910. Included in this figure were a remarkable variety of institutions. Governments, whether municipal, provincial, or national, supported many of them. Other museums belonged to learned societies, universities and colleges, or religious orders. Some traced their ancestry to the whims of an eccentric individual collector; others owed their birth to the great number of objects amassed by a geological survey or put together for an exposition. On occasion, museums sought comprehensiveness and embraced every known science, both pure and applied. For others, mastering a restricted domain, such as the natural history products of a particular locality, sufficed. Museums prospered in all kinds of quarters, some merely having been granted the use of a few showcases or rooms in libraries while others enjoyed independent buildings constructed specifically for them.

As museums were recast as educational institutions, questions of purpose, organization, and arrangement became of central concern. For the first time in the history of museums, curators were compelled to develop collections in directions that answered the needs of diverse social groups. Increasingly during the nineteenth century, museums were expected to serve a middle-class that had greater leisure, wealth, physical mobility and educational opportunity than ever before. This newly articulate audience sought out socially sanctioned activities that provided the perfect mixture of education and amusement. Conflicting ideals, for example, often had somehow to be reconciled between the expectations of the public and the requirements of a newly professionalized scientific elite.[6] As a result, museum curators frequently quibbled over seemingly minor issues of museum practice—such as the way to display specimens and how many objects to exhibit—but often underlying these technical debates were more fundamental differences of opinion concerning the function of museums.

Attitudes toward all these aspects of museum work varied according to local circumstances, institutional loyalties, and national allegiances. The plans and procedures adopted by a government repository located in a major metropolitan center, as a result, might seem to have limited relevance for a small collection developed in the hinterland. Yet, since museums, like other scientific institutions, were seldom started from scratch, successful models served as inspiration and example for less fortunate attempts elsewhere. Personnel were sometimes imported in order to put a particular plan or design into effect. With constant international imitation and cross-fertilization the norm, ideas and innovations freely traveled the museum circuit during the decades preceding 1900.

Metropolitan Inspiration:
The British Museum and the
"New Museum Idea"

Having abandoned its parent across town in Bloomsbury, the new
British Museum (Natural History) in South Kensington opened its
doors to the public in 1881, as the jubilee celebrations commemo-
rating Victoria's accession to the throne began. At the time, it was
the world's most remarkable natural history museum. A gothic
"temple to science," the museum exemplified Victorian architectural taste, which had been seen earlier in the ornate structure built
for the New Museum at Oxford. Following the Oxford plan, which
incorporated biological symbolism into the design of the museum,
the walls of the British Museum displayed bas-relief representations
of living and extinct species. The eclectic genius of the architect
Alfred Waterhouse combined Romanesque arcades and Baroque
staircases with cast-iron columns and a glass roof typical of contemporary railroad stations. A few observers found the terra-cotta facing
dingy and the high central hall empty, with adjacent side galleries
like factory rooms, but most visitors were delighted with the new
museum. It contained four acres of exhibition space in addition to
rooms for laboratories, workshops, and storage.[7]

It was not the dramatic building alone that placed the British Museum in the first rank, but the overall quality and extent of its collections. Especially in paleontology and mineralogy, the British Museum soon supplied the example to which every other national
museum aspired. Specimens were well displayed, without the
crowding that plagued so many government repositories that functioned more as storehouses. Exhibits made liberal use of descriptive
labels, drawings, photographs, diagrams, and models. Museum publications, too, won accolades from colleagues in sister institutions.
Descriptive catalogues provided more information than the typical
lists of museum holdings and served as definitive monographs,
while a series of elementary guidebooks provided fine examples of
lucid popularization. Because so many ideas had been tried and the
staff were eager to share their experience with others, the British
Museum became the most important institution in the world for
answering queries concerning museum practice.[8]

The British Museum also became an arena in which various

schemes for classifying and arranging natural history collections were championed. There, as elsewhere, the attractively simple "new museum idea"—the separation of study from exhibition materials—had triumphed by 1900.[9] According to its foremost exponent, British Museum director William Henry Flower, the essence of the idea involved organizing museums around a dual purpose: as research collections and as educational tools.[10] Flower insisted that numerous specimens representing specific and, especially, varietal forms had little significance for the average visitor and belonged in a segregated study section. He compared the practice of exhibiting every museum specimen to the framing and hanging of each page in every library book. The observer who confronted objects row upon row came away, if not bewildered, at least bored. Instead the liberal use of storage drawers and cabinets meant that specimens could be well preserved away from dust, pests, and light, housed in minimal space, and easily retrieved by researchers. To make the study collections truly useful, rooms adjacent to the exhibits needed to be furnished with adequate tables, reference books, proper lighting, and an accessible museum staff.[11]

The public exhibits, by contrast, were designed to impart a general understanding of the kingdoms of nature to the layman. Only the best specimens were displayed, in order to illustrate a particular principle or taxonomic category. No duplication of materials was permitted. Specimens were to be shown in uncrowded cases at a reasonable height, accompanied by informative labels. The curator might fashion his collection around such specific themes as geographical distribution or evolution. His resources could lead him to emphasize one group while excluding others. The function of guidebooks and catalogues was to increase the educational value of the public materials (see figure 1).[12]

Even before Flower's time, John Edward Gray, Keeper of Zoology during the middle decades of the nineteenth century, had tried to persuade his colleagues at the British Museum to follow a similar plan. Collections carefully pruned and arranged, he argued, gave the visitor "the greatest amount of instruction in the shortest most direct manner." Serious students, who could more easily compare and measure objects by using unmounted specimens in ordinary rooms away from onlookers, would also benefit, he insisted.[13] These ideas, however, threatened the designs of the formidable and influential

Figure 1. Plan for the Arrangement of an Ideal Museum of Natural History

WINDOWS		WINDOWS
STUDIES FOR OFFICERS	BOARD ROOM DIRECTOR AND SECRETARIAT	ASSISTANTS AND STUDENTS

R	E	S	E	R	V	E	REFRESHMENT ROOMS ETC	LIBRARY AND LECTURE ROOM	C	O	L	L	E	C	T	I	O	N	S

P	U	B	L	I	C	HALL	E	X	H	I	B	I	T	I	O	N

PUBLIC ENTRANCE

Source: W. H. Flower, *Essay on Museums and Other Subjects Connected with Natural History* (London, 1898), p. 50

Richard Owen, Flower's predecessor as superintendent, for a revitalized "national museum of natural history." The purpose of a national repository, insisted Owen, was to display every varietal and specific form, an aim that a lesser collection with limited resources could not hope to attain. Owen's plan called for a massive structure to accommodate the largest specimens of huge mammals such as elephants and whales that might be procured. Even duplicates could be exhibited in "distinct series with different aims," such as separate taxonomic, educational, and "British" collections.[14]

Owen's dream seemed to be materializing as he superintended the transfer of the British Museum's natural history collections from Bloomsbury to South Kensington during the early 1880s. Yet with the move completed and Flower succeeding to the directorship in 1884, the views of Gray began to enjoy a renaissance. Owen's plan for a synoptic index museum to the larger collection, for example, which had been incorporated into the new structure as "chapels" flanking the central hall, never found favor with his successors.[15] Such an abstract of the general collection became unnecessary because Flower and other curators mounted selective, not comprehensive, exhibits.

The new museum idea, ultimately triumphing at the British Museum, also helped to transform museum practice in other parts of the world. By 1900, most museum men accepted the principle that less was more instructive in museum displays and that the remaining inventory of specimens should be reserved for scientific investigators. Yet the acceptance of the new museum idea posed a quandary that confused the efforts of curators throughout the next several decades. Museums were intended to serve two distinct audiences simultaneously: a few scholars and the public at large. But it was difficult, if not impossible, to be equally responsive to the needs of both groups. As a result, any concentration on scientific requirements left directors open to the charge of neglecting their responsibilities to the public. Emphasizing the function of popular educator, on the other hand, suggested that museums were not performing their research function. The attempt to fulfill these two conflicting mandates provided a constant source of vexation to museum administrators all over the world.[16]

Other Metropolitan Museums: Europe and the United States

At the same time that the British Museum underwent expansion and reorganization, museums elsewhere in Europe experienced a surge of growth stimulated by the increasing size of their collections since midcentury. Many new buildings were constructed, and there were additions and renovations to old ones. The costly Vienna Naturhistorisches Hofmuseum opened to generous praise from "new museum" enthusiasts in 1889. In the same year, the zoological collections at the Paris museum were moved to new Romanesque-style quarters replete with animal carvings; anatomy and anthropology found improved accommodations a decade later. The Bohemian Museum opened in Prague in 1894 and the Royal Belgian Museum inaugurated new buildings in Brussels in 1903.[17]

Most Continental museums displayed every object in their possession and followed what Gray had contemptuously labeled the French plan. This called for "attaching each specimen to a separate stand, and marshalling them like soldiers on the shelves of a large open case."[18] A few museums, though, became converts to the views being articulated across the channel. The Prague museum, for example, displayed true-to-life exhibits incorporating natural surroundings. It also made liberal use of drawings and models. The new building in Vienna separated exhibits from an internal core of workrooms, storage, and study collections.[19]

Whereas by 1900 Germany could claim 150 museums of natural history, Britain 250, and France 300, the United States could count a respectable 250.[20] In America, where museums had grown in quantity and improved in quality during the latter half of the nineteenth century, successful institutions traditionally depended upon individual initiative. Examples include the Museum of Comparative Zoology, for which Louis Agassiz raised $300,000, and the museum at Yale University, which benefited from George Peabody's bequests. Public museums, directed by trustees and supported by government, gradually began to supplant the cabinets of natural history typically associated with private colleges, lyceums, and academies.[21] Foremost among these were the American Museum of Natural History in New York City and the National Museum, part of the Smithsonian Institution, in Washington, D.C. The American Museum, established in 1869, moved in 1877 into new quarters that occupied thirteen acres. Two years later the National Museum began to construct a new building to accommodate the donations received upon the dismantling of the Philadelphia Exposition of 1876. The museum's functional structure attracted worldwide attention for covering more than two acres while costing taxpayers only $250,000.[22]

Like the British, the Americans had become expert at mounting explanatory and true-to-life displays, incorporating habitats, charts, diagrams, and photographs. The National Museum perfected the use of plaster casts and models to make fragments of bone and fossils comprehensible, permit reconstruction of partial skeletons, or fill the gaps in a series of specimens. Along with the American Museum of Natural History, it established a reputation for portraying animal groups in their natural environments. This method, initiated with birds and mammals, was soon extended to fish, invertebrates, fossils, and finally to plants, through the use of finely worked glass and wax models. Both museums were the beneficiaries of techniques pioneered at the taxidermy firm run by Henry Ward in Rochester, New York, by hiring his trainees as curators.[23]

Colonial Museums:
F. A. Bather's Survey of 1893

Less widely known than the advances and expansion of natural history museums in European and American metropolitan centers during the latter half of the nineteenth century is their development in other parts of the world at about the same time. With the gradual accumulation of capital in the hinterland—coming from the development and exploitation of natural resources, made possible by improved transportation networks—colonial legislatures began to support public museums.[24] As part of a survey conducted under the auspices of the Museums Association in 1893, British Museum curator F. A. Bather evaluated museum resources throughout the Empire. What united these scattered and diverse institutions, according to Bather, was their colonial status. His paper is unique in its time for supplying cross-cultural comparisons and attempting general definitions of that little-studied category, colonial museums. Forty years would elapse before Bather's pioneering work was further elaborated and expanded in the series of directories and reports compiled by Henry A. Miers and Sydney F. Markham.

The first characteristic of a colonial museum, argued Bather, was a purview limited to materials garnered from the immediate environment. Visitors to the museum—whether fresh immigrants from abroad or inhabitants of the interior—came to see local products, which were best kept apart from any other collections the museum might possess. In order to encourage and sustain support by residents, Bather reasoned, all donations, no matter how common, should be accepted. The other special interest of most visitors involved applied science. Although Bather pitied the curator who had to spend his days analyzing ores, an emphasis on practical matters had to be tolerated; it was preferable to having no publicly recognized utilitarian function, and hence being under constant threat of closure. If separate technological museums were allowed to be established, support for museums of natural history might dwindle to nothing.[25]

The second characteristic used by Bather to distinguish colonial museums was their fundamental dependence upon foreign institutions. Bather argued that external support, particularly from the mother country, enabled colonial museums to weather day-to-day

problems, such as periodic financial depressions, invasion by insects, and deleterious climates. Assistance might take the form of
specimen exchanges or the advice of specialists whose skills were
unavailable in the colony.[26] Without the aid of metropolitan institutions, museums in the hinterland could neither function properly
nor, ultimately, survive.

It is questionable, however, whether Bather captured the essence
of the colonial museum despite the originality of his approach. Certainly curatorial staffs recognized the necessity of ministering to the
needs of the local populace, whose support they required in over-
coming daily tribulations. But colonial museums depended upon
metropolitan museums not only for materials, but also for architectural designs, organizational models, and qualified personnel. Bather's failure to discuss the actual political or economic status of the
countries in question is probably the result of his exclusive concern
with Britain's overseas possessions. The omission is significant
nevertheless, because colonialism in science need not follow the
mapmakers' or diplomats' charts. Historians now recognize that
new nations, such as Canada and Argentina, were as dependent
upon European traditions at the end of the nineteenth century as
were de facto colonies in Australia and New Zealand.[27]

Bather's shortcoming was that he underestimated and therefore
understated the frequency with which fine institutions were almost
literally carved out of the wilderness. Images of Victorian palaces of
science danced in the heads of museum men in the hinterland, and
they sought to forge reasonable facsimiles under adverse circumstances. Some of these museums emphasized local objects, including the artifacts of aboriginal peoples, technological implements,
and examples of native flora and fauna. But other museums, especially those associated with universities and colleges, sought to display broad collections selected to represent the diversity of animal,
vegetable, and mineral kingdoms on a worldwide scale. These were
intended principally for the instruction of students destined for careers in mining and medicine.

Like their metropolitan counterparts, colonial museums also set
out to educate or at least to "elevate" the middle and lower classes.
As one writer described the educational power of natural history
specimens: "Let each object represent so much knowledge, to
which the very mention of its name will immediately conjure up a

crowd of associations, relationships, and intimate acquaintances, and you will then see what a store of real knowledge may be represented in a carefully arranged cabinet." Advocates repeatedly explained that viewing properly organized natural history museums instilled, as well, a sense of order, method, and law. In addition to developing an individual's powers of observation and reflection, museums might stimulate healthy exercise. The resultant interest in some branch of natural history could lead the eager student into the "pure air and pleasant scenes" of the countryside, thus offering the "best antidote to habits of dissipation or immorality."[28]

Such frequent and enthusiastic endorsements of the multiple intellectual, social, and moral benefits conferred by museums of natural history would seem to indicate that their promoters were not simply mouthing empty rhetoric. As even one intensely critical historian of museums admits, the development of nonart museums has rested on foundations at least potentially more democratic than those supporting art museums.[29] Perhaps to a greater extent than for their colleagues elsewhere, colonial curators brought a missionary zeal to their work, which implied, as well, a concern with extending the frontiers of civilization.

Colonial and Metropolitan Museums: Some Comparisons

The museum movement traveled around the globe with remarkable speed as museums throughout North and South America, Africa, Asia, and the South Pacific were founded, renovated, or given new quarters during the last decades of the nineteenth century. Some colonial museums copied the general organization of European institutions. The Hobart Museum in Tasmania and the Museu Goeldi at Para, Brazil, for example, followed the Parisian pattern of supplementing specimen collections with "living" museums—botanical gardens, zoological parks, or aquariums.

As the *Encyclopedia Britannica* generalizes the case, the architecture of nineteenth-century museums was monumental and imposing, because of their association with national prestige and civic pride. Architects achieved a grandiose effect by incorporating "colonnades and arches, high vaulted interiors, [and] vast flights of stairs" into buildings often situated in parks or on broad avenues.[30] On occasion, colonial museums—such as the South African Museum at Cape Town and the National Museum of Victoria at Melbourne—imitated the architectural detail of the cathedrallike British Museum.[31]

More commonly, however, museums in the hinterland followed the design of a classical temple, which seemed to symbolize better the importance of the government that supported them. This architectural convention dictated a symmetrical facade, a commanding entrance, perhaps with a portico, and a multistoried central hall surrounded by balconies and staircases.[32] The third edition of James Fergusson's *History of the Modern Styles of Architecture* of 1891 singled out Montreal's Peter Redpath Museum as a particularly fine example of neoclassical style emerging from a rather conservative and provincial tradition of British colonial architecture.[33]

Some museums adapted their designs to the immediate environment or to local conditions. The Museum Paulista—also called the Museu Ypiranga, after the nearest trolley line—took advantage of São Paulo's mild climate by making its second floor rooms accessible only from an outdoor balcony. The Albert Museum at Jaipur followed Indo-Saracenic architectural style in its new building of 1876. Institutions without the money to erect their own facilities used structures intended for other purposes, such as public libraries and legislative buildings.

The size of the scientific staff and the resources of colonial museums differed widely from one location to another, as table 1 suggests. The first column shows the enormous range in budgets (converted to U.S. dollars), from $2,000 for the Canterbury Museum to $142,000 (including salaries) for the National Museum at Rio de Janeiro, the showcase of Brazilian museums. At this time, nationally supported museums in Paris and London each spent more than $200,000 a year. Another metropolitan institution, however, the Royal Belgian Museum in Brussels, had a budget of just over $30,000. A first-class British provincial museum spent not even $4,000 a year.[34] While colonial museums, then, failed to attract funding to equal that of perhaps the top five museums in the world, their budgets rank with those of the better European institutions.

The second column, showing the number of personnel in natural history in colonial museums, may likewise be compared to metro-

Table 1.   Colonial Museums, c. 1900

Name of museum and location	Annual budget or expenditure (U.S. dollars)	Size of natural history staff (number of persons)	Number of visitors a year
New Zealand			
Auckland	3,896	1	1890: 30,000
			1915: 100,000
Canterbury, Christchurch	2,000[a]	2	n.a.
Australia			
National, Melbourne	14,366.50	1893: 8	1889: 133,065
		1903: 10	1894: 93,395
			1898: >95,698
Australian, Sydney	1891–92: 38,600	1891–92: 15	>120,000
	1893: 23,678	1893: 13	
	1897: 31,460		
	1901: 25,220	1909–10: 7	
South Africa			
Rhodesian Bulawayo	5,357	1	n.a.
South African Cape Town	12,175	4	>88,000
Canada			
Provincial Halifax, N.S.	2,600	1	9,520[b]
Provincial Victoria, B.C.	5,500	2	34,640
India			
Indian Calcutta	19,181	5	1,000,000[b]
Government Central Madras	20,940[b]	n.a.	527,707[b]
South America			
La Plata	1898: 38,600	6	n.a.
	1909–10: 86,850		
National Rio de Janeiro	142,000[b]	7	150,000[ab]
		1909–10: 12	
National Santiago, Chile	32,140	8	>3,000[c]
	1909–10: 40,200		

Note: Statistics for 1900 are given in all instances in which they are available. Several numbers are shown for a single institution in instances in which amounts fluctuated significantly during even a short period.

Sources: F. A. Bather, "Some Colonial Museums," *Museum Association* (Dublin, 1895), pp. 193–239; L. V. Coleman, *Directory of Museums in South America* (Washington, D.C., 1929); "The Difficulties of the Australian Museum," *Natural Science* 15 (1899): 317–19; S. F. Markham and H. Hargreaves, *The Museums of India* (London, 1936); F. J. H. Merrill, "Natural History Museums of the United States and Canada," *New York State Museum Bulletin* 62 (1903): 3–233; H. A. Miers and S. F. Markham, *A Report on the Museums of Canada* (Edinburgh, 1932); *Minerva: Jahrbuch der gelehrten Welt* (Strassburg, 1891–); Paul Marshall Rea, "A Directory of American Museums of Art, History, and Science," *Bulletin of the Buffalo Society of Natural Sciences* 10 (October 1910): 1–347.

n.a. Not available.
a. Approximate figure.
b. Figures taken from the period 1929–36.
c. Sundays and holidays.

politan institutions such as the American Museum of Natural History, which housed fourteen scientists among its seventy-one employees in 1877. Many colonial museums accommodated scientific staffs of similar size.[35] In South America, where European traditions were especially strong because of the dominating influence of French, Italian, and German curators, departmental keepers often doubled as professors of natural science in the local university.

Statistics on the annual number of visitors suggest how well museums functioned as instruments of popular education. According to the third column, it was not unusual for a colonial institution to attract crowds of 100,000 people each year. Even illiterate natives, it was claimed, enjoyed exhibits showing local resources and the history of their people.[36] Although the British Museum and the American Museum were receiving about 400,000 visitors annually by 1900, another major institution, Prague's Bohemian Museum, boasted of an attendance of 90,000 in one year.[37] Colonial museums, then, ranked as popular institutions on a global level, especially given the more limited sizes of the local populations that they served.

A fourth quantitative indicator, omitted from table 1 because it was seldom recorded, is the size of collections—a figure that also reflects the significance of colonial museums. Ceylon's Colombo Museum contained more than 100,000 zoological specimens, the Indian Museum at Calcutta and the Geological Survey Museum in Ottawa each more than 350,000 items, emphasizing rocks and minerals, and the National Museum of Melbourne more than 500,000 objects. The world giant was perhaps the National Museum in Washington, D.C., with more than 3 million items during the early 1890s. By the 1930s it counted more than 12 million objects and was acquiring more at the rate of 500,000 a year.[38] Though trailing behind this leader, many colonial museums attempted to assemble reasonably large collections to be used for teaching or research purposes.

Tables 2 and 3 compensate for the fact that table 1 draws upon those institutions for which the fullest information is available, with the result that it portrays museums in India too favorably and those in New Zealand too critically. Tables 2 and 3 are more accurate for purposes of comparison. Table 2 shows Britain and the United States far ahead of even the most auspicious overseas environments.

Table 2. *Expenditure on Museums, by Country, c.1930*

Pence per capita

County	Approximate amount spent annually
Great Britain	8.0
South Africa	1.5
Canada	2.25
Australia	3.0
New Zealand	3.5
United States	8.0
India	0.033

Sources: S. F. Markham and W. R. B. Oliver, *A Report on the Museums and Art Galleries of New Zealand* (London, 1933), p. 72; S. F. Markham and H. C. Richards, *A Report on the Museums and Art Galleries of Australia* (London, 1933), p. 11; S. F. Markham and H. Hargreaves, *The Museums of India* (London, 1936), p. 4.

The low ranking of the colonies here is derived from the underdeveloped state of vast rural areas, which virtually canceled the advances of progressive cities and towns. Table 3 suggests, however, that moderate-sized colonial towns were in fact better endowed than their analogues in England. Ottawa, Dunedin, and Durban, for example, mustered considerable resources to support museums. Larger cities, too, such as Sydney and Cape Town, could be categorized as "good museum towns," according to the compilers' criterion of providing at least sixpence a year per capita for museums and art galleries.[39] In conclusion, then, all these figures support the view that colonial museums should be recognized as important loci of scientific activity.

Some Descriptions of Colonial Museums

Historians have largely failed to recognize the extent to which non-European museums of the late nineteenth century sought to assemble significant collections that included prime materials obtained abroad. Partial justification for museum building came from the desire to edify the local populace, especially younger colonists who had never seen the natural and artificial products that were commonplace in Europe. Apart from this public rationalization, though, somewhat different reasons inspired curators who found themselves working in places and under conditions that earlier they

Table 3. *Expenditure on Museums, by Town with Population between 100,000 and 160,000, c.1930*

Pence per capita

Town	Amount spent annually on museums and art galleries
England	
Birkenhead	5.0
Blackburn	4.9
Bournemouth	4.9
Brighton	6.4
Derby	4.0
Gateshead	2.4
Huddersfield	7.6
Middlesbrough	1.6
Norwich	12.0
Oldham	3.3
Preston	10.0
Southend-on-Sea	2.6
South Shields	0.7
Stockport	1.9
Tottenham	1.9
Wolverhampton	2.7
Woolwich	1.6
Canada	
Hamilton	1.7
Ottawa	54.0
Quebec	5.8
Australia	
Newcastle	0.5
New Zealand	
Christchurch	5.4
Dunedin	11.0
Wellington	6.5
South Africa	
Durban	9.5

Source: S. F. Markham and W. R. B. Oliver, *A Report on the Museums and Art Galleries of New Zealand* (London, 1933), p. 71.

would have thought inconceivable. Using European practices and methods as the measure, these men believed that a proper museum had to include objects of universal value as well as materials of local interest alone. The diversity of the natural world was to be shown at least through representative types, if not by a multiplicity of individual forms. Reputations, it was felt, depended upon the number of specimens amassed, with considerable cachet attached to the acquisition of exotic foreign materials.

If, then, a naturalist's fortune had led him to places like Cape Town or Bombay, the question became, as Melbourne's museum director Frederick McCoy put it, how to *grow* a museum in the hinterland. Growth required financial resources, which came to most museums at some stage in their development, but rarely at an adequate level or on a regular basis. Certain local objects, however, such as rare natural history specimens or archaeological implements, possessed commercial value and might be sold or exchanged for materials from abroad. International exhibitions, whether the huge fairs held in large metropolitan centers or the more modest colonial affairs, called attention to these treasures from all over the world.

Some colonial museums fared better than others in turning the accidents of nature to good account. In Africa, museums of natural history were concentrated in the extreme southern portion of the continent. South African and Rhodesian museums survived only in centers with large white populations, such as Cape Town, Durban, Pietermaritzburg, and Grahamstown. Although some institutions attempted to attract native inhabitants, more typical, perhaps, was the museum that excluded blacks every day but Thursday, when admittance depended on wearing boots or shoes. Even the descendants of English and Dutch settlers failed to attend museums in great numbers, because, according to one observer, the magnificent climate vastly outshone their rather unappealing contents. Despite above-average architecture and design, the meager financial resources of most South African institutions left them impoverished by North American or European standards.[40]

African museums were well endowed, however, in comparison with their neighbors farther east. The impetus for the creation of a number of India's principal museums—in Bombay, Calcutta, and Madras—came from the donation of the collection of a local philosophical or scientific society. Following the initial bequest, though, specimens often fell prey to the forays of insects and extremes of temperature and humidity. These adverse conditions were exacerbated by poor curatorial care. Hindu workers, reluctant to take life, tended to tolerate pests, while the caste system encouraged a rigid adherence to assigned duties only. In the words of one observer, there was no other place in the world where museums "count[ed] for so little . . .[were] so meagerly supported, or . . .[were] so few and far between." Many large towns lacked museums altogether; those that existed appeared as "gingerbread palaces, fantastic and bizarre, or gloomly prisonlike edifices . . .[with] galleries more suited to be mausoleums." Some still exhibited freaks and monsters to the curious; others showed obsolete maps and charts occasionally displayed upside down. As in Africa, widespread illiteracy, extreme poverty, and patterns of rural settlement made museums irrelevant to the vast majority of the populace.[41]

Museums in Canada, South America, and Australasia

Elsewhere, however, the museum movement was more successful. The leading museums of postconfederation Canada, for example, were concentrated within a 150-mile radius of Montreal. The largest collections belonged to the museum of the Geological Survey of Canada, later christened the National Museum, which was transferred to Ottawa after nearly forty years in Montreal, along with Survey headquarters, in 1881. Ottawa's huge reserve of 150,000 paleontological specimens placed it far ahead of any other Canadian museum. Nevertheless, the large number of items displayed in Montreal's Natural History Society museum, together with McGill University's Peter Redpath Museum, made the city a close second to the national capital. Quebec City, which for a time had alternated with Toronto as capital of the young colony of Canada, placed third because of the excellent collections assembled in the museum of Laval University and in the provincially supported Museum of Public Instruction of Quebec.

Although the balance of power would shift with the creation of the Royal Ontario Museum several decades later, Ontario's best museums, those in Kingston and Toronto, housed only a fraction as much material as was exhibited in the threes main museum centers of Ottawa, Montreal, and Quebec City.[42] The typical Canadian museum of the day—crowded into several rooms and controlled by Catholic educational institutions in Quebec or by some other kind of organization or university in Ontario—usually contained around 5,000 natural history specimens. Municipal museums, more closely associated with the needs of the local citizenry than with those of a learned society or school, were almost entirely absent. As one authority who surveyed Canadian museums in the 1930s summed up the case, Canada had long been active in collecting, but its educational museums were embryonic and its museum endowments negligible.[43]

South American museums, by contrast, tried to function both as research institutions and as instruments of popular enlightenment. Supported by national or provincial governments, important museums could be found in every capital city. Rio de Janeiro, Buenos Aires, Santiago de Chile, and Montevideo erected autonomous museums of natural history. Bogotá and Caracas combined natural history with art and other subjects and housed all the collections under one roof. Most of these institutions enjoyed comfortable annual appropriations and spacious quarters that accommodated extensive staffs. They undertook considerable scientific fieldwork and issued a variety of research publications. According to one observer, the larger museums in South America attracted 100,000 to 150,000 visitors annually, which amounted to about 5–10 percent of the local population.[44]

Museums' resources in Australia and New Zealand compared favorably with those of cities of similar size elsewhere. In New Zealand, as a result of the action of public-spirited and energetic citizens, five towns—Nelson, Christchurch, Wellington, Auckland, and Dunedin—possessed museums by 1877. Australian museums ranged from excellent to mediocre, despite the strong hand of individual state governments in funding. Following settlement patterns, museums tended to be concentrated in the southeastern corner of the mainland. By the 1870s about a dozen museums, emphasizing geology or zoology, had been created; another dozen had opened, through the efforts of local learned societies, by 1900.[45]

The success of the museum movement is especially surprising given the adverse circumstances under which colonial institutions, even in these most favorable situations, had to function. Political and financial problems daily tried the patience of curators such as McCoy, whose museum fell victim to ministerial whims as internal government reorganizations shuffled it from one department to another. The acute depression of the early 1890s caused the dismissal of the entire scientific staff at the Queensland Museum in Brisbane, and the acquisitions budget for the Australian Museum in Sydney was reduced more than 80 percent at that time.[46]

Unanticipated acts of God seemed to strike even well-endowed and long-established colonial museums with distressing frequency. Fires destroyed collections at Sydney and at New Westminster (British Columbia), where the entire town went up in flames. Earthquakes wrecked museums in South America and New Zealand. Less extreme circumstances—fluctuations in humidity and temperature, as well as an excessive amount of sunlight—caused mounted specimens to shrink, crack, and fade. In tropical areas, damage was also caused by the incursions of moths, mites, birds, and monkeys. By contrast, John William Dawson in Montreal had to contend with a snow blockade that delayed a shipment of materials from New York.[47]

The attitude of the local populace was sometimes just as intractable as the environment. One sympathetic Australian politican complained to McCoy that "it is difficult to indoctrinate people with ideas altogether foreign to those which have already occupied their minds and it is still more difficult to implant a new idea in a mind hitherto fallow."[48] Several curators worried about vandalism and the theft of coins and revolvers because "visitors were too indiscriminately admitted." Indeed, when thieves stole gold from several Australian museums such fears proved to be well founded.[49]

According to those who surveyed colonial museums, themselves museum professionals, the skill and energy of the curatorial staff spelled the difference between success and failure. In South America, a number of Europeans, such as Hermann von Ihering at São Paulo and Rudolph Amandus Philippi at Santiago, transformed the drudgery of museum building into magnificent testimony to a life's work. In Australia, the sons of prominent metropolitan keepers such as Henry Woodward and Robert Etheridge served their scientific apprenticeships as curators at the outposts of the British Empire. In many instances, however, the problems encountered and the low salaries paid made colonial positions somewhat less than attractive to ambitious naturalists.

A handful of active, enthusiastic men directed museums located in the principal urban centers of Canada, Argentina, Australia, and New Zealand. One historian, commenting on the striking similarities among these societies, noted that "obviously the formal political status of a territory must not be confused with the reality of its positions within a wider economic system." Nevertheless, with the exception of Argentina, all these countries were colonies or former colonies of Great Britain. Argentina, because of its strong economic ties to Britain, has been seen as part of that country's "informal empire."[50]

It is not surprising that museums of natural history enjoyed particular success in these environments where "outposts of transplanted European society" were created. Unlike most other colonial or neocolonial situations, these countries had been recently settled by European immigrants. Native peoples, already decimated or otherwise subjugated, posed no threat to establishing young societies that would eventually surpass the standards of living of their European progenitors. From around 1870 to 1914, these "Dominion capitalist societies" became the "centre pieces of British imperial strategy." Wool, meat, dairy products, cereals, and lumber were exported from their sparsely settled interiors in return for industrial products, loan capital, and immigrants. Among the European powers engaged in this process, Great Britain was especially active in financing railroads and other means of transport which helped to get food and raw materials to market.[51] It is not accidental that this period of especially rapid economic growth coincided with the time of museum expansion and that the British impress on their development was especially strong.[52]

Enterprising directors found in Canada, South America, and Australasia, then, fertile environments for museum building. Most curators emphasized local products and resources initially, subscribing to the theory that residents were most interested in "the things they can find about home."[53] Attracting as many visitors as possible helped to justify their position as educational institutions. In order to do so, they often displayed coins, ancient relics, and ethnological materials alongside natural history specimens of every description. It was not unusual to show technological apparatus, such as agricultural machinery. The particular interests of the keeper or the contents of a bequest also influenced the composition of collections. As Bather remarked during his survey of colonial institutions, "the first lesson a curator has to learn is to cut his coat according to his cloth."

Australian museums, for example, perfected geological and mineralogical exhibits, while the Canterbury Museum displayed specimens of wool accompanied by practical hints from sheepbreeders.[54]

Soon colonial curators in these favorable circumstances no longer remained content to direct museums of merely local significance and sought prime foreign specimens as well. They aimed to assemble collections to rival those at the great metropolitan museums, to be "to the naturalist what a dictionary is to the scholar."[55] Julius Haast at Christchurch and McCoy at Melbourne mastered the mechanisms by which excellent collections could be built, by exchanging or purchasing specimens abroad. Museums affiliated with local universities, such as the La Plata Museum and the Peter Redpath Museum, strove to display comprehensive and instructive series of natural objects for those who wished to master the rudiments of systematic biology or mineralogy. Several other institutions, including the public museum of Buenos Aires, established solid reputations for scientific research.[56]

Colonial museums, however, tended to exhibit specimens row upon row and for the most part neglected to incorporate up-to-date techniques such as explanatory labels and habitat cases. A few institutions, nevertheless, adopted the new museum idea and accordingly separated research from exhibition collections.[57] McCoy set out with a master plan to develop teaching collections based on the exhibition of generic types and representative geographical groups of species. This "index" to organic life required no subsequent rearrangement while new specimens could be continuously integrated into the general collection.[58] Making a distinction between local and foreign objects gave other museums some rudimentary arrangement. The remarkable collection of moa skeletons at the Canterbury Museum, for example, distinguished New Zealand specimens from the other exhibits strikingly.[59]

Conclusion

In his pioneering survey of 1893, F. A. Bather distinguished colonial museums by their adaptation to local circumstances and their dependence upon foreign institutions. Certainly museums in the hinterland did follow metropolitan models and did import European staff. But the personnel often stayed on to shape a new kind of scientific career in the hinterland, while metropolitan patterns worked only insofar as the immediate environment permitted. The very existence of these indigenous museums overseas, in fact, militated against extreme scientific dependence, because they protected local treasures against plunder by European institutions. They may be seen as encouraging the first steps toward scientific independence, in the sense that they provided a haven for natural resources and products of the locality. They came to offer, in addition, scientific training and a variety of positions to native-born talents inclined toward natural history.

Colonial museums flourished, however, only so long as the museum movement prospered elsewhere. By the time the youngsters who flocked to the British Museum in South Kensington during the 1880s and 1890s had grown to maturity, museums of natural history had begun to lose their disciplinary centrality. The natural sciences—now pursued by specialist geologists, zoologists, and botanists—presented more promising vistas from the microscopic, rather than the macroscopic, level. Even Darwinian evolution, which at first had accelerated the zeal to collect by rationalizing taxonomy and giving new scientific significance to varieties, seemed to offer greater inducements to the geneticist in the laboratory than to the ornithologist or mammalogist in the field. Those who remained in the field found that such new techniques as photography provided better data about ecology and behavior than copious museum specimens.[60]

Developments external to biological discourse, such as the rise of public and private research institutes and the remarkable expansion of universities, also diverted resources and interest from museums to other scientific endeavors. Nevertheless, until the second or third decade of the twentieth century, the history of natural history was largely caught up with the development of museums. To look at these museums today is to glimpse magnificent monuments to a remarkably hardy and adaptable tradition.

1. David Murray, *Museums: Their History and Their Use* (Glasgow, 1904), vol. 1, p. 224.

2. Exceptions to this general rule are the interesting study by Camille Limoges, "The Development of the Muséum d'Histoire Naturelle of Paris, c. 1800–1914," in *The Organisation of Science and Technology in France 1808–1914*, ed. Robert Fox and George Weisz (Paris: Maison des Sciences de l'Homme; Cambridge: Cambridge University Press, 1980), pp. 211–40; and "Papers Presented at the International Conference on the History of Museums and Collections in Natural History," *Journal of the Society for the Bibliography of Natural History* 9 (1980): 365–670. S. Bedini traces the development of museums devoted to physical sciences and technology in "The Evolution of Science Museums," *Technology and Culture* 6 (1965): 1–29. See also the recent study by Sally Gregory Kohlstedt, "Australian Museums of Natural History: Public Priorities and Scientific Initiatives in the 19th Century," *Historical Records of Australian Science* 6 (1983): 1–29.

3. Murray, *Museums*, vol. 1, pp. 5–19, 186–208.

4. Carl von Linné, *Reflections on the Study of Nature*, trans. James E. Smith (London, 1785), pp. 24ff.

5. See, for example, A. R. Wallace, "Museums for the People," *MacMillan's Magazine* 19 (1869): 244–50.

6. For a discussion of these issues see Duncan F. Cameron, "The Museum: a Temple or a Forum," in the special issue of UNESCO's *Journal of World History* 14 (1972): 189–202, especially p. 194, and Kenneth Hudson, *A Social History of Museums: What the Visitors Thought* (London: Macmillan & Company, 1975, especially p. 13 and chap. 2.

7. J. Mordaunt Crook, *The British Museum* (London: Allen Lane, 1972), pp. 200, 204ff., 207. See Henry Scadding's description of the interior of the New Museum in his "On Museums and Other Classified Collections, Temporary or Permanent, as Instruments of Education in Natural Science," *Canadian Journal of Industry*, n.s.13 (1871): 15–16. Superb illustrations and discussion of the architecture of the South Kensington Museum appear in *Alfred Waterhouse and the Natural History Museum*, by Mark Girouard (London: British Museum, 1981), p. 38 and elsewhere. See Richard Owen, *The Life of Richard Owen* (London, 1895), vol. 2, p. 53; A. B. Meyer, *Studies of the Museums and Kindred Institutions of New York City, Albany, Buffalo, and Chicago, with Notes on Some European Institutions* (Washington, D.C.: Government Printing Office, 1905), pp. 523–24; A. E. Gunther, *A Century of Zoology at the British Museum through the Lives of Two Keepers, 1815–1914* (London: Dawsons, 1975), p. 349; William T. Stearn, *The Natural History Museum at South Kensington* (London: Heinemann, 1981), pp. 44–45, 49–52, which describes the interior embellishments.

8. Meyer, *Museums and Kindred Institutions*, p. 526. See also E. O. Hovey, "Notes on Some European Museums," *American Naturalist* 32 (1898): 708–10, 712; O. C. Farrington, "Notes on European Museums," ibid. 33 (1899): 775–76.

9. Among the various suggestions for organizing museums of natural history are those of Louis Agassiz, "On the Arrangement of Natural History Collections," *Annals and Magazine of Natural History*, n.s. 3, 9 (1862): 415–19; W. A. Herdman, "An Ideal Natural History Museum," *Proceedings of the Literary and Philosophical Society of Liverpool* 12 (1887): 61–81; and A. L. Herrera, "Les Musées de l'avenir," *Memorias de la Sociedad Científica "Antonion Alzale"* 9 (1896): 221–51.

10. Meyer, *Museums and Kindred Institutions*, p. 325, credits Louis Agassiz with being the first to articulate this principle, which was later adopted in Europe.

11. William Henry Flower, *Essays on Museums and Other Subjects Connected with Natural History* (London, 1898), pp. 13, 15–19.

12. Ibid., pp. 16–19, 38.

13. J. E. Gray, "Museums, Their Use and Improvement," presidential address before the British Association for the Advancement of Science, Botany and Zoology Section, Bath, 1864, pp. 77–78; on Gray's view of museums see also Stearn, *Natural History Museum at South Kensington*, pp. 35–36.

14. Richard Owen, *On the Extent and Aims of a National Museum of Natural History* (London, 1862), pp. 54, 14, 22–23, 27, 70–73, 63; see also Stearn, *Natural History Museum at South Kensington*, pp. 36–37, 57.

15. Gunther, *Century of Zoology*, pp. 349, 352.

16. Indeed, Flower himself had insisted that only in national museums could research be as important as instruction; elsewhere, education should be the primary function. Flower, *Museums and Other Subjects*, p. 38.

17. A. C. L. Günther, "Objects and Uses of Museums," presidential address before the British Association for the Advancement of Science, Section D, Biology, Swansea, 1880, pp. 591–92; Flower, *Museums and Other Subjects*, pp. 43–45; Farrington, "Notes on European Museums," p. 764; Anton Fritsch, "The Natural History Departments of the Bohemian Museum," *Natural Science* 8 (1896): 168–71; Meyer, *Museums and Kindred Institutions*, pp. 584–85, 597; Jean Schopfer, "Natural History Museum at Paris," *Architectural Record* 10 (1900): 55–75.

18. Anton Fritsch, "The Museum Question in Europe and America," *Museums Journal* 3 (February 1904): 247–56, especially p. 252; Gray, "Museums, Their Use and Improvement," p. 77.

19. Fritsch, "Bohemian Museum," pp. 170–71; Farrington, "Notes on European Museums," p. 764.

20. Meyer, *Museums and Kindred Institutions*, p. 324.

21. L. V. Coleman, *The Museum in America* (Washington, D.C.: American Association of Museums, 1939), vol. 2, pp. 13, 14.

22. Meyer, *Museums and Kindred Institutions*, p. 329; see George Brown Goode, ed., *The Smithsonian Institution, 1846–1896: The History of Its First Half Century* (Washington, D.C., 1897), pp. 328–30.

23. The success of the habitat idea depended upon the remarkable advances in taxidermy realized at Ward's Natural Science Establishment. Ward's students—including W. T. Hornaday, F. S. Webster, and F. A. Lucas—took up posts at the National and American museums and at lesser U.S. institutions as well; see G. B. Goode, "Recent Advances in Museum Method," *Report of the U.S. National Museum for the Year Ending 30 June 1893* (1895), pp. 42–45.

24. Suggested by A. E. Gunther, who also introduces the category of colonial museums; see his *Century of Zoology*, pp. 152–53, 320–22, and his *Founders of Science at the British Museum, 1753–1900* (Suffolk: Halesworth Press, 1980), p. 125.

25. F. A. Bather, "Some Colonial Museums," in *Museums Association* (Dublin, 1895), pp. 231–33.

26. Ibid., p. 235.

27. For a masterly discussion of the problems associated with terms such as *colonial* and *empire*, as well as a new framework for their use by historians of science, see Roy MacLeod, "On Visiting the 'Moving Metropolis': Reflections on the Architecture of Imperial Science," *Historical Records of Australian Science* 5 (1982): 1–16, reprinted in the present volume.

28. Wallace, "Museums for the People," p. 245; John Ellor Taylor, ed., *Notes on Collecting and Preserving Natural History Objects* (London, 1876), p. 2; Scadding, "Museums as Instruments," p. 23; Günther, "Objects of Museums," p. 593.

29. Hudson, *History of Museums*, p. 24.

30. *Encyclopedia Britannica* (1970), s. v. "Museum Architecture."

31. *Nature* 56 (1897): 32; R. T. M. Pescott, *Collections of a Century: The History*

of the First Hundred Years of the National Museum of Victoria (Melbourne, National Museum of Victoria, 1954), p. 54.

32. Mark Girouard, *Alfred Waterhouse*, pp. 25–26.

33. Fergusson, *History of the Modern Styles of Architecture*, 3d ed. (London, 1891), vol. 2, p. 170.

34. Meyer, *Museums and Kindred Institutions*, pp. 526, 590, 602; H. Bolton, "Provincial Museums and the Museums Association," *Museums Association* (Sheffield, 1899), p. 91.

35. I count as natural history staff the general director or keeper, curators of sections or departments, taxidermists, osteologists, collectors, and preparators.

36. Henry A. Miers and S. F. Markham, *Reports on the Museums of Ceylon, British Malaya, the West Indies, etc.* (London: Museums Association, 1933), p. 9.

37. Meyer, *Museums and Kindred Institutions*, pp. 521, 330; Fritsch, "Bohemian Museum," p. 171.

38. S. F. Markham and H. Hargreaves, *The Museums of India* (London: Museums Association, 1936), pp. 130, 221; Henry Marc Ami, *Report on the State of the Principal Museums in Canada and Newfoundland* (London, 1897), pp. 67–68; Pescott, *Collections of a Century*, p. 86.

39. S. F. Markham and W. R. B. Oliver, *A Report on the Museums and Art Galleries of New Zealand* (London: Museums Association, 1933), p. 69.

40. H. A. Miers and S. F. Markham, *A Report on the Museums and Art Galleries of British Africa* (Edinburgh: T. & A. Constable, 1932), pp. viii, 3, 10, 17, 31–32.

41. Markham and Hargreaves, *Museums of India*, pp. 3, 21, 49, 57–58.

42. Ami, *Museums in Canada*. Even in 1937, educational institutions in Quebec housed 116 museum collections, 100 of which were devoted to natural sciences, while the other eight provinces could claim only 64 similar collections among them; see also Archie F. Key, *Beyond Four Walls: The Origins and Development of Canadian Museums* (Toronto: McClelland & Stewart, 1973), p. 110.

43. Miers and Markham, *Museums of Canada*, pp. 8–9, 27.

44. L. V. Coleman, *Directory of Museums in South America* (Washington, D.C.: American Association of Museums, 1929), pp. 3–4, 8.

45. Markham and Oliver, *Museums of New Zealand*, pp. 75–81; S. F. Markham and H. C. Richards, *A Report on the Museums and Art Galleries of Australia* (London: Museums Association, 1933), pp. 2–6, 27–28. On Australian museums see also Kohlstedt, "Australian Museums of Natural History."

46. Pescott, *Collections of a Century*, p. 37; Bather, "Some Colonial Museums," pp. 226–29; "The Difficulties of the Australian Museum," *Natural Science* 15 (November 1899): 318.

47. J. W. Dawson to H. A. Ward, 25 January 1881, Henry A. Ward papers, Special Collections, University of Rochester Library, Rochester, N.Y.

48. William Denison to Frederick McCoy, 29 June [1859?], McCoy correspondence, Mitchell Library, Sydney, Australia.

49. Markham and Richards, *Museums of Australia*, p. 29; Gerard Krefft, "The Improvements Effected in Modern Museums in Europe and Australia," *Transactions of the Royal Society of New South Wales* (1868), p. 20; Ronald Strahan, *Rare and Curious Specimens: An Illustrated History of the Australian Museum, 1827–1979* (Sydney: Australian Museum, 1979), p. 31; Herbert M. Hale, "The First Hundred Years of the Museum, 1856–1956," *Records of the South Australian Museum* 12 (1956): 73.

50. H. S. Ferns, *Britain and Argentina in the Nineteenth Century* (Oxford: Clarendon Press, 1960), p. 487.

51. Barrie Dyster, "Argentine and Australian Development Compared," *Past and Present* 84 (August 1979): 91–110, especially p. 91; Warwick Armstrong and John

Bradbury, "Industrialisation and Class Structure in Australia, Canada, and Argentina: 1870 to 1980," in *Political Economy of Australia*, ed. E. L. Wheelwright and K. Buckley, vol. 5 (Sydney, 1983), pp. 43–74; see also E. L. Wheelwright, "Australia and Argentina: A Comparative Study," in his *Radical Political Economy*, (Sydney, 1974), pp. 270–96, especially pp. 270–71.

52. A recent comparative study of science funding suggests that the extraordinarily high level of British support for museums during this period reveals a "passion for museums" in that country; see Nathan Reingold and Joel N. Bodansky, "The Sciences, 1850–1900, a North Atlantic Perspective," *Biological Bulletin* 168 (1985): 44–61, on p. 48.

53. H. H. Howorth, "Some Casual Thoughts on Museums," *Natural Science* 7 (1895): 98.

54. Bather, "Some Colonial Museums," p. 207; Julius Haast to Edward Jollie, 6 September 1870, p. 7, Canterbury Museum Records, Canterbury Public Library, Christchurch, New Zealand.

55. William Swainson, *Taxidermy, Bibliography, and Biography* (London, 1840), pp. 77–78.

56. Markham and Richards, *Museums of Australia*, p. 23. According to Coleman, *Museums in South America*, p. 26, the Buenos Aires museum possessed about 90 percent of all known species of tertiary mollusca.

57. Gerard Krefft, curator of the Australian Museum in Sydney, claimed that J. E. Gray's views had been realized in Australia; see his "Improvements Effected in Modern Museums," pp. 15–25, especially p. 21.

58. As proposed by J. S. Henslow, one of McCoy's mentors, in "On a Typical Series of Objects in Natural History Best Adapted to Local Museums," paper presented before the British Association for the Advancement of Science, Glasgow, 1855, p. 111.

59. Bather, "Some Colonial Museums," p. 199.

60. Coleman, *Museum in America*, vol. 3, pp. 225, 226.

The Influence of New Zealand
on Rutherford's Scientific
Development

LAWRENCE BADASH

Rutherford's official biographer, A. S. Eve, tells a story about dinner at Trinity College high table in the 1930s. A visitor was seated next to a portly, probably rumpled-looking man with a shaggy mustache. The latter talked genially and knowingly about the dominions and then agriculture. Afterward the stranger asked someone the identity of "that Australian farmer," for the man certainly did not match the popular image of a university intellectual. He was told that his dinner companion was the great physicist Lord Rutherford.[1]

Here, on a superficial level, we can see part of the influence New Zealand (not Australia) had on Ernest Rutherford, for throughout his life abroad he remained alert to conditions in his native country and retained the enthusiasm, heartiness, bluffness, and clearheadedness that might be associated with a rural, colonial upbringing. In many ways he remained an unpolished gem.

My remarks here deal, however, not with New Zealand's influence on Rutherford's personality, but its effect on his scientific career. Since the achievements of exceptional people are based in no small measure upon their *internal* qualities, I cannot in good conscience credit New Zealand with having produced a genius. I can, however, attempt to describe the environment of this country that permitted, or even encouraged, a youth to become a scientist. My focus is thus on New Zealand and its characteristics in the latter part of the nineteenth century, and my use of Rutherford for illustration means not that he was typically able, but that the opportunities open to him were typically available.

First, allow me to suggest that New Zealand *does* deserve a significant measure of credit for Rutherford's success, or, conversely, to dispose of the view that a man of Rutherford's abilities would have succeeded in any environment. Had he been an Eskimo or an Aborigine, I think you will agree that tribal leadership might have been conceivable, but not scientific leadership. A scientific career, then, requires a literate person. I take this to be fundamental. A developed educational system offers much support, although self-educated men who become great scientists—Benjamin Franklin and Michael Faraday come to mind—show that it is not vital. Further, an industrial society is presumably more likely to spawn scientists than an agrarian society, but the difference is one of degree. Only when *agrarian* becomes *primitive*, with its concomitant illiteracy, would I draw the line of likelihood.

Thus the fact that New Zealand produced a young Ernest Rutherford who could read and write was crucial. Many countries, however, educate their youngsters, so can we find any unique circumstances or combination of them that proved particularly advantageous? I cannot point to any single feature of New Zealand society that provides the key. I believe it was rather a fortuitous constellation of circumstances that provided the fertile soil for Rutherford's intellectual development.

New Zealand was discovered in 1642 by Abel Tasman, but not populated by Europeans until after James Cook's circumnavigation of the islands in 1769. During the decades that followed, settlements of former convicts and escaped prisoners from Australia, as well as shipwrecked and runaway sailors, subsisted by trading muskets and rum with the native Maoris for flax and food, while a contingent of missionaries tried to save the Maoris' souls. Until the late 1830s a state of near anarchy prevailed, in part because the British government in London was otherwise occupied and reluctant to acquire another colony. By 1840 the situation had changed sufficiently for London to accept the responsibility of protecting the islands. This change, brought about by the achievement of a critical mass of population, was, of course, destined to encourage yet further increases in the number of settlers.[2]

In the mother country the Industrial Revolution had brought a striking alteration to the traditional agrarian life. To escape crowded and unsanitary mill towns, the danger of imprisonment from the poor laws, and increased social controls from an upper class that sought to smother the kind of political unrest that could lead to an anglicized model of the French Revolution, many chose to emigrate. The majority went to America and Canada, but a hardy handful voluntarily selected the distant shores of New Zealand.[3]

In large measure, this unlikely locale proved attractive because it provided the opportunity for social experiment. Edward Gibbon Wakefield, a reformer of conservative cast, sought to relieve the pressure cooker of discontent in Great Britain by providing the safety valve of emigration. He envisaged not rough, frontier communities, however, but pastoral models of English towns. These would attract finance and financiers, as well as a working class; and land prices would deliberately be maintained at a high level to discourage laborers from deserting the entrepreneurs and establishing their own farms. This policy would also tend to concentrate the settlers in communities, rather than allow them to disperse over the countryside, and thus be more productive of "civilization." Wakefield's agents bought much land from the Maoris and the British government during the 1840s and established several settlements, most notably in New Plymouth, Wellington, and Nelson. The project ultimately failed for economic and political reasons, but while it functioned it succeeded in assisting shiploads of pioneers to establish themselves in New Zealand. In general, these were of a better class than those who went to other colonies, with more education, better skills, and higher principles.[4]

Rutherford's paternal grandfather was a wheelwright and joiner from Scotland, while his father, who had been brought to New Zealand as a young child, pursued a number of technological activities, in addition to farming. His maternal grandmother was a widowed teacher from the south of England; when she remarried, her post was filled by her daughter, Ernest's mother.[5] Rutherford's parents were vigorous, intelligent, hard-working, sober pioneers, trying to tame the land and offer their large, close-knit family whatever benefits were attainable. Self-reliance was tempered by a sense of community; chores and other responsibilities were interspersed with an active outdoor life; formal education was supplemented by spelling bees and discussions of current events at home, while the older children took responsibility for certain lessons of their younger siblings. However poor they were—and for their state during Rutherford's youth the term is accurate—his mother found means to help others less fortunate than they. If this seems almost a storybook portrayal, it is because of the character of Rutherford's parents and a country in which honest labor was rewarded more often than not.

The young Rutherford was an avid reader, enjoying several popular Victorian authors and wild West adventure stories as well. He also exhibited an interest in the way mechanical objects worked, thereby reflecting a widespread nineteenth-century phenomenon: the success of the how-to book.[6] To the respect for education and the technological curiosity he derived from his parents, Rutherford added an enthusiasm he retained throughout his life.

New Zealand's economy fluctuated during the nineteenth century, according to changing local and international conditions. Aside from variable exports of wood and flax and periodic gold rushes, the islands were too remote and too little endowed with resources for much foreign trade. The flax, which grew abundantly, was not the type from which linen is made but was used for cordage instead. Competition from other fibers for the manufacture of ships' ropes prevented this from becoming a major industry. From about 1870 onward the national government pursued a policy of spending on internal improvements in an effort to mitigate the effects of a regional depression. The facilities for transport and communication thus established served to tie the country together, but the overseas borrowing and the lack of increased productivity and exports to match the investment boom had dragged New Zealand into a worldwide depression by about 1880. Not until refrigerator ships were introduced a decade later did the export of meat and dairy products to Europe, coupled with rising prices for other items such as wheat, bring prosperity to New Zealand. This variable economy prevented the country from developing as rapidly as it might have under affluent conditions, and it precluded striking rises in the standard of living of the population. Yet inexorable progress toward the trappings of civilization was nevertheless made, and the people's closeness to the soil spared them the housing and hunger problems common in depressed economic conditions in more industrial or urban societies. It is hard to imagine that Rutherford's home or school life during his youth would have been significantly different had his family been considerably more affluent.[7]

At the age of five, Rutherford entered the village primary school. In the same year the Education Act of 1877 was passed, requiring free schooling for children between the ages of seven and thirteen. Until the government took over the schools, they were run by private individuals or the church. Rutherford's first enrollment, at an age younger than the country's new minimum (which itself was an enlightened development on the world scene), reflects the circumstances that schools had formerly been privately run, parents were often willing to pay for what they felt was a proper education, and the one-room schoolhouses were inclined to accommodate them, for financial if not pedagogic reasons. Public education was not an unmixed blessing, by twentieth-century standards, for repetitive drill and inflexible discipline, copied from the mother country's practices, were far more common than individual attention to the student's intellectual capability.[8]

Young Rutherford's easygoing personality took both the uncreative disciplinarian and the unpredictable iconoclast in stride, and it was his good fortune that his teachers of both types took an interest in him. A primary-school instructor is credited with awakening Rutherford's interest in science shortly before his eleventh birthday. The vehicle was a small popular science text called *Physics*, written by Balfour Stewart of Manchester. It was part of another nineteenth-century publishing tradition of books that avoided dull, dry, formal presentations of their materials, stressing instead how much could be learned of nature from simple observation and experiment. That such a book was imported to New Zealand and came into the hands of an elementary-school teacher not known for a strong personal interest in science speaks well for the quality of rural schoolmasters and for the country's trade in books.

Another instructor offered an optional class in Latin each morning before school opened. To be sure, it increased his income; the significant point, however, is that Rutherford's parents and others were quite willing to undertake the expense. With such support at home and in school, Rutherford received an education sufficiently broad and deep for him to compete successfully for a provincial scholarship to Nelson College. This was a private institution (called public in the British world), modeled after Eton and Harrow, but largely devoid of their social snobbery. As the only secondary school in the region, and with secondary education not compulsory, Nelson College catered to those academically acceptable boys whose parents were willing to pay the tuition. In addition to sons of the so-called better classes, children of men who worked with their hands were enrolled. Since all the Rutherford boys attended Nelson (one other on a scholarship), expenses were apparently not beyond the means of a modest and frugal farm family.[9]

At Nelson, Rutherford was exposed to instructors of greater intellectual achievement than he had met earlier. Perhaps without exception, they had received typical university educations in classics in the British Isles and as classicists were deemed competent to teach not only Latin and Greek, but mathematics and science, and even, with Victorian hubris, to run the Foreign Office or the Exchequer. Thanks to W. S. Littlejohn in particular, Rutherford developed a primary concern with mathematics and secondary interest in science, reflecting his instructor's own abilities, but Rutherford won honors almost every year in Latin, French, English literature, and history as well and was top student in the school. Secondary education of the quality offered here, less than fifty years from the establishment of the country, is impressive testimony of the determination and desires of the settlers of New Zealand.[10]

If primary and secondary schooling show nineteenth-century New Zealand to have been an enlightened country, then its colleges do nothing to diminish that impression. These institutions of higher education were founded in the period around 1875, in such cities as Auckland, Wellington, Dunedin, and Christchurch. A minority looked to them to elevate the standards of mankind—in contemporary terms, knowledge for its own sake—but most supported the colleges for their practical goal of preparing students for social-Darwinian survival of the fittest in the mercantile battle of life.[11] Canterbury College, in Christchurch, which Rutherford entered in 1890, had at that time five professors, three lecturers, and about one hundred seventy students. While it was certainly not a large institution, its staff were able to provide high-quality education because the pedagogical tradition demanded a rather well defined curriculum, limited to the basics and with few choices. Thus Rutherford, who carried with him a coveted Junior Scholarship—one of only ten awarded throughout the country—selected his courses from the standard fare of Latin, Greek, English, French, philosophy, general history and political economy, pure mathematics, applied mathematics, physics, chemistry, and natural science.[12]

The two professors who influenced him most were Charles H. H. Cook, sixth wrangler in the Cambridge University mathematical tripos and a former fellow of St. John's College, Cambridge, and Alexander W. Bickerton, a graduate of the Royal School of Mines in London and a public lecturer on science of considerable ability. Cook, much like Littlejohn, and in the tripos tradition, was a methodical, rigorous problem solver, but not a creative mathematician. For Rutherford this was perfectly adequate, for his own contributions were based on physical insight, not mathematical inventiveness. Bickerton was quite different. His entry in the *Encyclopedia of New Zealand* (1966 edition) lists him as "scientist and eccentric." The former he was by profession, and the latter—well, he was hardly an amateur. His was probably the first original scientific mind that Rutherford had encountered, and if Bickerton's brilliance bordered on the bizarre, no one could deny the stimulation of his restless presence and the wide range of his interests. In his own major research he contrived grandiose cosmological theories without providing supporting evidence, then criticized the scientific establishment for ignoring him. Rutherford, happily, caught the spark of creative inquiry from Bickerton, but the solid, level-headed, proof-required mathematical molding of Littlejohn and Cook kept him close enough to the facts before him that Bickerton's spark ignited not a momentary flare but a long-burning flame.[13]

His association with Bickerton was crucial to Rutherford's scientific development, for he communicated a thirst for knowledge and a feeling that the book of nature could be read. The importance of that idea—that one *could* make original contributions—cannot be overstated. He transmitted to Rutherford, moreover, a laboratory style of sealing-wax-and-string improvisation, of back-of-the-envelope calculations, of preference for the joy of discovery rather than the satisfaction of measuring a known quantity to another decimal place of accuracy.[14]

Beyond that, Rutherford was introduced to the life of science through Bickerton. He revised proofs of one of his professor's papers, he witnessed the transitory fame Bickerton enjoyed when a newly discovered star seemed to confirm his cosmological theory, and he was encouraged to participate in the activities of the local scientific society. The last was more than socialization to the profession. The college library was pitifully small and could not affort the important scientific periodicals; a few blocks away in Christchurch, however, in the rooms of the Philosophical Institute of Canterbury, Rutherford was able to consult such journals as the *Philosophical Magazine* and the *Proceedings* and *Philosophical Transactions of the Royal Society of London*. Also at hand were the basic monographs on electricity and magnetism, such that Rutherford could feel reasonably confident that he was abreast of the field in which he chose to do his research.[15] Thus, despite the general poverty of Canterbury College in laboratory equipment and library holdings, the means for reaching the research front nevertheless existed.

The most vigorous scientific organization in the South Pacific was the Australasian Association for the Advancement of Science (AAAS). It held its third annual meeting in Christchurch in January 1891, but there is no evidence that Rutherford returned early from the summer vacation after his freshman year to attend. The AAAS next visited New Zealand in 1904, long after Rutherford's departure. His debut as a scholar therefore occurred before the student-run Canterbury College Science Society; he appeared before the Philosophical Institute of Canterbury, a constituent society of the New Zealand Institute, and, of course, the proprietor of the library Rutherford used. His paper appeared on a crowded program, on 7 November 1894, and only its title was read, but its content of physical science differed so much from the natural history subjects most commonly presented that it was accepted for publication in the New Zealand Institute's *Transactions*.[16]

In this fashion Rutherford not only became socialized to the professional behavior of preparing papers, reading and publishing them, and attending meetings, but he also benefited from the mutually supportive atmosphere generated by people with common interests. Beyond this somewhat general encouragement, he was fortunate to have a fellow student, John Angus Erskine, whose interests and abilities were so close to Rutherford's own that the two interacted and collaborated; Erskine, for example, used Rutherford's magnetic detector for radio waves in his own work.[17] Though there did not exist in New Zealand what might be called a flourishing research community, the local enthusiasts nevertheless offered relief from the uncertainties of scientific isolation and the ability to consult with well-informed contemporaries and elders. This made Christchurch, New Zealand, at that moment a sufficiently nourishing environment for the fledgling scientist.

But the social, cultural, and economic forces that had made it possible to dream of a career in scientific research now proved to have been exhausted. While the New Zealand characteristics of common sense, getting on with the job, and hard, honest work, which led to upward mobility for many, had indeed functioned in Rutherford's case, the problem was that the colony simply had no scientific research positions to offer someone so talented. In fact, with the exception of university posts and some comparatively uncreative jobs in industrial and government laboratories, there were few opportunities anywhere in the world for the pursuit of basic scientific research.

The choices open to Rutherford were limited. He could perhaps become a science master in a secondary school, and if he were able to save enough of his earnings to finance a year or more in Europe, he could obtain further scientific training and possibly a university post. Teaching positions, however, were not readily available, and the prospect of accumulating such a sum was unlikely. He could aim for a medical career, one of the few professions in which advanced work was not uncommon and one in which scholarships to European institutions were relatively plentiful. Despite the encouragement he received from a Canterbury friend studying at Edinburgh, Rutherford could not see himself as a physician.[18] We come finally to the only singular opportunity open to him, and it must be noted that its uniqueness has nothing to do with New Zealand, but lies with ideas of education, philanthropy, and empire in Great Britain.

In 1851 a great exhibition was held in the Crystal Palace, an enormous iron-and-glass structure erected in London's Hyde Park. Industrial and commercial arts of many countries were exhibited, and the financial success of this fair left its managers looking for uses for the profits. In 1891 they inaugurated the annual award of almost twenty research scholarships in science, the recipients coming from within the British Empire but permitted to attend any university at which their programs were approved. The University of New Zealand, of which Canterbury College was a part, was entitled to nominate a candidate every other year and in 1895 nominated Rutherford.[19] In this manner he was able to travel to Cambridge University, where he was welcomed as a research student in J. J. Thomson's Cavendish Laboratory. New Zealand had, without question, provided the multiplicity of factors that enabled Rutherford to aspire to a research career, but it took the greater wealth and philanthropy of the mother country to make the dream a reality.[20]

The research for this essay was supported by a travel grant from the American Philosophical Society.

1. A. S. Eve, *Rutherford* (Cambridge: Cambridge University Press, 1939), p. 364.

2. J. C. Beaglehole, "Discovery and Exploration," pp. 3–4, 7–12; I. L. G. Sutherland, "Maori and Pakeha," pp. 48, 54–57; both in *New Zealand*, ed. Horace Belshaw (Berkeley and Los Angeles: University of California Press, 1947).

3. W. T. G. Airey, "New Zealand in Evolution (1840–1880)," in *New Zealand*, ed. Belshaw, pp. 73–75.

4. Ibid. pp. 75–81, 86–87; Samuel Leathem, "Industry and Industrial policy," in *New Zealand*, ed. Belshaw, pp. 164–65. Muriel F. Lloyd Prichard, *An Economic History of New Zealand to 1939* (Auckland and London: Collins, 1970), p. 25.

5. Eve, *Rutherford*, p. 1.

6. Richard Kuczkowski, "This Way to the Egress," *The Nation* 226 (21 June 1975): 758–60.

7. Lloyd Prichard, *Economic History of New Zealand*, pp. 4, 8–9, 22, 130–79 passim, 203; Airey, "New Zealand in Evolution," pp. 87–88.

8. Lloyd Prichard, *Economic History of New Zealand*, pp. 76, 153; A. E. Campbell, "Education," in *New Zealand*, ed. Belshaw, p. 223.

9. Eve, *Rutherford*, p. 3; Campbell, "Education," p. 230; Lloyd Prichard, *Economic History of New Zealand*, p. 145.

10. "Nelson College in Rutherford's day, 1887–1888–1889," 3 pp. of mimeographed notes sent to the author by Headmaster B. H. Wakelin; "Notes on W. S. Littlejohn," 6 pp. of typewritten notes prepared by Rutherford in 1934, Add. 7653/PA305, Rutherford collection, Cambridge University Library (hereafter RCC); C. H. Broad (a classmate and later headmaster of Nelson College), "Lord Rutherford's Early Years: Some Reminiscences," *The Times* (London), 22 October 1937, p. 18.

11. R. A. Loughnan, *New Zealand at Home* (London: Newnes, [1908]), p. 96; Lloyd Prichard, *Economic History of New Zealand*, p. 226.

12. *Weekly Press* (Christchurch), 7 February 1890, p. 35; *Canterbury College Calendar for the Year, 1895*, p. 74; James Hight and Alice Candy, *A Short History of the Canterbury College* (Auckland: Whitcombe and Tombs, 1927), pp. 32, 117, 142, 164.

13. Hight and Candy, *Short History of Canterbury College*, pp. 117–21; R. M. Burdon, *Scholar-Errant: A Biography of Professor A. W. Bickerton* (Christchurch: Pegasus, 1956), passim.

14. Burdon, *Scholar-Errant*, pp. 47, 55, 118–21; J.S.S.C., "Professor E. Rutherford, M.A., B.Sc.," *Canterbury College Review*, no. 4 (October 1898): 21–22.

15. Burdon, *Scholar-Errant*, pp. 56–58, 61–62. Information regarding the periodicals comes from the Canterbury University Library, while the monographs are cited by Rutherford in "Magnetization of Iron by High-Frequency Discharges," *Transac-*

tions of the *New Zealand Institute* 27 (1894): 481–513; reprinted in *The Collected Papers of Lord Rutherford of Nelson*, ed. James Chadwick, 3 vols. (London: Allen & Unwin; New York: Interscience, 1962, 1963, 1965), vol. 1, pp. 25–26.

16. Rutherford, "Magnetization of Iron."

17. J. A. Erskine, "A Comparison of the Magnetic Screening Produced by Different Metals," *Transactions of the New Zealand Institute* 28 (1895): 178–82; "On the Screening of Electric-Motive Force in the Fields Produced by Leyden-Jar Discharges," ibid., 30 (1897): 459–62.

18. John Stevenson to Rutherford, 7 September 1893, RCC.

19. The University of New Zealand actually submitted two names to London and the other candidate was selected. He chose to get married instead, and Rutherford was awarded the 1851 Exhibition.

20. "The Great Exhibition," *The Times* (London), 1 May 1851, pp. 8–9; *Record of the Science Research Scholars of the Royal Commission for the Exhibition of 1851* (London: The Commissioners, 1961); University of New Zealand, Minutes of Proceedings of the Senate, 1891 to 1895, pp. 3, 14, 99–101; this volume is preserved in the University of Canterbury Registry; typewritten notes entitled "Lord Rutherford," sent by the Commission's secretary, Sir Evelyn Shaw, to Rutherford's biographer, A. S. Eve, Add. 7653/PA312.8, RCC.

Notes on Contributors

Lawrence Badash is Professor of History of Science at the University of California, Santa Barbara. His research interests include the history of nuclear physics and the nuclear arms race.

David Wade Chambers, Chairman of Social Studies of Science, Deakin University, Victoria, Australia, is the author of a number of university-level textbooks, including *Red and Expert, Imagining Nature, Beasts and Other Illusions,* and *The Lessons of Lysenko.*

Kathleen G. Dugan received her Ph.D. from the University of Kansas and is now at the Graduate School, Academia Sinica, Beijing, People's Republic of China.

R. W. Home, Professor of History and Philosophy of Science at the University of Melbourne and editor of *Historical Records of Australian Science,* is at work on a general history of the Australian physics community to 1945.

Richard A. Jarrell, who holds an A.B. from Indiana University and the M.A. and Ph.D. degrees from the University of Toronto, is Associate Professor and Chairman of Natural Science, Atkinson College, York University in Toronto.

Sally Gregory Kohlstedt, who teaches at Syracuse University and is now a fellow at the Woodrow Wilson International Center for Scholars, writes on the cultural and institutional context of science in nineteenth-century America and has recently edited, with Margaret Rossiter, *Historical Writing on American Science* (1985).

Edwin T. Layton, Jr., is a historian of technology at the University of Minnesota with a special interest in the interaction of science and technology, particularly as it pertains to the hydraulic turbine. He is President of the Society for the History of Technology.

Roy MacLeod was educated at Harvard, Cambridge, and the London School of Economics and is Professor of History at the University of Sydney, where he teaches imperial history, military studies, and the social history of science.

Ann Moyal is a historian of Australian science and technology and an analyst of science and technology. Her latest book, *"A Bright and Savage Land": Scientists in Colonial Australia,* was published by Collins in October 1986.

D. J. Mulvaney, C M G, is Professor Emeritus of Prehistory and an honorary fellow of the Department of History at the Australian National University.

Ronald L. Numbers is Professor of the History of Medicine and the History of Science at the University of Wisconsin–Madison, where he teaches the history of American science and medicine.

Lewis Pyenson is Professeur Titulaire in the Département d'Histoire, Université de Montreal. He received the 1986 Herbert C. Pollock Award of the Dudley Observatory.

Nathan Reingold is a Senior Historian at the National Museum of American History, Smithsonian Institution.

Marc Rothenberg is Editor of the Joseph Henry Papers, Smithsonian Institution.

C. B. Schedvin is professor of Economic History, University of Melbourne and is completing a history of the Commonwealth Scientific and Industrial Research Organization

Susan Sheets-Pyenson coordinates the program in Science and Human Affairs at Concordia University in Montreal.

John Harley Warner is Assistant Professor of the History of Medicine at Yale University and is the author of *The Therapeutic Perspective: Medical Practice, Knowledge, and Identity in America, 1820–1885* (Harvard University Press, 1986).

Index of Names